£3

INDUSTRIAL LOCOMOTIVES 1982

INDUSTRIAL LOCOMOTIVES
of GREAT BRITAIN
1982

INDUSTRIAL RAILWAY SOCIETY

Published by the INDUSTRIAL RAILWAY SOCIETY
at 422, Birchfield Lane, Oldbury, Warley, West Midlands, B69 1AF.

© INDUSTRIAL RAILWAY SOCIETY 1982.

ISBN 0 901096 43 1 (hardbound)
ISBN 0 901096 44 X (softbound)

Distributed by IRS Publications, 47, Waverley Gardens, London, NW10 7EE.

Printed in Great Britain by A B Printers Ltd., Leicester.

CONTENTS

SECTION ONE   ENGLAND

SECTION TWO   SCOTLAND

SECTION THREE   WALES

SECTION FOUR   NATIONAL COAL BOARD

SECTION FIVE   MINISTRY OF DEFENCE, ARMY DEPARTMENT

SECTION SIX   IRELAND

SECTION SEVEN   BRITISH RAILWAYS

The text of this book incorporates all amendments notified on or before 1st January, 1982.

FOREWORD
to the sixth edition

This is the sixth in a line of EL (existing locomotive information as opposed to historical) books which commenced in 1968. The lists now include all known locomotives of gauge 1'3" and above excluding only British Railways Capital Stock and N.C.B. underground stock.

The lists are in accordance with our records for January, 1982 and we would stress again that the information all derives from the observations and researches of members and others. We shall rely on similar enthusiastic support to keep the records up to date, so please send your observations (and your queries too) to the Hon. Records Officer and his Assistants, as listed below. All such information will be distributed to members through the bi-monthly bulletin.

Current observations - N.C.B.
> R.Darvill,(Assistant Records Officer), 20, Oakfield Road, Rugby, Warwickshire, CV22 6AU.

Current observations - M.O.D.,A.D.
> G.P.Roberts (Assistant Records Officer), 2, Saxon Close, Hillingdon, Middlesex.

Current observations - other owners
> J.A.Foster (Assistant Records Officer), 30A, Pinner Road, Hunters Bar, Sheffield, S11 8UH.

Historical matters - N.C.B.
> A.R.Etherington (Assistant Records Officer), 23, Moira Road, Woodville, Burton-on-Trent, Staffs.

Historical matters - other owners
> W.K.Williams (Hon. Records Officer), 60, Westhill Road, Kings Norton, Birmingham, West Midlands, B38 8TH.

INTRODUCTION

The format is the same as in previous editions, with the locomotives listed under owners arranged alphabetically in Counties. The categories of the entries (industrial, preserved, etc) will be obvious from the titles in the great majority of cases, and only those entries which may be subject to ambiguity are noted 'Pvd'.

Coal Concentration Depots, although owned by N.C.B., are listed under the name of the merchants operating them, under the relevant County. Locomotives at scrapyards purely for scrapping are not generally listed, only if they have been 'resident' for a year or more. Those at dealers yards are however listed in order that members can keep a full picture of their movements.

Contractors locos can be found on sites in all parts of the country, working on sewer and tunnel schemes, etc. The locomotive details are shown in the usual way in the lists for the county in which the plant depot, or main plant depot if more than one is in use, is situated. A list of contractors and the relevant counties is shown after the Locomotive Builders tables.

When a works is closed, this is indicated after the title or subtitle, thus: PENTEWAN SANDS CARAVAN LTD, PENTEWAN. (Closed)

When a works is still in operation but no longer uses a certain type of rail traffic, this is indicated by a note after the gauge concerned, thus: Gauge : 3'0".   (TL 186352)   RTC.   RTC = Rail Traffic Ceased.

Regular steam working is noted by RSW in the heading; this does not necessarily imply daily steam operation, but that the normal motive power is steam.

GAUGE. The gauge of the railway system is given at the head of the locomotive list. At preservation sites and museums, etc, where several gauges differ by only a fraction they are usually all listed under the nominal gauge of the majority.

GRID REFERENCE. An indexed six-figure grid reference is given in brackets after the gauge, and denotes the location of the loco shed or stabling point. Most references are known but details of any missing references are welcome.

NUMBER AND NAME. The title of the locomotive - number, name or both - is given in the first two columns. A name unofficially bestowed and used by staff but not carried on the loco is indicated by inverted comas, "   ". Locomotives under renovation at preservation sites, etc, may not carry their intended title but these are shown, unless it is definitely known that the name will not be retained. Ex B.R. locomotives are further identified by the inclusion of the B.R. number in brackets, if not carried now.

TYPE. The type of locomotive is given in column three. The Whyte system of wheel classification is used in the main, but when driving wheels are not connected by outside rods but by chains or other means (as in various 'Sentinel' steam locos and diesel locos) they are shown as 4w (Four-wheeled), 6w (Six-wheeled), or if only one axle is motorised it is shown as 2w-2. The following abbreviations are used

T	Side tank or similar - a tank positioned externally and fastened to the frame.
CT	Crane tank - a T type loco fitted with load lifting apparatus.
PT	Pannier tank - side tanks not fastened to the frame.
ST	Saddle tank.
STT	Saddle tank with tender.
WT	Well tank - a tank located between the frames under the boiler.
VB	Vertical boilered locomotive.
F	Fireless steam locomotive.
D	Diesel locomotive - unknown transmission.
DC	Diesel locomotive - compressed air transmission.
DM	Diesel locomotive - mechanical transmission.
DE	Diesel locomotive - electrical transmission.
DH	Diesel locomotive - hydraulic transmission.
P	Petrol or Paraffin locomotive - unknown transmission.

8

```
PM    Petrol or Paraffin locomotive - mechanical transmission.
PE    Petrol or Paraffin locomotive - electrical transmission.
R     Railcar - a vehicle primarily designed to carry passengers.
BE    Battery powered electric locomotive.
CE    Conduit powered electric locomotive.
RE    Third rail powered electric locomotive.
WE    Overhead wire powered electric locomotive.
```

CYLINDER POSITION. Shown in column four; this column is blank in the case of non-steam locomotives apart from those with a steam outline - see below.

```
IC    Inside Cylinders.
OC    Outside Cylinders.
VC    Vertical Cylinders.
G     Geared transmission - coupled to IC, OC or VC.
```

STEAM OUTLINE. Diesel or Petrol locomotives with a Steam locomotive appearance added are shown by S/O in column four.

MAKERS. The builder of the locomotive is shown in column five; the abbreviations used are listed on a later page.

MAKERS NUMBER AND DATE. The sixth column shows the works number, the next column the date which appears on the plate, or the date the loco was built if none appears on the plate.

```
c.    denotes circa, i.e. about the time of the date quoted.
reb.  denotes Rebuilt.
Pvd.  denotes Preserved on site.
```

DOUBTFUL INFORMATION. Information known to be doubtful is denoted as such by the wording, or printed in brackets with a question mark. Thus, P (1234?) 1910 denotes that the loco is a Peckett of 1910 vintage and possibly of works number 1234.

Locomotives which have ceased work are indicated by OOU (Out of Use), or as Dsm (Dismantled) if incomplete. These are only shown when the condition is permanent, not of a temporary nature. Wkm motorised trollies which have been converted to engineless trailers are shown as DsmT.

There are cases where all locomotives are OOU but the tramway is still in use; with a road tractor or man-power for example. Also some owners use locos on their own internal system after the B.R. connection has been severed.

Electric locomotives at Steelworks and N.C.B. sites are usually to be found working at Coke Ovens, the grid reference for the Ovens being shown in the usual place, when known.

Certain locomotives are capable of working on either rail or on normal road surfaces (most are of either Unilok or Unimog manufacture). These are indicated by R/R in column four.

Hints on recording observations

As already explained, the Society is almost entirely reliant on the observations of enthusiasts to keep the records up to date, and we are always pleased to receive reports of visits, which should be sent to the Hon. Records Officer or his assistants. The following will, we hope, be of assistance to those sending in reports.

Report anything you see; on the locomotive side this means reporting locos present, even if there is no change from the published list. Someone else might go in six months and find a change, the date of which can then be narrowed down to this six months. In addition to locos, details of rolling stock, layouts and items of historical information gleaned, are all welcome and will be filed for future reference. It is surprising how often these items are required later by other enthusiasts. There is nothing to beat a note made on the spot; a note now is far better than trying to get the information from the recollection of employees in five years time.

On the other hand, care should be taken to distinguish observation from inference. If you see a 0-4-0ST OC with worksplate P 1704 and named FRANCIS, all well and good; but if it does not carry a worksplate, that fact should be stated. You may infer the loco is P 1704 from the name, by the number stamped on the motion, or by reference to the pocket book. If you say 'number 1704 on motion' or 'loco assumed to be P 1704' as the case may be, we will know the position precisely, which is important when names and tanks are exchanged, for example. If a loco carries no identification, you will probably be able to guess its make from design features, and a note of livery may help.

A thorough search of all premises is worthwhile, as locos (particularly if OOU) are frequently hidden away - and how often have surprises turned up in this way. Further, if you search diligently, and a loco is missing it may be assumed to have gone and enquiries can be made, as to where it has gone or if it has been scrapped. Please try to ascertain such information from the staff. Similarly, in the case of new arrivals, dates should be obtained if possible.

Firms' titles change from time to time, so please check from the office or board at the entrance. Subsidiary companies frequently display the title of the parent company, but we always use the name under which the company trades, i.e. the subsidiary where such exists.

The necessity for obtaining permission to see locomotives applies with equal force to 'preserved' locations other than public parks and museums. Established systems which operate trains or have 'Open Days' will usually permit inspection of locos by arrangement. Locomotives shown as at private homes or farms are very often in storage or in course of restoration and as a general rule members are advised not to embarrass the owners by writing for permission but to wait until they announce that their machines are available for inspection.

Finally, try to establish friendly relations with the firms visited, as we are only allowed access by their courtesy. Don't be a nuisance or hold up production, nor expose yourself to danger, but show a healthy interest in the processes being carried out. In this way, not only you, but other enthusiasts will be welcome later.

LOCOMOTIVE BUILDERS

AB	Andrew Barclay, Sons & Co Ltd, Caledonia Works, Kilmarnock.
AB/MV	Built jointly by AB and MV.
A.C.Cars	A.C.Cars Ltd, Thames Ditton, Surrey.
Acton	Acton Works, London Transport Executive, Acton, London.
AE	Avonside Engine Co Ltd, Bristol.
AEC	Associated Equipment Co Ltd, Southall.
AEG	Allgemeine Elektricitaets Gesellschaft, Berlin, Germany.
AH	A.Horlock & Co, North Fleet Iron Works, Kent.
AL	American Locomotive Co, U.S.A./Canada.
Alan Keef	Alan Keef Ltd, Cote Farm, Cote, near Bampton, Oxon.
Alldays & Onions	Alldays & Onions Ltd, Birmingham.
AP	Aveling & Porter Ltd, Invicta Works, Canterbury, Kent.
Afd	S.E. & C.R., Ashford Works, Kent.
AT	Les Ateliers Métallurgiques Nivelles, Tubize & La Sambre Works, Tubize, Belgium.
Atlas	Atlas Loco & Mfg. Co Ltd, Cleveland, Ohio, U.S.A.
Atw	Atkinson Walker Wagons Ltd, Preston, Lancs.
AW	Sir W.G.Armstrong, Whitworth & Co (Engineers) Ltd, Newcastle-Upon-Tyne.
Barlow	H.N.Barlow, Southport, Lancs.
BD	Baguley-Drewry Ltd, Burton-on-Trent.
BE	Brush Electrical Engineering Co Ltd, Loughborough, Leics.
Belliss & Seekings	Belliss & Seekings Ltd, Birmingham.
Berwyn	Berwyn Engineering, Thickwood, Chippenham.
Bg	E.E.Baguley Ltd, Burton-on-Trent.
BGB	Becorit (Mining) Ltd, Grove Street, Mansfield Woodhouse, Nottingham.
BgC	Baguley Cars Ltd, Burton-on-Trent.
Bg/DC	Built by Bg for DC; makers numbers identical.
BH	Black, Hawthorn & Co Ltd, Gateshead.
Bilsthorpe	Stanton & Staveley Ltd, Bilsthorpe, Notts.
BL	W.J.Bassett Lowke Ltd, Northampton.
BLW	Baldwin Locomotive Works, Philadelphia, Pennsylvania, U.S.A.
Booth	J.Booth & Bros, Rodley, Leeds.
Borsig	A.Borsig, G.m.b.H., Berlin, Germany.
Boston Lodge	Boston Lodge Workshops, Festiniog Railway.
Bow	Bow Locomotive Works, North London Railway.
BP	Beyer, Peacock & Co Ltd, Gorton, Manchester.
BPH	Beyer, Peacock (Hymek) Ltd, Gorton, Manchester.
BRC	Birmingham Railway Carriage & Wagon Co Ltd, Smethwick.
Bredbury	Bredbury & Romiley U.D.C., Cheshire.
Bredonvale	Bredonvale Products Ltd, Defford, Worcs.
Brown & Tawse	Brown & Tawse Ltd, West Horndon, Essex.
BT	B.E., Traction Division, Falcon Works, Loughborough.
BT/WB	Built jointly by BT & WB.
BTH	British Thomson-Houston Co Ltd, Rugby.
Bton	Brighton Works, British Railways. (Previously S.R. and L.B.S.R.)
Bury	Bury, Curtis & Kennedy, Liverpool.
BV	Brook Victor Electric Vehicles Ltd, Burscough Bridge, near Ormskirk.
Cail	Société Francaise de Constructions Mécaniques, Denain, France.
Cdf	Cardiff Works, Taff Vale Railway.
CE	Clarke Chapman Ltd, International Combustion Division, Clayton Works, Hatton, near Derby.
Chance	
Chaplin	Alexander Chaplin & Co Ltd, Cranston Hill Works, Glasgow.
CL	Cammell Laird & Co Ltd.

Clarkson	H.Clarkson & Son, York.
Cleveland	Dorman Long (Steel) Ltd, Cleveland Works, Yorks N.R.
Corpet	Corpet Louvet & Cie, La Courneuve, Seine, France.
CoSi	Coleby-Simkins Engineering, Melton Mowbray, Leics.
Couillet	S.A. des Usines Métallurgiques du Hainaut, Couillet, Belgium.
Cowlairs	North British Railway, Cowlairs Works, Glasgow.
Cravens	Cravens Ltd, Darnall, Sheffield.
Crewe	Crewe Works, Cheshire; British Railways. (Previously L.M.S.R. and L.N.W.R.)
CS	Christoph Schöttler Maschinenfabrik, G.m.b.H., Diepholz, Germany.
Curwen	A.Curwen, All Cannings, near Devizes, Wilts.
D	Dubs & CO, Glasgow Locomotive Works, Glasgow.
Dar	British Railways, Darlington Works, (and predecessors).
DB	Sir Arthur P. Heywood, Duffield Bank Works.
DC	Drewry Car Co Ltd, London. (Suppliers only).
Decauville	Société Nouvelle Des Establissements Decauville Ainé, Petit Bourg, Corbeil, S & O, France.
Derby	Derby Works; British Railways. (Previously L.M.S.R. and Midland Railway).
Derby C. & W.	Litchurch Lane Carriage Works, Derby; British Railways. (Previously L.M.S.R. and Midland Railway).
DeW	DeWinton & Co, Caernarvon.
Diema	Diepholzer Maschinenfabrik (Fr. Schöttler G.m.b.H.) Diepholz, Germany.
DK	Dick Kerr & Co Ltd, Preston, Lancs.
DM	Davies & Metcalfe Ltd, Romiley, Stockport, Cheshire.
Dodman	Alfred Dodman & Co, Highgate Works, Kings Lynn.
Don	Doncaster Works; British Railways. (Previously L.N.E.R. and G.N.R.)
Donelli	F.L.Donelli, Reggio, Italy.
Dowty	Dowty Group Ltd, Ashchurch, Glos.
DP	Davey Paxman & Co Ltd, Colchester, Essex.
Dtz	Motorenfabrik Deutz, A.G., Cologne, Germany.
Dundalk	Dundalk Works; Great Northern Railway Of Ireland.
EB	E.Borrows & Sons, St. Helens.
Eclipse	Eclipse Peat Co Ltd, Ashcott, Somerset.
EE	English Electric Co Ltd, Preston, Lancs.
EES	E.E., Stephenson Works, Darlington. (Successors to R.S.H.D.)
EEV	E.E., Vulcan Works, Newton-Le-Willows, Lancs. (Successors to V.F.)
Elh	Eastleigh Works, Hants; British Railways. (Previously S.R. and L.S.W.R.)
EV	Ebbw Vale Steel, Iron & Coal Co Ltd, Ebbw Vale, Mon.
EW	E.B.Wilson & Co.
FB	Société Franco-Belge de Matériel de Chemins de Fer, La Croyère, Belgium.
FH	F.C.Hibberd & Co Ltd, Park Royal, London; later at Butterley Works, Derbyshire.
FJ	Fletcher Jennings & Co, Lowca Engine Works, Whitehaven.
Foster Rastrick	Foster, Rastrick & Co, Stourbridge, Worcs.
Freud	Stahlbahnwerke Freudenstein & Co, Tempelhof, Dortmund, Berlin, Germany.
Frichs	A/S Frichs Maskinfabrik & Kedelsmedie, Arhus, Denmark.
B.J.Fry	B.J.Fry Ltd, Dorchester, Dorset.
FW	Fox, Walker & Co, Atlas Engine Works, Bristol.
GB	Greenwood & Batley Ltd, Leeds.
GE	George England & Co Ltd, Hatcham Ironworks, London.
GEC	General Electric Co Ltd, Witton, Birmingham.
GECT	G.E.C. Traction Ltd, Newton-Le-Willows, Lancs.
Geo Stephenson	George Stephenson, Hetton, Durham.
GEU	General Electric Co, Erie, Pennsylvania, U.S.A.
GH	Gibb & Hogg, Airdrie.
Ghd	Gateshead Works; North Eastern Railway.

GM	General Motors Ltd, Electro-Motive Division, La Grange, Illinois, U.S.A.
GMT	Gyro Mining Transport Ltd, Victoria Road, Barnsley, South Yorks.
Gorton	Gorton Works; Manchester; Great Central Railway.
Grazebrook	M. & W. Grazebrook Ltd, Dudley.
GR	Grant, Ritchie & Co, Kilmarnock.
GRC	Gloucester Railway Carriage & Wagon Co Ltd, Gloucester.
G & S	G. & S. Light Engineering Co Ltd, Stourbridge, Worcs.
Guest	Guest Engineering & Maintenance Co Ltd, Stourbridge, Worcs.
H	James & Fredk. Howard Ltd, Britannia Ironworks, Bedford.
Hackworth	Timothy Hackworth, Soho Works, Shildon, Co. Durham.
Hano	Hannoversche Maschinenbau - A.G. (vormals Georg Egestorff), Hannover, Germany.
Hartmann	Sächsische Maschinenfabrik, vormals Richard Hartmann A.G., Chemnitz, Germany.
HB	Hudswell Badger Ltd, Hunslet, Leeds.
HC	Hudswell, Clarke & Co Ltd, Railway Foundry, Leeds.
HE	Hunslet Engine Co Ltd, Hunslet, Leeds.
Wm.Hedley	William Hedley, Wylam Colliery, Northumberland.
Hen	Henschel & Sohn G.m.b.H., Kassel, Germany.
HF	Haydock Foundry Co Ltd, Haydock, Lancs.
H(L)	Hawthorn & Co, Leith, Edinburgh.
HL	R. & W. Hawthorn, Leslie & Co Ltd, Forth Bank Works, Newcastle-Upon-Tyne.
HLT	Hughes Locomotive & Tramway Engine Works Ltd, Loughborough, Leics.
Hor	Horwich Works, Lancs; British Railways. (Previously L.M.S.R. and L.Y.R.)
HU	Robert Hudson Ltd, Leeds.
HW	Head, Wrightson & Co Ltd.
Inchicore	Inchicore Works, Dublin; Great Southern Railways. (Previously G.S. & W.R.)
Iso	Iso Speedic Co Ltd, Fabrications & Electric Vehicles, Charles Street, Warwick.
Jaywick Rly	Jaywick Light Railway, near Clacton, Essex.
JF	John Fowler & Co (Leeds) Ltd, Hunslet, Leeds.
Jung	Arn. Jung Lokomotivfabrik G.m.b.h., Jungenthal, Germany.
K	Kitson & Co, Airedale Foundry, Leeds.
KC	Kent Construction & Engineering Co Ltd, Ashford, Kent.
T.Kennon.	Thos Kennon & Son, Dublin.
Kershaw	John Kershaw Ltd, Mytholmroyd.
Kitching	A.Kitching, Hope Town Foundry, Darlington, Co. Durham.
KKM	Knecht Krohnke Maschinenfabrik, Hamburg, Germany.
Krauss	Lokomotivfabrik Krauss & Co, Munich, Germany & Linz, Austria.
KrMaf	Krauss-Maffei A.G., Munich, Germany.
Krupp	Friedrich Krupp, Maschinenfabriken Essen, Abt. Lokomotivbau, Essen, Germany.
KS	Kerr, Stuart & Co Ltd, California Works, Stoke-on-Trent.
KTH	Kitson, Thompson & Hewittson, Leeds.
L	R. & A. Lister & Co Ltd, Dursley, Glos.
Lake & Elliot	Lake & Elliot Ltd, Braintree, Essex.
Lancs Tanning	Lancashire Tanning Co Ltd, Littleborough, Lancs.
Lane	Charles Lane, Liphook, Hants.
LB	Lister Blackstone Traction Ltd, Dursley, Glos.
LBC	London Brick Co Ltd.
Lewin	Stephen Lewin, Dorset Foundry, Poole.
Lima	Lima Locomotive Works Incorporated, Lima, Ohio, U.S.A.
LMM	Logan Mining & Machinery Co Ltd, Dundee.
Loco Ent	Locomotion Enterprises (1975) Ltd, Bowes Railway, Springwell, Gateshead, Tyne & Wear.

Locospoor	
Longhedge	Longhedge Works, London; S.E. & C.R.
M	Metropolitan Carriage & Wagon Co Ltd, Birmingham.
MAK	Maschinenbau Kiel G.m.b.H., Kiel-Friedrichsort, Germany.
Mather & Platt	Mather & Platt Ltd, Park Works, Manchester.
MC	Metropolitan Cammell Carriage & Wagon Co Ltd, Saltley, Birmingham.
Mercury	The Mercury Truck & Tractor Co, Glos.
MH	Muir Hill Engineering Ltd, Trafford Park, Manchester.
Minirail	Minirail Ltd, Frampton Cotterell, Bristol, Glos.
Mkm	Markham & Co Ltd, Chesterfield, Derbyshire.
Motala	A.B. Motala Verkstad, Motala, Sweden.
Moyse	Locotracteurs Gaston Moyse, La Courneuve, Seine, France.
MR	Motor Rail Ltd, Simplex Works, Bedford.
MV	Metropolitan-Vickers Electrical Co Ltd, Trafford Park, Manchester.
MW	Manning, Wardle & Co Ltd, Boyne Engine Works, Hunslet, Leeds.
N	Neilson & Co, Springburn Works, Glasgow.
NB	North British Locomotive Co Ltd, Glasgow.
NBQ	N.B., Queens Park Works, Glasgow.
NCC	Northern Counties Committee.
Nea	Neasden Works, London; Metropolitan Railway.
NNM	Noord Nederlandsche Machinefabriek B.V. Winschoten, Holland.
Nohab	Nydquist & Holm A.B., Trollhättan, Sweden.
NR	Neilson Reid & Co, Glasgow.
NW	Nasmyth, Wilson & Co Ltd, Bridgewater Foundry, Patricroft, Manchester.
Oakeley	Oakeley Slate Quarries Co Ltd, Blaenau Ffestiniog, Merion.
Oerlikon M.C.	
OK	Orenstein & Koppel A.G., Berlin, Germany.
Oldbury	Oldbury Carriage & Wagon Co Ltd, Birmingham.
P	Peckett & Sons Ltd, Atlas Locomotive Works, St. George, Bristol.
Permaquip	The Permanent Way Equipment Co Ltd, Pweco Works, Lillington Road North, Bulwell, Nottingham.
PR	Park Royal Vehicles, Park Royal, London.
R & R	Ransomes & Rapier Ltd, Riverside Works, Ipswich, Suffolk.
Redstone	Redstone, Penmaenmawr.
Regent St.	Regent Street Polytechnic, London.
RH	Ruston & Hornsby Ltd, Lincoln.
R.Heath	Robert Heath & Sons Ltd, Norton Ironworks, Stoke-on-Trent.
Riordan	Riordan Engineering Ltd, Surbiton, Surrey.
RM	Road Motors Ltd, West Drayton, London.
Robel	Robel & Co, Maschinenfabrik, Munchen, 25, Germany.
RP	Ruston, Proctor & Co Ltd, Lincoln.
RR	Rolls Royce Ltd, Sentinel Works, Shrewsbury. (Successors to S).
RS	Robert Stephenson & Co Ltd, Forth Street, Newcastle-Upon-Tyne and Darlington.
RSH	Robert Stephenson & Hawthorns Ltd.
RSHD	R.S.H., Darlington Works.
RSHD/WB	Built by RSHD but ordered by WB.
RSHN	R.S.H., Newcastle-Upon-Tyne Works. (Successors to HL).
R.Thomas	Richard, Thomas & Co Ltd, Crowle Brickworks, Lincs; constructed from parts supplied by FH.
Ruhrtaler	Ruhrtaler Maschinenfabrik Schwarz & Dyckerhoff, Mülheim, Germany.
RWH	R. & W. Hawthorn & Co, Newcastle-Upon-Tyne. (Later HL)
RYP	R.Y.Pickering & Co Ltd, Wishaw, Scotland.

S	Sentinel (Shrewsbury) Ltd, Battlefield, Shrewsbury.
Sabero	Hulleras de Sabero Y Anexas S.A., Sabero, Spain.
Sara	Sara & Burgess, Penryn, Cornwall.
St.Rollox	St.Rollox Works, Glasgow; Caledonian Railway.
S & H	Strachan & Henshaw Ltd, Ashton, Bristol.
Schalker	Schalker Eisenhette Maschinenfabrik G.m.b.H., 465, Gelsenkirken, Magdeburger Strasse 37, West Germany.
Schichau	F.Schichau, Maschinen-und Lokomotivfabrik, Elbing, Germany. (Now Elbtag, Poland).
SCW	I.C.I. Ltd, South Central Workshops, Tunstead, Derbyshire.
Sdn	Swindon Works, Wilts; British Railways. (Previously G.W.R.)
Siemens	Siemens Bros Ltd.
SIG	Schweizerische Industriegesellschaft, Neuhausen am Rheinfall, Switzerland.
SL	Severn-Lamb Ltd, Stratford-Upon-Avon.
SLM	Schweizerische Lokomotiv-and Maschinenfabrik, Winterthur, Switzerland.
SMH	Simplex Mechanical Handling Ltd, Elstow Road, Bedford.
Spence	W.Spence, Cork Street Foundry, Dublin.
SS	Sharp, Stewart & Co Ltd, Atlas Works, Manchester and Atlas Works, Glasgow. (Latter from 1888).
Stoke	Stoke-Works; North Staffordshire Railway.
Str	Stratford Works, London; Great Eastern Railway.
Syl	Sylvester Steel Co, Lindsay, Ontario, Canada.
TG	T.Green & Son Ltd, Leeds.
TH	Thomas Hill (Rotherham) Ltd, Vanguard Works, Hooton Road, Kilnhurst, South Yorkshire.
TH/S	Built by TH, utilising frame of S steam loco.
Thakeham	Thakeham Tiles Ltd, Thakeham, Sussex.
Thornton	Thorntons Ltd, Huddersfield.
Todd, Kitson & Laird	Todd, Kitson & Laird, Leeds.
UMM	Underground Mining Machinery Ltd, Aycliffe, Co.Durham.
Unilok	Hugo Aeckerle & Co, Hamburg, Germany; also constructed under licence by Engineering Products Ltd, Frances Street, Dublin, Eire.
Unimog	Mercedes Benz Ltd, (Unimog), West Germany.
VE	Victor Electrics Ltd, Burscough Bridge, Lancs.
VF	Vulcan Foundry Ltd, Newton-Le-Willows, Lancs.
VIW	Vulcan Iron Works, Wilkes-Barre, Penna., U.S.A.
VL	Vickers Ltd, Barrow-In-Furness.
Vollert	Hermann Vollert K.G. Maschinenfabrik, 7102 Weinsberg/Wurtt, West Germany.
VW	Vulcan-Werke Hamburg & Stettin A.G., Stettin, Germany.
WB	W.G.Bagnall Ltd, Castle Engine Works, Stafford.
WC	Wellman Cranes Ltd, Darlaston, Staffs.
WCB	Whitcomb Locomotive Co, Wilkes-Barre, Rochelle, Illinois, U.S.A.
WCI	Wigan Coal & Iron Co Ltd, Kirkless, Lancs.
WhC	Whiting Corporation, Harvey, Illinois, U.S.A.
WkB	Walker Bros (Wigan) Ltd, Wigan, Lancs.
Wkm	D.Wickham & Co Ltd, Ware, Herts.
WMD	Waggon & Maschinenbau G.m.b.H., Donauwörth, Germany.
WN	Wm. Neill & Son (St. Helens) Ltd, Bold Ironworks, St. Helens, Lancs.
Wolf	R.Wolf A.G., Abteilung Lokomotivfabrik Hagans, Erfurt, Germany.
Woolwich.	Woolwich Arsenal, London.
WR	Wingrove & Rogers Ltd, Kirkby, Liverpool.
WSO	Wellman, Smith, Owen Engineering Corporation Ltd, Darlaston, Staffs.
YE	Yorkshire Engine Co Ltd, Meadow Hall Works, Sheffield.
9E	Nine Elms Works, London; London & South Western Railway.

Listed below are the firms who use locos on tunnel and sewer contracts, etc. The locos are to be found in all parts of the Country but the details of the loco fleets are listed under the firm's main depot in the County shown below :

TITLE OF FIRM	COUNTY
Associated tunnelling Co Ltd.	Greater Manchester.
Baillie Contracting Co Ltd.	West Midlands.
Cementation Mining Ltd.	South Yorkshire.
Clugston Construction.	Humberside.
Delta Construction Ltd.	Surrey.
J.F.Donelan & Co Ltd.	Greater Manchester.
Fairclough Civil Engineering Ltd.	Staffordshire.
Frénch Kier Construction Ltd.	Norfolk.
J.J.Gallagher & Co Ltd.	West Midlands.
M.J.Gleesons (Contractors) Ltd.	Greater London.
Grant Lyon Eagre Ltd.	Humberside.
J.H.Tractors Ltd.	South Yorkshire.
Johnston Construction Ltd.	Warwickshire.
T.Kilroe & Sons (1975) Ltd.	Greater Manchester.
Lilley/Waddington Ltd.	Essex.
Martin & Co (Contractors) Ltd.	West Midlands.
Sir Robert McAlpine & Sons Ltd.	Northamptonshire.
Miller Buckley Plant Ltd.	Warwickshire.
John Mowlem & Co Ltd.	Hertfordshire.
J.Murphy & Sons Ltd.	Staffordshire.
Norwest Holst.	Merseyside.
Edmund Nuttall Ltd.	Strathclyde.
Raynesway Plant Ltd.	Derbyshire.
Sheridan Contractors Ltd.	West Midlands.
Stepney Contractors Ltd.	Humberside.
A.Streeter & Co Ltd.	Surrey.
Tarmac Construction Ltd.	Cambridgeshire.
Taylor Woodrow Plant Ltd.	Greater London.
Thyssen (Great Britain) Ltd.	Dyfed.
Tyne-Tees Tunnelling Great Britain.	Durham.
Wilson Kinmond & Marr Ltd.	Strathclyde.

SECTION 1 ENGLAND

AVON	18	ISLE OF WIGHT	103
BEDFORDSHIRE	19	KENT	103
BERKSHIRE	22	LANCASHIRE	108
BUCKINGHAMSHIRE	22	LEICESTERSHIRE	113
CAMBRIDGESHIRE	24	LINCOLNSHIRE	117
CHANNEL ISLANDS	28	MERSEYSIDE	119
CHESHIRE	28	NORFOLK	122
CLEVELAND	31	NORTHAMPTONSHIRE	126
CORNWALL	37	NORTHUMBERLAND	129
CUMBRIA	42	NORTH YORKSHIRE	130
DERBYSHIRE	48	NOTTINGHAMSHIRE	135
DEVONSHIRE	57	OXFORDSHIRE	137
DORSET	59	SALOP	140
DURHAM	60	SOMERSET	143
EAST SUSSEX	65	SOUTH YORKSHIRE	145
ESSEX	66	STAFFORDSHIRE	153
GLOUCESTERSHIRE	70	SUFFOLK	158
GREATER LONDON	72	SURREY	159
GREATER MANCHESTER	81	TYNE & WEAR	163
HAMPSHIRE	87	WARWICKSHIRE	166
HEREFORD & WORCESTER	90	WEST MIDLANDS	169
HERTFORDSHIRE	93	WEST SUSSEX	175
HUMBERSIDE	95	WEST YORKSHIRE	177
ISLE OF MAN	101	WILTSHIRE	181

AVON

BIRNBECK PIER RAILWAY, WESTON-SUPER-MARE.
Gauge : 1'3".   (   )

	THE CUB	4w-4wDM		Minirail	1954	OOU

BRISTOL INDUSTRIAL MUSEUM, PRINCES WHARF, CITY DOCK, BRISTOL.
Gauge : 4'8½".   (   )

| S3 | PORTBURY | 0-6-0ST | OC | AE | 1764 | 1917 | + |
| S9 | HENBURY | 0-6-0ST | OC | P | 1940 | 1937 | |

+ Currently under renovation elsewhere.

BRISTOL SUBURBAN RAILWAY SOCIETY, BITTON STATION.
Gauge : 4'8½".   (ST 670705)

45379		4-6-0	OC	AW	1434	1937
3		0-6-0ST	OC	FW	242	1874
	EDWIN HULSE	0-6-0ST	OC	AE	1798	1918
	LITTLETON No.5	0-6-0ST	IC	MW	2018	1922
	FONMON	0-6-0ST	OC	P	1636	1924
No.9		0-6-0T	OC	RSHN	7151	1944
		0-4-0DM		Bg/DC	2158	1941
		4wDM		RH	235519	1945
PWM 3769	B8W	2w-2PMR		Wkm	6648	1953

BRITISH GAS CORPORATION, SOUTH WESTERN REGION, SEABANK WORKS, HALLEN, near AVONMOUTH.
Gauge : 4'8½".   (ST 536820)

0-4-0DM	RH	418792	1959

COMMONWEALTH SMELTING LTD, AVONMOUTH.
Gauge : 4'8½".   (ST 522797)

5		4wDH	S	10005	1959
6		4wDH	S	10048	1960
7		4wDH	S	10023	1960

DODINGTON DEVELOPMENTS LTD, DODINGTON CARRIAGE MUSEUM & LIGHT RAILWAY,
CHIPPING SODBURY.
Gauge : 2'0".   (ST 750799)

| | DODINGTON DRAGON | 4wDM | HE | 4394 | 1952 |
| | | 4wDM | HE | 4395 | 1952 |

FISONS LTD, AVONMOUTH.
Gauge : 4'8½",   (ST 519792)

0-4-0DH	EEV	D1124	1966

IMPERIAL CHEMICAL INDUSTRIES LTD, SEVERNSIDE WORKS, SEVERN BEACH.
Gauge : 4'8½".   (ST 539828)

	IBURNDALE	0-6-0DE	YE	2725 1958
	KILDALE	0-6-0DE	YE	2741 1959

PORT OF BRISTOL AUTHORITY, AVONMOUTH DOCKS.
Gauge : 4'8½".   (ST 513784, 514787)

34	0-6-0DH	S	10148 1963
35	0-6-0DH	S	10149 1964
36	0-6-0DH	S	10150 1964
37	0-6-0DH	S	10151 1964
39	0-6-0DH	RR	10218 1965
40	0-6-0DH	RR	10219 1965
41	0-6-0DH	RR	10220 1965
42	0-6-0DH	RR	10221 1965

SOMERSET RAILWAY MUSEUM, BLEADON & UPHILL STATION. (Closed)
Gauge : 4'8½".   (ST 325578)

1338	0-4-0ST OC	K	3799 1898
W 79976	4wDMR	A.C.Cars	1958
	0-4-0T OC	HE	1684 1931
1	4wVBT VCG	S	9374 1947
	4wDM	FH	3057 1946

WESTERN FUEL CO LTD,
Filton Coal Concentration Depot.
Gauge : 4'8½".   (ST 611788)

	0-4-0DM	Bg	3410 1955

Princess Road, Wapping Wharf, Bristol.
Gauge : 4'8½".   (ST 586722)

	WESTERN PRIDE	0-6-0DM	HC	D1171 1951

BEDFORDSHIRE

JOSEPH ARNOLD & SONS LTD, SILICA SAND QUARRIES, LEIGHTON BUZZARD.
Gauge : 2'0".   RTC.   Locos are kept at :-

	Billington Road Workshops.	(SP 927241)		
	Double Arches Quarry Shed & Works.	(SP 942287, 943285)		
	4wDM	MR	4701 1936	Dsm
	4wDM	MR	4707 1936	Dsm
24	4wDM	MR	4805 1936	OOU
	4wDM	MR	7215 1938	Dsm

BEDFORDSHIRE COUNTY COUNCIL, DOVERY DOWN COUNTY PRIMARY SCHOOL, HEATH ROAD,
LEIGHTON BUZZARD.
Gauge : 2'0".   (   )

30	4wDM	MR	8695 1941	Pvd

BLUE CIRCLE INDUSTRIES LTD, DUNSTABLE DEPOT, HOUGHTON REGIS, DUNSTABLE.
Gauge : 4'8½".   (TL 015233, 016232)

		0-4-0DE		RH	425477	1959	
	PATRICIA	0-6-0DH		JF	4240017	1966	

CENTRAL ELECTRICITY GENERATING BOARD, LITTLE BARFORD POWER STATION, near ST. NEOTS.
Gauge : 4'8½".   (TL 183576)   RTC.

E.D.4	EDMUNDSONS	0-4-0ST	OC	AB	2168	1943	
No.11		0-4-0DM		AB	413	1957	

CHETTLE AND SONS, PODINGTON SCRAPYARD.
Gauge : 4'8½".   (   )

		4wDH		NB	27544	1959	Dsm

GEORGE GARSIDE (SAND) LTD, SILICA QUARRIES, DOUBLE ARCHES QUARRY SHED, LEIGHTON BUZZARD.
Gauge : 2'0".   (SP 944285, 943284)   RTC

No.13	ARKLE	4wDM		MR	7108	1937	OOU
No.31	MILL REEF	4wDM		MR	7371	1939	OOU
No.34	RED RUM	4wDM		MR	7105	1936	OOU

LEIGHTON BUZZARD NARROW GAUGE RAILWAY SOCIETY.
Gauge : 2'0".   Locos are kept at :-

Pages Park Shed.	(SP 929242)	
Stonehenge Workshops.	(SP 941275)	

No.2	PIXIE	0-4-0ST	OC	KS	4260	1922	
3	RISHRA	0-4-0T	OC	BgC	2007	1921	e
No.4	THE DOLL	0-6-0T	OC	AB	1641	1919	
No.5	ELF	0-6-0WT	OC	OK	12740	1936	
6	CARAVAN	4wDM		MR	7129	1938	
8986	PAM	4wDM		OK	8986		
8		4wDM		RH	217999	1942	
No.10	HAYDN TAYLOR	4wDM		MR	7956	1945	
11	P.C.ALLEN	0-4-0WT	OC	OK	5834	1912	
12	CARBON	4wPM		MR	6012	1930	
		4wDM		HE	3646	1946	
15	TOM BOMBADIL	4wDM		FH	2514	1941	
		4wDM		L	11221	1939	e
No.17	DAMREDUB	4wDM		MR	7036	1936	
No.21	FESTOON	4wPM		MR	4570	1929	e
22		4wDM		MR	11003	1954	
23		4wDM		MR	11298	1965	
24		4wDM		MR	11297	1965	
25		4wDM		MR	60S317	1966	
26		4wDM		MR	60S318	1966	+
43		4wDM		MR	10409	1954	
No.44	KESTREL	4wDM		MR	7933	1941	
		4wDM		MR	7214	1938	
	REDLANDS	4wDM		MR	5603	1931	a
"No.5"		4wDM		MR	5608	1931	b
R8		4wDM		MR	5612	1931	c
R7	No.131	4wDM		MR	5613	1931	d
No.6		4wDM		MR	5875	1935	b

    a - For use as mobile engine test-bed.
    b - Converted to a brake-van.
    c - In use as a weed-killer wagon.
    d - Converted to a base for a hedge-cutter.
    e - Currently elsewhere for renovation.
    + - Brake column is stamped with number 11164 in error.

LONDON BRICK CO LTD, ARLESEY WORKS.
Gauge : 2'0".   (TL 184353)   RTC.

| | | | 4wDM | | MR | 8927 | 1944 | Dsm |
| | | | 4wDM | | MR | 11312 | 1966 | Dsm |

P.J.MACKINNON, WALNUT LODGE SAWMILLS, 36, LUTON ROAD, WILSTEAD.
Gauge : 1'8".   (TL 063433)

| | | | 4wDM | | OK | 6703 | | Pvd |

R.PEARMAN, BIGGLESWADE.
Gauge : 2'0".   (   )

| | | | 4wDM | | MR | 11001 | 1956 | |

      Loco at a temporary, private, location.

PLEASURE-RAIL LTD, WHIPSNADE & UMFOLOZI RAILWAY, WHIPSNADE ZOO.
Gauge : 3'6".   (TL 004172)

| 390 | | | 4-8-0 | OC | SS | 4150 | 1896 | |
| | | | 2w-2PM | | Ford | | 1938 | |

Gauge : 2'6".

No.1		CHEVALLIER	0-6-2T	OC	MW	1877	1915	
No.2		EXCELSIOR	0-4-2ST	OC	KS	1049	1908	
No.3		CONQUEROR	0-6-2T	OC	WB	2192	1922	
No.4		SUPERIOR	0-6-2T	OC	KS	4034	1920	
	L116		4wDM		MR	5606	1931	
			0-6-0DM		JF	4160004	1951	
9		VICTOR	0-6-0DM		JF	4160005	1951	
			4wBE		WR	1393	1939	
11			4wBE		WR	1616	1940	Dsm
20			4wBE		WR	1801	1940	Dsm

SIMPLEX MECHANICAL HANDLING LTD, SIMPLEX WORKS, ELSTOW ROAD, BEDFORD.
Gauge : 3'0".   (TL 053488)

			4wDM			MR	3797	1926
				Rebuild of	MR	1363		
No.2			4wDM			MR	3965	1939

      Locos are privately preserved.
      New locos and locos under repair usually present.

VAUXHALL MOTORS LTD.
Dunstable Works.
Gauge : 4'8½".   (TL 024224)   RTC.

| | | | 4wDM | | RH | 394012 | 1956 | OOU |

Luton Works.
Gauge : 4'8½".   (TL 105205)

| 29368 | | | 4wDM | | RH | 338419 | 1954 | |

WOBURN ABBEY (NARROW GAUGE) RAILWAY, WOBURN PARK.
Gauge : 2'0".   (SP 968328)

| | DUCHESS | | 4wDM | S/O | RH | 223749 | 1944 |
| | | | 4wDM | S/O | RH | 239381 | 1946 |

BERKSHIRE

MINISTRY OF DEFENCE, ARMY DEPARTMENT, THATCHAM DEPOT.
       See Section Five for full details.

F.STAPLETON, near NEWBURY.
Gauge : 2'0".   (   )

	4wDM	HE	4556	1954	
	4wDM	RH	235729	1944	

BUCKINGHAMSHIRE

M.CAPRON, 8, MOON STREET, WOLVERTON.
Gauge : 1'11".   (SP 818409)

	4wDM	RH	183773	1937

A.COCKLIN & J.THOMAS, 40, WYE CLOSE & SHENLEY BROOK ROAD, BLETCHLEY.
Gauge : 2'0".   (SP 852345, 852341)

	4wDM	L	4228	1931
	4wDM	OK	4805	
1	4wDM	OK	6705	
	4wDM	OK	7371	
2	4wDM	OK	7600	

V.GOLDBERG, HIGH WYCOMBE.
Gauge : 1'3".   (   )

	BL	

GOODMAN BROS LTD, SCRAP DEALERS, NEW BRADWELL, WOLVERTON.
Gauge : 4'8½".   (SP 832413)

0-6-0ST OC	HL	3138	1915	OOU

MARLOW SAND & GRAVEL CO LTD, WESTHORPE PITS, LITTLE MARLOW.
Gauge : 2'0".   (SU 867870)

3	4wDM	MR	5867	1934	
4	4wDM	MR	8790	1943	
5	4wDM	MR	21283	1965	
	4wDM	MR	7176	1937	Dsm

W.H.McALPINE, FAWLEY HILL, FAWLEY GREEN, near HENLEY-ON-THAMES.
Gauge : 4'8½".   (SU 755861)

No.31		0-6-0ST IC	HC	1026 1913
29	ELIZABETH	0-4-0ST OC	AE	1865 1922
	JOYCE	4wVBT VCG	S	7109 1927 +
No.5		0-4-0DM	Bg	3027 1939

+ Currently under renovation at B.Turners, Ripley, Surrey.

QUAINTON RAILWAY SOCIETY LTD, QUAINTON ROAD STATION, near AYLESBURY.
Gauge : 4'8½".   (SP 736189, 739190)

6024	KING EDWARD II	4-6-0 4C	Sdn	1930
6989	WIGHTWICK HALL	4-6-0 OC	Sdn	1948
7200		2-8-2T OC	Sdn	1934
7715		0-6-0PT IC	KS	4450 1930
9466		0-6-0PT IC	RSH	7617 1952
(30585)	E 0314	2-4-0WT OC	BP	1414 1874
41298		2-6-2T OC	Crewe	1951
41313		2-6-2T OC	Crewe	1952
46447		2-6-0 OC	Crewe	1950
	TRYM	0-4-0ST OC	HE	287 1883
	SWANSCOMBE	0-4-0ST OC	AB	699 1891
	SYDENHAM	4wWT G	AP	3567 1895
	PUNCH HULL	0-4-0ST OC	AB	776 1896
L44		0-4-4T IC	Nea	1896
4		0-4-0F OC	AB	1477 1916
No.1	SIR THOMAS	0-6-0T OC	HC	1334 1918
	ALEXANDER	0-4-0ST OC	AB	1865 1926
No.2	ISEBROOK			
	L.N.E.R. 49	4wVBT VCG	S	6515 1926
	HORNPIPE	0-4-0ST OC	P	1756 1928
3		0-4-0ST OC	HL	3717 1928
	SCOTT	0-4-0ST OC	WB	2469 1932
	TOM PARRY	0-4-0ST OC	AB	2015 1935
	JILL	0-4-0T OC	P	1900 1936
No.1	COVENTRY No.1	0-6-0T IC	NB	24564 1939
	SWORDFISH	0-6-0ST OC	AB	2138 1941
11		4wVBT VCG	S	9366 1945
1	MILLOM	0-4-0ST OC	HC	1742 1946
7		4wVBT VCG	S	9376 1947
	LAPORTE	0-4-0F OC	AB	2243 1948
		0-4-0ST OC	P	2104 1950 +
		0-4-0ST OC	P	2105 1950 +
11	CHISLET	0-6-0ST OC	YE	2498 1951
	JAMES I	0-4-0ST OC	P	2129 1952
		0-6-0ST IC	HE	3782 1953
	JUNO	0-6-0ST IC	HE	3850 1958
	SPARTACUS	0-6-0ST IC	HE	3890 1964
	REDLAND	4wDM	KS	4428 1929
	OSRAM	0-4-0DM	JF	20067 1933
T1		4wDM	FH	2102 1937
		4wDM	FH	3765 1955
54233		4w-4wRER	GRC	1939
9040		2w-2PMR	Wkm	6963 1955
9037		2w-2PMR	Wkm	8197 1958
		2w-2PMR	Wkm	8263 1959 DsmT

+ Actually built in 1948 but plates are dated as shown.

SOUTHERN DEPOT CO. N.C.B. COAL CONCENTRATION DEPOT, GRIFFIN LANE, AYLESBURY.
Gauge : 4'8½".   (SP 804143)

(D2324)		0-6-0DM	(RSH	8183	1961
			(DC	2705	1961

TRACK SUPPLIES & SERVICES LTD, OLD WOLVERTON ROAD, OLD WOLVERTON, MILTON KEYNES.
Gauge : 4'8½".   (SP 818416)

No.1		4wDM	RH	200796	1941	OOU
		4wDM	RH	221644	1943	OOU
101	A1	4wDM	RH	224341	1944	OOU
No.1		4wDM	RH	275886	1949	OOU
		0-4-0DM	RH	319286	1953	OOU
	ARMY 401	0-4-0DH	NBL	27422	1955	OOU
		0-4-0DH	NBL	27424	1955	OOU

Gauge : 3'0".

L1		4wDM	RH	170200	1934	Dsm
L3		4wDM	RH			OOU +
L5		4wDM	RH			OOU +
		4wDM	RH	244574	1947	OOU
L7	20	4wDM	RH	244575	1947	Dsm
L2	21	4wDM	RH	252798	1947	OOU

+ These are two of 182146 1936; 186304 1937; 187058 1937.

Gauge : 2'0".

	A.M.W. 166	4wDM	RH	193987	1939	OOU
		4wDM	RH	202969	1940	Dsm
		4wDM	RH	200802	1941	OOU
	A.M.W. No.224	4wDM	RH	203020	1941	OOU
		4wDM	RH	432652	1959	OOU
		4wBE	WR	M7556	1972	OOU

Also other locos are occasionally present for repairs & resale.

TUNNEL CEMENT LTD, PITSTONE WORKS, near IVINGHOE.
Gauge : 4'8½".   (SP 932153)

		0-6-0DM	WB	3160	1959
2		4wDH	S	10159	1963
3		4wDH	RR	10264	1966

CAMBRIDGESHIRE

BLUE CIRCLE INDUSTRIES LTD, NORMAN WORKS, CHERRY HINTON, CAMBRIDGE.
Gauge : 4'8½".   (TL 481573)

4wDM		FH	3990 1962

A.B.M.BRAITHWAITE, CALDREES MANOR, ICKLETON.
Gauge : 1'6".   (TL 493438)

4-2-2	OC	Regent St.	1898

BRITISH SUGAR CORPORATION LTD, WOODSTON FACTORY, PETERBOROUGH.
Gauge : 4'8½".   TL 175976)

98						
		0-4-0DM		RH	243080	1947
		0-4-0DM		RH	390776	1956
		0-6-0DM		RH	347750	1959

CIBA-GEIGY LTD, DUXFORD.
Gauge : 4'8½".   (   )

03030					
	4wDM R/R		Unilok	2109	1980

Mr.DRAGE, SOUTH CAMBS RAILWAY, NEW BUILDINGS FARM, HEYDON, near ROYSTON.
Gauge : 4'8½".   (TL 419409)

11	13	NEWCASTLE	0-6-0ST	IC	MW	1532	1901
			0-4-0ST	OC	AB	1219	1910
44		CONWAY	0-6-0ST	IC	K	5469	1933
960236			2w-2PMR		Wkm	1519	1934

E.HAMPTON, CHURCH FARM, CHURCH LANE, FENSTANTON, ST. IVES.
Gauge : 3'0".   (TL 321687)

(ED 10)	E 9				
		4wDM		RH	411322	1958

IMPERIAL WAR MUSEUM, DUXFORD AERODROME.
Gauge : 4'8½".   (   )

10756				
	4wPM		MR	1364	1918

A.KING & SONS (METAL MERCHANTS), KNAPPETTS SCRAPYARD, SNAILWELL, near NEWMARKET.
Gauge : 4'8½".   (TL 638678)

03020	F 134 L	0-6-0DM		Sdn		1958
03012	F 135 L	0-6-0DM		Sdn		1958
	F 136 L	4wDM		RH	305315	1952

M.KNIGHT, Location Unknown.
Gauge : 75cm.   (TF 393079) ?

	2-6-2T	OC	AT	2369	1948

LONDON BRICK CO LTD.
No.2 Works, Kings Dyke, Whittlesey.
Gauge : 3'0".   (TL 239974)   RTC.

4				
	4wDM		RH		+ OOU

+ One of 182146 1936; 186304 1937; 187058 1937.

Warboys Works,
Gauge : 2'11".   (TL 311818)

	4wWE		Allens		1950

Gauge : 2'0".

3		4wDM		MR	7474	1940	OOU
		4wDM		MR	8936	1944	Dsm
2	L 114	4wDM		MR	9792	1955	
		4wDM		MR	11264	1964	
	L 115	4wDM		MR	20585	1955	OOU
		4wDM		MR	22045	1959	

PETERBOROUGH RAILWAY SOCIETY LTD. NENE VALLEY INTERNATIONAL STEAM RAILWAY.
Gauge : 4'8½".   Locos are kept at :-

| | | Wansford Steam Centre. | | | (TL 093979) | |
| | | Woodston Sugar Factory, Peterborough. | | | (TL 184979) | |

34081	92 SQUADRON		4-6-2	3C	Bton		1948
70000	BRITANNIA		4-6-2	OC	Crewe		1951
73050	CITY OF PETERBOROUGH		4-6-0	OC	Derby		1954
3052	CAR No.90		4w-4wRER				
3142	11161, 11201	4w-4w+4w-4wRER					
	11187		4w-4wRER				
3.628			4-6-0	4CC	Hen	10745	1911
1178			2-6-2T	OC	Motala	516	1914
Nr.740	D.F.D.S. DANISH SEAWAYS		2-6-4T	3C	Frichs	86	1928
80 014			0-6-0T	OC	Wolf	1228	1927
64-305			2-6-2T	OC	KrMaf	11535	1935
1697			4-6-0	OC	Nohab	2082	1944
Nr.656			0-6-0T	OC	Frichs	360	1949
1928			2-6-4T	OC	Nohab	2229	1953
	PITSFORD		0-6-0ST	OC	AE	1917	1923
	DEREK CROUCH		0-6-0ST	IC	HC	1539	1924
			0-6-0ST	OC	AE	1945	1926
	JACKS GREEN		0-6-0ST	IC	HE	1953	1939
			0-6-0ST	OC	P	2000	1942
75006	L1		0-6-0ST	IC	HE	2855	1943
1	THOMAS		0-6-0T	OC	HC	1800	1947
90432			0-4-0ST	OC	AB	2248	1948
	PERCY		4wPM		FH	2894	1944
	FRANK		4wDM		FH	2896	1944
			4wDM		RH	294268	1951
			4wDM		RH	321734	1952
	HORSA		0-4-0DH		RSHD/WB	8368	1962
VR 001			2w-2PMR		Wkm		1944

C.ROADS, HILLSIDE, WIMPOLE.
Gauge : 4'8½".   (TL 364510)

0-6-0ST OC	HC	1208	1916

RUGBY PORTLAND CEMENT CO LTD. BARRINGTON CEMENT WORKS.
Gauge : 4'8½".   (TL 396504)

7	0-4-0DE	RH	499435	1963
8	0-4-0DE	RH	499436	1963
9	4wDH	TH	186V	1967
14	4wDM	FH	3716	1955
15	4wDH	TH	127V	1963
		Rebuilt	TH	240V 1972

SEADYKE LTD. ALUMINIUM PANEL DEALERS. WISBECH.
Gauge : 4'8½".   (   )

0-4-0DM	HE	2068	1940 + Dsm

+ Retained for use as an emergency generator.

TARMAC CONSTRUCTION LTD, PLANT DEPOTS, FENGATE, PETERBOROUGH & WHARF ROAD,
off OUNDLE ROAD, PETERBOROUGH.
Gauge : 2'0".   (TL 207987,   )

430/34			4wBE	CE	5239	1967
432/24	420298		4wBE	CE	5378	1967
432/25			4wBE	CE	5481/2	1968
432/26	420299		4wBE	CE	5238/2	1966
432/29	420301		4wBE	CE	5446	1968
432/30			4wBE	CE	5481/5	1968
432/32	420302	8	4wBE	CE	5481/1	1968
432/33	420303		4wBE	CE	5481	1968
432/34	420304		4wBE	CE	5481/4	1968
432/35			4wBE	CE	5481/3	1968
432/36		2	4wBE	CE	5628	1969
432/37			4wBE	CE	5628	1969
432/38			4wBE	CE	5640	1969
432/39			4wBE	CE	5640	1969
432/40	420310		4wBE	CE	5640	1969
432/43			4wBE	CE	5866A	1971
432/44			4wBE	CE	5866B	1971
432/45	420836		4wBE	CE	5868A	1971
432/46	420837		4wBE	CE	5868B	1971
432/47	421195		4wBE	CE	5868C	1971
432/48	420314		4wBE	CE	5378	1967
432/49			4wBE	CE	5238/1	1966
432/50			4wBE	CE	5378	1967
	428001		4wBE	CE	B0156	1973
	428002		4wBE	CE	B0176	1974
	428003		4wBE	CE	B0182A	1974
	428004		4wBE	CE	B0182B	1974
	428005		4wBE	CE	B0182C	1974
	428006		4wBE	CE	B0428	1974
	428007		4wBE	CE	B0428	1974

Gauge : 1'6".

432/28			4wBE	CE	5431	1968
432/41	421194		4wBE	CE	5645	1969
432/42	420311		4wBE	CE	5783/2	1970
			4wBE	CE	5783	1970

Locos present in yards between contracts.

Mr.THORNHILL, BARNWELL JUNCTION STATION, near CAMBRIDGE.
Gauge : 4'8½".   (TL 472596)

960225	2w-2PMR	Wkm	1308	1933

E.M.WILCOX LTD, ROYCE ROAD, PETERBOROUGH.
Gauge : 4'8½".   (TL 207994)

	0-4-0DM	JF	4210143	1958	Dsm

In use as a standby generator.

CHANNEL ISLANDS

ALDERNEY RAILWAY SOCIETY, MANNEZ QUARRY, ALDERNEY.
Gauge : 4'8½".

PWM 3776	CADENZA		2w-2PMR	Wkm	6655 1953 +
PWM 3954	MARY LOU		2w-2PMR	Wkm	6939 1955
No.1	GEORGE	RLC/009025	2w-2PMR	Wkm	7091 1955
9028			2w-2PMR	Wkm	7094 1955
No.2	SHIRLEY	RLC/009029	2w-2PMR	Wkm	7095 1955
9022			2w-2PMR	Wkm	8086 1958

+ Currently under renovation at Riduna Bus Garage, Newtown.

DEPARTMENT OF THE ENVIRONMENT, PROPERTY SERVICES AGENCY, GROSNEZ FORT, ALDERNEY.
Gauge : 4'8½".

| | | 4wVBT VCG | S | 6909 1927 + |
| HO5 | MOLLY 2 | 4wDM | RH | 425481 1958 |

+ Converted to a Sand Blaster & Crane Match Truck.

MR. PALLOT, JERSEY.
Gauge : 2'0".

| 2 | 4wDM | MR | 11143 1960 | |
| | 4wDM | MR | 60S383 1969 | Dsm |

CHESHIRE

ASSOCIATED OCTEL CO LTD, STANLOW.
Gauge : 4'8½". (SJ 415767)

2	0-4-0DM	RH	313394 1952
3	0-4-0DM	RH	313396 1952
4	0-4-0DM	RH	319284 1952

BOWATERS UNITED KINGDOM PAPER CO LTD, NORTH ROAD, ELLESMERE PORT.
Gauge : 4'8½". (SJ 391788)

| 15 | 4wDH | S | 10174 1963 |
| C 930 | 4wDH | S | 10175 1964 |

B.P. CHEMICALS (U.K.) LTD, ELWORTH WORKS, SANDBACH.
Gauge : 4'8½". (SJ 729633)

| | 0-4-0DE | RH | 544997 1969 |

JOSEPH BRIERLEY & SONS LTD, SCRAP DEALERS, WILDERSPOOL CAUSEWAY, HOWLEY, WARRINGTON.
Gauge : 4'8½". (SJ 609876) RTC.

| | 0-4-0DM | JF | 22898 1940 | OOU |

BRITISH OXYGEN CO LTD, MARSH WORKS, WIDNES.
Gauge : 4'8½".   (SJ 503846)

| | ENTERPRISE | 0-6-0DH | HE | 7189 | 1970 | |

BRITISH SALT CO LTD, MIDDLEWICH.
Gauge : 4'8½".   (SJ 718643)

| (D2150) | | 0-6-0DM | Sdn | | 1960 | |
| | | 4wDH | RR | 10243 | 1966 | |

BRITISH STEEL CORPORATION, SCUNTHORPE DIVISION, LANCASHIRE WORKS,
Bewsey Road Mills, Warrington.
Gauge : 4'8½".   (SJ 602898)

	LEOPARD					
37		0-4-0DH	NB	27652	1956	
39		0-4-0DE	YE	2615	1957	Dsm
42		0-4-0DE	YE	2636	1957	
		0-4-0DE	YE	2639	1957	
	PANTHER	0-4-0DH	NB	27737	1958	

Monks Hall Mills, Athertons Quay, Warrington.
Gauge : 4'8½".   (SJ 593876)

MH6						
	LION	0-4-0DM	JF	4210129	1957	OOU
No.9		0-4-0DH	NB	27657	1957	
		0-4-0DE	YE	2683	1958	

BURMAH OIL TRADING LTD, BURMAH REFINERY, STANLOW,
Gauge : 4'8½".   (SJ 423758)   RTC.

| (D2767) | | 0-4-0DH | NB | 28020 | 1960 | |
| | | Rebuilt by | AB | | 1968 | OOU |

COUNTRY KITCHEN FOODS LTD, LINDOW MOSS, MOOR LANE, WILMSLOW.
Gauge : 2'0".   (SJ 823803)

	4wDM	L	38296	1952	Dsm
No.2	4wDM	L	40009	1954	
	4wDM	LB	52528	1961	
	4wDM	Alan Keef No.4	1979		

B.CROFT, PEAT PRODUCTS, WHITE MOSS, NURSERY ROAD, ALSAGER.
Gauge : 2'0".   (SJ 775553)   RTC.

| 10 | 4wPM | MR | 9104 | 1942 | OOU |

P.A.HANTON, FOUR WINDS, CRAUFORD ROAD, EATON, CONGLETON.
Gauge : 1'3".   (NY 871656)

| ERMINTRUDE | 4wPM | T.Stanhope |

IMPERIAL CHEMICAL INDUSTRIES LTD, DIVISIONAL ENGINEERING WORKS, RUNCORN.
Gauge : 4'8½".   (SJ 498816)

0401/1	BRUNEL	0-4-0DE	YE	2613	1956
0401/2	CAVENDISH	0-4-0DE	YE	2611	1956
0401/3	DANBYDALE	0-6-0DE	YE	2714	1958
0401/4	MARDALE	0-6-0DE	YE	2743	1959
0401/5	ESKDALE	0-6-0DE	YE	2718	1958

NEWTON	0-4-ODE	YE	2612	1956
	0-6-ODE	YE	2669	1958

Serving Castner-Kellner Works.

IMPERIAL CHEMICAL INDUSTRIES LTD, MOND DIVISION, WINNINGTON, WALLERSCOTE & LOSTOCK WORKS, NORTHWICH.
Gauge : 4'8½".   (SJ 641746, 642737 - Workshops, 649747, 682744, 684745)

	0-6-ODE	EE	1552	1948
JOHN BRUNNER	0-6-ODE	EE	1901	1951
SOLVAY	0-6-ODE	EE	1902	1951
JOULE	0-6-ODE	EE	1903	1951
PERKIN	0-6-ODE	EE	1904	1951
FARADAY	0-4-ODE	RH	402803	1957
KELVIN	0-4-ODE	RH	402807	1957
RUTHERFORD	0-4-ODE	RH	412710	1957
NEWCOMEN	0-4-ODE	RH	416207	1957
F.A.FREETH	0-4-ODE	RH	416213	1957
LUDWIG MOND	6wDE	GECT	5578	1980
	4wBE	Vollert		1980
	4wBE	Vollert		1980

IMPERIAL CHEMICAL INDUSTRIES LTD, MOND DIVISION, WINNINGTON WORKS CRYSTAL PLANT.
Gauge : 2'6".   (SJ 643749)

	4wBE	WR	c1948	
	4wBE	WR	K7070	1970

MACCLESFIELD CORPORATION, ENGINEERS DEPARTMENT STORE, MACCLESFIELD.
Gauge : 2'0".   (SJ 920739)

	4wPM	MR	7033 1936	OOU

MANCHESTER SHIP CANAL CO LTD, ELLESMERE PORT & STANLOW.
Gauge : 4'8½".   (SJ 399777, 419763)

3001	0-6-ODH	S	10144	1963
3002	0-6-ODH	S	10145	1963
3003	0-6-ODH	S	10146	1963
3004	0-6-ODH	S	10147	1963
3005	0-6-ODH	S	10162	1963
CE 9604	4wDM	Robel		

Locos return to Mode Wheel Workshops, Greater Manchester for repairs, etc.

MIDLAND ROLLMAKERS LTD, CREWE.
Gauge : 4'8½".   (SJ 715543)

GRAZEBROOK	4wDM	RH	279597	1949
	4wDM	RH	495994	1963

MINISTRY OF DEFENCE, ROYAL ORDNANCE FACTORY, RADWAY GREEN, ALSAGER.
Gauge : 4'8½".   (SJ 784545)

61/50011	TOAD	0-4-ODH	JF	4220015 1962

Also uses M.O.D.,A.D. locos; for details see Section Five.

SHELL U.K. OIL LTD, STANLOW & THORNTON-LE-MOORS.
Gauge : 4'8½".   (SJ 425766, 432761, 440758)

1		0-4-ODH	JF	4220045	1967
2		0-4-ODH	S	10065	1961
3		4wDH	TH	235V	1971
4		0-4-ODH	TH	160V	1966
5		4wDH	TH	220V	1970
6		0-4-ODH	JF	4220044	1967
7		4wDH	TH	236V	1971
8		4wDH	TH	288V	1980
9		4wDH	TH	287V	1980
10		4wDH	TH	234V	1971

T.A.C. CONSTRUCTIONAL MATERIALS LTD, BOLD WORKS, DERBY ROAD, WIDNES.
Gauge : 4'8½".   (SJ 522875)

29467		4wDM	FH	3866	1958
28076	POILITE II	4wDM	FH	3953	1961

U.K.F. FERTILISERS LTD, FERTILISER FACTORY, INCE MARSHES.
Gauge : 4'8½".   (SJ 472765)

12082		0-6-ODE	Derby		1950
		0-6-ODH	EEV	D1233	1968	OOU
		0-6-ODH	EEV	D3986	1970

CLEVELAND

ACCREDITED PROCESSED METALS LTD, SCRAP MERCHANTS, PARADINE WORKS, CARGO FLEET, MIDDLESBROUGH.
Gauge : 4'8½".   (NZ 506205)

	4wDM	RH	279599	1950
	4wDM	RH	411318	1957

Also other locos for scrap occasionally present.

BLACKETT, HUTTON & CO LTD, STEELFOUNDERS, GUISBOROUGH.
Gauge : 4'8½".   (NZ 612155)

-	4wDM	RH	265617	1948

BRITISH CHROME & CHEMICALS LTD, INDUSTRIAL CHEMICAL DIVISION, EAGLESCLIFFE CHEMICAL WORKS, URLAY NOOK.
Gauge : 4'8½".   (NZ 402146)

No.1	FD 62/1	4wDM	FH	3822	1956
No.2	FD 62/20	4wDM	FH	3808	1956

BRITISH STEEL CORPORATION, REDPATH ENGINEERING LTD, STRUCTURAL DIVISION, TEESSIDE BRANCH, BRITANNIA CONSTRUCTIONAL WORKS, MIDDLESBROUGH.   (Closed)
Gauge : 4'8½".   (NZ 482210, 484219, 486209)

3	PR No.003	MSC 31	4wDM	RH	299107	1950	Dsm
			4wDM	FH	3942	1960	OOU

BRITISH STEEL CORPORATION, REDPATH DORMAN LONG LTD, MANUFACTURING DIVISION, TEESSIDE BRANCH, TEESSIDE ENGINEERING WORKS, NORTH ORMESBY, MIDDLESBROUGH.
Gauge : 4'8½".   (NZ 508203)

TEES-SIDE No.6	4wDM	FH	3933	1960	
TEES-SIDE No.7	4wDH	TH	105V	1961	
TEES-SIDE No.9	4wDM	FH	3935	1960	

BRITISH STEEL CORPORATION, TEESSIDE DIVISION, TEESSIDE WORKS.
Apprentice Training Centre, Middlesbrough.
Gauge : 4'8½".   (NZ 546211)

42		0-4-0DH	JF	4220020	1961	OOU

Cargo Fleet Works, Middlesbrough.
Gauge : 4'8½".   (NZ 521203)

No.47	0-4-0DH	JF	4220040	1967	Dsm
No.50	0-4-0DH	JF	4220043	1968	
51	4wDH	TH	225V	1970	
52	4wDH	TH	226V	1970	

Cleveland Works, South Bank and Lackenby Works, Middlesbrough.
Gauge : 4'8½".   (NZ 546211, 563223)

	MC 1	0-4-0DC	Cleveland		1958	OOU a
	MC 2	0-4-0DC	Cleveland		1958	OOU a
	MC 3	0-4-0DC	Cleveland		1958	OOU a
		4wWE	AB/MV		1948	OOU
No.2		4wWE	AB/MV		1948	Dsm
1		4wWE	Cleveland		c1967	OOU b
2		4wWE	Cleveland		1974	b
T 3640	BARRIER WAGON No.1	4wDH	Cleveland		1972	Dsm c
T 3585	BARRIER WAGON No.2	4wDH	Cleveland		1972	Dsm c
9111/161	STANTON No.51	0-4-0DE	YE	2886	1961	OOU
32	CHRIS MOODY	0-6-0DH	S	10032	1960	
54		0-6-0DH	S	10054	1961	
55		0-6-0DH	TH/S	109C	1961	
56		0-6-0DH	S	10056	1961	
64		0-4-0DH	S	10064	1961	
68		0-6-0DH	S	10068	1961	
78		0-6-0DH	S	10078	1961	
101		0-6-0DH	S	10101	1962	
102		0-6-0DH	S	10102	1962	
105		0-4-0DH	S	10105	1962	
155		0-8-0DH	S	10136	1962	Dsm
166		0-6-0DH	S	10166	1963	
167		0-6-0DH	S	10167	1964	
168		0-6-0DH	S	10168	1964	
201		4wDH	TH	201V	1969	OOU
202		4wDH	TH	202V	1969	OOU
203		4wDH	TH	203V	1969	OOU
205		4wDH	TH	205V	1969	OOU d
206		4wDH	TH	206V	1969	OOU d
(207)		4wDH	TH	207V	1969	OOU d
(209)		4wDH	TH	209V	1969	OOU d
210		4wDH	TH	210V	1969	OOU d
211		4wDH	TH	211V	1969	OOU d
210		0-6-0DH	RR	10210	1964	
211		0-6-0DH	RR	10211	1964	
224		0-6-0DH	RR	10224	1965	
225		0-6-0DH	RR	10225	1965	
234		0-6-0DH	RR	10234	1965	
292		0-6-0DH	RR	10292	1971	

251	WALTER URWIN	6wDE		GECT	5414	1976
252	BOULBY	6wDE		GECT	5415	1976
253	ESTON	6wDE		GECT	5416	1976
254	BROTTON	6wDE		GECT	5417	1976
255	LIVERTON	6wDE		GECT	5418	1976
256	LINGDALE	6wDE		GECT	5425	1977
257	NORTH SKELTON	6wDE		GECT	5426	1977
258	GRINKLE	6wDE		GECT	5427	1977
259	CARLIN HOW	6wDE		GECT	5428	1977
260	ROSEDALE	6wDE		GECT	5429	1977
261	STAITHES	6wDE		GECT	5430	1977
262	LOFTUS	6wDE		GECT	5431	1977
263	LUMPSEY	6wDE		GECT	5432	1977
264	PORT MULGRAVE	6wDE		GECT	5461	1977
265	ROSEBERRY	6wDE		GECT	5462	1977
266	SHERRIFFS	6wDE		GECT	5463	1977
267	SLAPEWATH	6wDE		GECT	5464	1977
268	KIRKLEATHAM	6wDE		GECT	5465	1977
269	LONGACRES	6wDE		GECT	5466	1977
270	CHALONER	6wDE		GECT	5467	1977
271	GLAISDALE	6wDE		GECT	5469	1978
272	GROSMONT	6wDE		GECT	5470	1978
273	KILTON	6wDE		GECT	5471	1978
274	ESKDALESIDE	6wDE		GECT	5472	1978
275	RAITHWAITE	6wDE		GECT	5473	1978
276	SPAWOOD	6wDE		GECT	5474	1978
277	WATERFALL	6wDE		GECT	5475	1978
1		4wDM		Robel		1980
LM21	REDCAR 2	4wDM		Robel		1980

a) Locos built incorporating the frames of 0-4-0ST/OC ex Dorman Long Ltd.
b) Rebuilds of 4wDH TH 128V/1963 and S 10004/1959 respectively; used at
   Cleveland No.3 Billet Mill. (Closed)
c) Rebuilds of 4wDH S 10027/1960 and S 10080/1960 respectively.
d) Slave units formerly used with locos TH 201V to 203V.

Hartlepool Works, Hartlepool.
Gauge : 4'8½".   (NZ 511286)

36	4wDH		TH	222V	1970
37	4wDH		TH	223V	1970
38	4wDH		TH	224V	1970
450	4wDH		TH	231V	1971
451	4wDH		TH	232V	1971
452	4wDH		TH	233V	1971
453	4wDH		TH	221V	1970
455	4wDH		TH	258C	1975
		Rebuild of	S	10001	1959
456	4wDH		TH	259C	1975
		Rebuild of	S	10041	1960
457	4wDH		TH	260C	1975
		Rebuild of	S	10011	1959
	0-4-0WE		GB	2937	1960
	4wWE		GB	420306	1972

Redcar Coke Ovens, Redcar.
Gauge : 4'8½".   (NZ 562257)

1	4wWE		GB	420355/1	1976
2	4wWE		GB	420355/2	1976
3	4wWE		GB	420408	1977

<u>Skinningrove Works, Carlin How, Saltburn-By-The-Sea.</u>
Gauge : 4'8½".   (NZ 708194)

1	0-4-0DH	S	10125 1963	
2	0-4-0DH	S	10126 1963	
3	0-4-0DH	S	10127 1963	
7	0-4-0DH	S	10131 1963	
9	0-4-0DH	S	10133 1963	Dsm
44	0-6-0DH	S	10081 1961	
45	0-6-0DH	S	10080 1961	

<u>South Bank Coke Ovens.</u>
Gauge : 4'8½".   (NZ 536214)

1	4wWE	WC	1504 1968
	4wWE	WC	1505 1968

<u>CENTRAL ELECTRICITY GENERATING BOARD.</u>
<u>Hartlepool Power Station, Seaton Carew.</u>
Gauge : 4'8½".   (   )

4wDH	RH	544996 1968

<u>North Tees Power Station, Haverton Hill.</u>
Gauge : 4'8½".   (NZ 478214)

2	4wDH	S	10003 1959

<u>CLEVELAND COUNTY COUNCIL, PRESTON PARK & MUSEUM, EAGLESCLIFFE.</u>
Gauge : 4'8½".   (NZ 430158)

0-4-0VBT VCG	HW	21 1870

<u>CLEVELAND POTASH LTD, BOULBY MINE, LOFTUS.</u>
Gauge : 4'8½".   (NZ 763183)

HARWOODDALE	0-6-0DE	YE	2724 1958

<u>GEORGE COHEN, SONS & CO LTD, COBORN WORKS, CARGO FLEET.</u>
Gauge : 4'8½".   (NZ 510206)   RTC.

4wDH	S	10017 1960	OOU
0-4-0DM	JF	4210016 1949	OOU
4wDM	FH	4011 1966	OOU

Also other locos for scrap occasionally present.

<u>C.HERRINGS & SONS LTD, SCRAP MERCHANTS, HARTLEPOOL.</u>
Yard (NZ 515307) with locos for scrap occasionally present.

<u>IMPERIAL CHEMICAL INDUSTRIES LTD, AGRICULTURAL DIVISION, BILLINGHAM.</u>
Gauge : 4'8½".   (NZ 475228)

3	COMMONDALE	0-6-0DE	YE	2666 1957
6	FARNDALE	0-6-0DE	YE	2719 1958
11	LINGDALE	0-6-0DE	YE	2742 1959
		6wDH	TH	296V 1981

IMPERIAL CHEMICAL INDUSTRIES LTD, WILTON WORKS, MIDDLESBROUGH.
Gauge : 4'8½".   (NZ 564218)

	GUISBOROUGH	0-6-0DE	EE	1553	1948	
(07005)	LANGBAURGH	0-6-0DE	RH	480690	1962	
(07011)		0-6-0DE	RH	480696	1962	
		2w-2DMR	Wkm	7591	1957	
		2w-2PMR	Wkm	7603	1957	DsmT

MINISTRY OF DEFENCE, NAVY DEPARTMENT, EAGLESCLIFFE SPARE PARTS CENTRE.
Gauge : 4'8½".   (NZ 410148)

	4wDM R/R	Unimog	2660	1978

MONSANTO CHEMICALS LTD, SEAL SANDS, near GREATHAM.
Gauge : 4'8½".   (NZ 535241)

GM 245	0-6-0DH	EEV	3870	1969

NORTHERN MACHINE TOOL CO, NORTH ORMESBY, MIDDLESBROUGH.
Gauge : 4'8½".   (NZ 487226)   RTC.

	4wDM	FH	3896	1959	OOU

E.PEARSON & SONS LTD, MIDDLESBROUGH.
Yard (NZ 492214) with locos for scrap or resale occasionally present.

W.G.READMAN LTD, IRON & SCRAP MERCHANTS, REDSTEEL WORKS, MIDDLESBROUGH.
Gauge : 4'8½".   (NZ 511207)

	4wDM	RH	425482	1958	OOU

RIBBLESDALE CEMENT LTD, FORTY FOOT ROAD, MIDDLESBROUGH.
Gauge : 4'8½".   (NZ 487207)

	0-4-0DE	RH	312989	1952

SALTBURN MINIATURE RAILWAY, VALLEY GARDENS, SALTBURN.
Gauge : 1'3".   (NZ 666215)

PRINCE OF WALES	0-6-0DE S/O	H.N.Barlow	1953

SHELL U.K. OIL, TEESPORT REFINERY, GRANGETOWN, MIDDLESBROUGH.
Gauge : 4'8½".   (NZ 556231)

2	BN 1364	4wDE	Moyse	1364	1976
3	BN 1365	4wDE	Moyse	1365	1976
		4wDE	Moyse	1464	1979

Also locos hired from B.R. Capital Stock.

T.SMITH & SON, SCRAP METAL PROCESSORS, LONGHILL INDUSTRIAL ESTATE, HARTLEPOOL.
Gauge : 4'8½".   (NZ 513313)

3	M.S.C. 31	P.R. No.003	4wDM	FH	3942	1960	OOU

SMITHS DOCK CO LTD, SOUTH BANK, CLEVELAND.
Gauge : 4'8½".   (NZ 530216)

	0-4-0DM	JF	4200018	1947

STEETLEY REFRACTORIES LTD. MAGNESIA DIVISION. PALLISER WORKS, HARTLEPOOL.
Gauge : 4'8½".   (NZ 509352)

DL2		0-4-0DM	HC	D1346	1965	
		0-4-0DH	HE	7425	1981	

TEES & HARTLEPOOL PORT AUTHORITY.
Conservancy Commission, Grangetown.
Gauge : 4'8½".   (NZ 546232)

No.1		0-4-0DH	S	10137	1962	
No.2		0-4-0DH	S	10170	1964	
No.3		0-4-0DH	S	10208	1965	OOU
4	(D2024)	0-6-0DM	Sdn		1958	
5	(D2023)	0-6-0DM	Sdn		1958	
No.6		0-6-0DH	RR	10215	1965	
	(D2205)	0-6-0DM	(VF	D212	1953	
			(DC	2486	1953	
		2w-2PMR	Wkm	(6607	1953?)	

Engineers Department, Middlesbrough Dock.
Gauge : 4'8½".   (NZ 497206, 505206)

2		0-4-0DM	(RSH	7925	1959	
			(DC	2592	1959	

Tees Dock Potash Terminal, Lackenby Dock, Grangetown.
Gauge : 4'8½".   (NZ 549235)

7	GLAISDALE	0-6-0DE	YE	2723	1958	

TEES STORAGE CO LTD, CLEVELAND DOCKYARD, MIDDLESBROUGH.
Gauge : 4'8½".   (NZ 508207)

		4wDM	FH	3958	1961	

T.J.THOMSON & SON LTD, MILLFIELD SCRAP WORKS, STOCKTON.
Gauge : 4'8½".   (NZ 438193)

2		4wDM	RH	321735	1952	
56		4wDM	RH	338424	1955	OOU
34		0-4-0DE	RH	381757	1955	OOU
		0-6-0DM	RH	395303	1956	
33	BOYLE	0-4-0DE	RH	408309	1957	OOU
No.198	ELIZABETH	0-4-0DE	RH	421436	1958	
82		4wDM	RH	425485	1959	OOU
87		4wDM	RH	463152	1961	OOU
	HELEN	4wDH	TH	264V	1976	

Also other locos for scrap occasionally present.

THOMAS TURNBULL & SONS LTD, METAL MERCHANT, VULCAN IRONWORKS,
THORNABY PLACE, THORNABY.
Gauge : 4'8½".   (NZ 449181)

	4wDM	RH	312434	1951	

Also other locos for scrap occasionally present.

VULCAN MATERIALS COMPANY (U.K.) LTD. HARTLEPOOL TINPLATE WORKS, LONGHILL, HARTLEPOOL.
Gauge : 4'8½".   (NZ 513308)

	0-4-0DM		JF	4210136	1958	
7900	0-4-0DM		RSH	7900	1958	

R.G.WALL, 9, SERPENTINE ROAD, HARTLEPOOL.
Gauge : 1'6".   (NZ 494331)

	0-4-0VBT	Rebuilt by R.G.Wall

CORNWALL

A.R.C. (SOUTH WESTERN) LTD.
Penlee Quarries, Newlyn.
Gauge : 2'0".   (SW 467281)   RTC.

	0-4-0WT	OC	Freud	73	1901	Pvd

West Of England Quarry, St. Keverne.
Gauge : 2'0".   (SW 809217)   RTC.

	4wDM		L	30947	1947	Dsm
	4wDM		LB	51509	1960	Dsm

BIRD PARADISE, HAYLE.
Gauge : 1'3".   (SW 555365)

	CHOUGH	0-4-0WTT	OC	W.V.O.Heiden	1968
No.3	ZEBEDEE	4wDM		L 10180	1938
6	PRINCESS ANNE	4-6-2DE	S/O	H.N.Barlow	1962

N.BOWMAN, EGLOSKERRY.
Gauge : 60cm.   (   )

	LILIAN	0-4-0ST	OC	HE	317	1883
2		4wDM		MR	5646	1933

CORNISH STEAM LOCOMOTIVE SOCIETY LTD, IMPERIAL KILN, BUGLE, near ST.AUSTELL.
Gauge : 4'8½".   (SX 013583)

		0-4-0ST	OC	P	1611	1923
	DOCKYARD No.19	0-4-0ST	OC	WB	2962	1950
79815	ALFRED	0-4-0ST	OC	WB	3058	1953
		0-4-0F	OC	WB	3121	1957

CORNWALL MINING SERVICES LTD, POOL INDUSTRIAL ESTATE, REDRUTH.
Gauge : 2'0".   (   )

	0-4-0BE		WR	7849	1975

DELABOLE SLATE (1977) LTD, LOWER PENGELLY, DELABOLE, CAMELFORD.
Gauge : 1'11".   (SX 075835)

		4wDM		MR	3739	1925	Pvd

E.C.C. PORTS LTD, FOWEY JETTIES.
Gauge : 4'8½".   (SX 125530)

D3452	73601	0-6-0DE		Dar		1957	
3476	73603	0-6-0DE		Dar		1957	
D3497		0-6-0DE		Don		1957	Dsm

ENGLISH CLAYS LOVERIN POCHIN & CO LTD, ROCKS WORKS, near BUGLE.
Gauge : 4'8½".   (   )

		4wDH		S	10029	1960	

G.J.A.EVANS, 'THE INNY VALLEY RAILWAY', TRECARRELL MILL, TREBULLET, LAUNCESTON.
Gauge : 1'10 3/4".   (SX 320772)

	VELINHELI	0-4-0ST	OC	HE	409	1886	
4	SAN JUSTO	0-4-2ST	OC	HC	639	1903	
	SYBIL	0-4-0ST	OC	WB	1760	1906	
		4wDM		FH	1896	1935	DsmT
No.3		4wDM		MR	9546	1950	

FALMOUTH SHIPREPAIR LTD, FALMOUTH DOCKS.
Gauge : 4'8½".   (SW 822324)   RSW.

No.3		0-4-0ST	OC	HL	3597	1926	
5		0-4-0ST	OC	HC	1632	1929	Dsm
129		0-4-0DH		S	10129	1963	

GEEVOR TIN MINES LTD, PENDEEN, near ST. JUST.
Gauge : 1'6".   (SW 375346)   (Locos are used mainly underground.)

1	0-4-0BE	Geevor		1973
2	0-4-0BE	Geevor		1950
3	0-4-0BE	Geevor		1950
4	0-4-0BE	Geevor		1952
5	0-4-0BE	Geevor		
6	0-4-0BE	Geevor		
7	0-4-0BE	Geevor		1954
8	0-4-0BE	Geevor		
9	0-4-0BE	Geevor		
10	0-4-0BE	Geevor		
11	0-4-0BE	Geevor		
12	0-4-0BE	Geevor		
13	0-4-0BE	Geevor		1973
14	0-4-0BE	Geevor		
15	0-4-0BE	Geevor		1973
	0-4-0BE	WR	6130	1959
	0-4-0BE	WR	6402	1961
1	4wBE	WR	H6583	1968
2	4wBE	WR	K6916	1970
3	4wBE	WR	L7496	1971
4	4wBE	WR	K6915	1970
5	4wBE	WR	L7495	1971
	0-4-0BE	WR		
	0-4-0BE	WR		

		4wBE			CE	5514	1968	
		4wBE			CE	5623	1969	
		4wBE			CE	5712	1969	
No.2		4wBE			CE	5739	1970	
		4wBE			CE	5764	1970	
		4wBE			CE	BO485	1975	
		4wBE			CE	BO485	1975	
		4wBE			CE	BO1592A	1978	
		4wBE			CE	BO1592B	1978	

LAPPA VALLEY RAILWAY, ST.NEWLYN EAST, near NEWQUAY.
Gauge : 1'3". (SW 839564)

No.4	MUFFIN	0-6-0	OC	Berwyn		1967
	ZEBEDEE	0-6-2T	OC	SL	34	1974
	POOH	4wDM		L	20698	1942
	LAPPA LADY	4w-4wDER		Minirail		1965
		4w-4PM		E.Booth		1975

PENTEWAN SANDS CARAVAN LTD, PENTEWAN. (Closed)
Gauge : 2'6". (SX 020471)

		4wDM		RH	189992	1938	OOU
		4wDM		RH	244558	1946	OOU

POLDARK MINING CO LTD, WENDRON FORGE MUSEUM, WENDRON, near HELSTON.
Gauge : 4'8½". (SW 683315)

	0-4-0ST	OC	P	1530	1919
POLDARK MINING CO LTD 6	0-4-0ST	OC	P	1530	1919

D.PREECE, PENGALLY FARM, TAVISTOCK ROAD, CALLINGTON,
Gauge : 2'0". (SX 364698)

No.1	SAMSON	4wDM		FH	1887	1934
No.1		4wDM		RH	186318	1937
		0-4-0DM	S/O	Bg	3235	1947

L.M.ROBERTSON, TRAGO MILLS, TWO WATERS FOOT, DOBWALLS, near LISKEARD.
Gauge : 1'3". (SX 177647)

1	PIONEER	4wPM		c1939

ST. AUSTELL CHINA CLAY MUSEUM LTD, WHEAL MARTYN MUSEUM, CARTHEW, near ST. AUSTELL.
Gauge : 4'8½". (SX 004555)

	JUDY	0-4-0ST	OC	WB	2572	1937

Gauge : 4'6".

	LEE MOOR No.1	0-4-0ST	OC	P	783	1899

ST. PIRRAN LTD, PENDARVES MINE, LITTLE PENDARVES, CAMBORNE.
Gauge : 60cm.   (SW 647385)   (Locos used mainly underground.)

1		4wBE	CE	5554/1	1968
2		4wBE	CE	5554/2	1968
3		4wBE	CE	5554/3	1968
5		4wBE	CE	5728	1969
6		4wBE	CE	5876	1971
7		4wBE	CE	5876	1971
8		4wBE	CE	5923	1972
9		4wBE	CE	5932	1972
10		4wBE	CE	B0163B	1973
11		4wBE	CE	B0163A	1973

SOUTH CROFTY LTD, POOL, near CAMBORNE.
Gauge : 1'10".   (SW 664409, 668413)

1		0-4-0BE	WR	Rebuilt South Crofty	
2		0-4-0BE	WR	Rebuilt South Crofty	
3		0-4-0BE	WR	Rebuilt South Crofty	
4		0-4-0BE	WR	Rebuilt South Crofty	
5		0-4-0BE	WR	Rebuilt South Crofty	
6		0-4-0BE	WR	Rebuilt South Crofty	
7		0-4-0BE	WR	Rebuilt South Crofty	
8		0-4-0BE	WR	Rebuilt South Crofty	
9		0-4-0BE	WR	Rebuilt South Crofty	
10		0-4-0BE	WR	Rebuilt South Crofty	
11		0-4-0BE	WR	Rebuilt South Crofty	
12		0-4-0BE	WR	Rebuilt South Crofty	
13		0-4-0BE	WR	Rebuilt South Crofty	
14	171501	0-4-0BE	WR	Rebuilt South Crofty	
15		0-4-0BE	WR	Rebuilt South Crofty	
16		0-4-0BE	WR	Rebuilt South Crofty	
17		0-4-0BE	WR	Rebuilt South Crofty	
18		0-4-0BE	WR	Rebuilt South Crofty	
19		0-4-0BE	WR	Rebuilt South Crofty	
20		0-4-0BE	WR	Rebuilt South Crofty	
21		0-4-0BE	WR	Rebuilt South Crofty	
22		0-4-0BE	WR	Rebuilt South Crofty	
23		0-4-0BE	WR	Rebuilt South Crofty	
24		0-4-0BE	WR	Rebuilt South Crofty	
25		0-4-0BE	WR	Rebuilt South Crofty	
26		0-4-0BE	WR	Rebuilt South Crofty	
27		0-4-0BE	WR	Rebuilt South Crofty	
28		0-4-0BE	WR	Rebuilt South Crofty	
29		0-4-0BE	WR	Rebuilt South Crofty	
30		0-4-0BE	WR	Rebuilt South Crofty	
31		0-4-0BE	WR	Rebuilt South Crofty	
32		0-4-0BE	WR	Rebuilt South Crofty	
33		0-4-0BE	WR	Rebuilt South Crofty	
34		0-4-0BE	WR	Rebuilt South Crofty	
35		0-4-0BE	WR	Rebuilt South Crofty	
36		0-4-0BE	WR	Rebuilt South Crofty	
37		0-4-0BE	WR	Rebuilt South Crofty	
38		0-4-0BE	WR	Rebuilt South Crofty	
39		0-4-0BE	WR	Rebuilt South Crofty	
40		0-4-0BE	WR	Rebuilt South Crofty	
01		4wBE	CE	B0960	1976
02		4wBE	CE	1524	1977
03		4wBE	CE	1524/2	1977
04		4wBE	CE	B01524	1977
05		4wBE	CE	1557A	1977
06		4wBE	CE	B1557C	1977

07		4wBE	CE	1557B	1977
08		4wBE	CE	1810A	1978
09		4wBE	CE	1810B	1978
010		4wBE	CE	B1827	1978
		4wBE	CE	B2930	1981
		4wBE	CE	B2930	1981
		4wBE	CE	B2930	1981
		4wBE	CE	B2930	1981
		4wBE	CE	B2930	1981
		4wBE	CE	B2930	1981
		4wDM	HE	6342	1970
		4wDM	HE	7083	1971
		4wDM	HE	7084	1972
		4wDM	HE	7087	1972
		4wDM	HE	7273	1972
		4wDM	HE	7320	1973
		4wDM	HE	7516	1977

Principally underground; battery locos are brought to the surface at 1500 hours daily for battery charging.

The WR locos are continually being rebuilt with both original and newly fabricated parts. The original WR locos were :-
4309-10 1950, 4820 1952, 4879 1952, 5314 1955, 5843 1956, 5927 1957, 5928-30 1957, 6303 1960, 6393 1961, C6712 1963, 6880 1964, F7029-30 1966, F7113-15 1966, J7293 1970, J7373-76 1969, L7526-31 1971, L7532 1972, L7533 1971, 7719-21 1974, N7833-39 1974.

TOLGUS TIN (CORNWALL) LTD, TOLGUS TIN MINE MUSEUM, PORTREATH ROAD, REDRUTH.
Gauge : 2'2".   (SW 689443)

		4wDM	RH	371547	1954

C.WARE, SCRAP DEALER, CARHARROCK, near ST. DAY.
Gauge : 4'8½".   (SW 741417)

TO 9362		4wDM	RH	349041	1953	OOU

WHEAL CONCORD LTD, BLACKWATER, near TRURO.
Gauge : 2'0".

	0-4-0BE	WR		Dsm
	BE	WR		
	BE	WR		
	BE	WR		

WHEAL JANE LTD, CLEMO'S SHAFT, BALDHU, near TRURO.
Gauge : 2'0".   (SW 772427)   Used mainly underground.

1		4wBE	CE	5512/1	1968
2		4wBE	CE	5512/2	1968
4		4wBE	CE	5688/2	1969
5	SPITFIRE	4wBE	CE	5766	1970
7		4wBE	CE	5839A	1971
		4wBE	CE	5839B	1971
10		4wBE	CE	5839D	1971
11		4wBE	CE	5918	1972
		4wBE	CE	5946	1972
		4wBE	CE	5946/1	1972
		4wBE	CE	5957/1	1972
		4wBE	CE	5957/2	1972
17		4wBE	CE	B0139A	1973
18		4wBE	CE	B0139B	1973
		4wBE	CE	B0174	1974

```
      20                        4wBE              CE   BO466/1 1976
    No.21                       4wBE              CE   BO466/2 1976
                                4wBE              CE    BO466  1976
```

CUMBRIA

ALBRIGHT & WILSON LTD, MARCHON CHEMICAL WORKS, KELLS, WHITEHAVEN.
Gauge : 4'8½".   (NX 966160)
```
                                4wDH              S    10085 1961  OOU
    10086                       4wDH              S    10086 1961
                                0-4-ODH           RR   10206 1965
```

A.R.C.(NORTHERN) LTD, SHAP BECK LIMEWORKS.
Gauge : 4'8½".   (NY 554181)   RTC.
```
              ORMSBY            0-4-ODM           JF   22077 1938  Pvd
```

AYLE COLLIERY CO LTD, AYLE EAST DRIFT, ALSTON.
Gauge : 2'0".   (NY 728498, 731499)
```
                                4wDM              HE   3496 1947
                                4wBE              GB   2382 1951  OOU
                                0-4-ODM           HE   4979 1955  Dsm
                                4wDM              HE   4991 1955
```

BARROW-IN-FURNESS EDUCATION DEPARTMENT, GEORGE HASTWELL TRAINING CENTRE,
ABBEY ROAD, BARROW-IN-FURNESS.
Gauge : 4'8½".   (SD 208711)
```
    No.7      "CHLOE"           0-4-OST  IC       SS   1435 1863
```

BRITISH FUEL CO, CARLISLE COAL CONCENTRATION DEPOT, LONDON ROAD, CARLISLE.
Gauge : 4'8½".   (NY 417550)
```
      24  D2004  LANCELOT       0-6-ODM           HC   D851 1955
```

BRITISH GYPSUM LTD.
Cocklakes Works, near Cumwhinton.
Gauge : 4'8½".   (NY 459510)   RTC.
```
              F.G.F.            0-4-ODH           AB    552 1968  OOU
```

Thistle Plaster Works, Kirkby Thore.
Gauge : 4'8½".   (NY 646268)
```
              J.W.H.           0-4-ODH            AB    558 1970
              B.G.            0-4-ODH             AB    559 1970
```

BRITISH NUCLEAR FUELS LTD, WINDSCALE FACTORY, SELLAFIELD.
Gauge : 4'8½".   (NY 025034)

| 1 | 4300/B/0001 | 0-4-0ST OC | P | 2027 1942 |

BRITISH STEEL CORPORATION, B.S.C.REFRACTORIES GROUP, MICKLAM WORKS,
LOWCA, near WHITEHAVEN. (Closed)
Gauge : 2.'6".   (NX 986227)   (Underground)

| | 4wBE | WR | |

BRITISH STEEL CORPORATION, CUMBRIA DIVISION.
Barrow Works, Barrow-In-Furness.
Gauge : 4'8½".   (SD 191700)

| 1 | 507/34 | 0-4-0DE | YE | 2802 1960 |
| 2 | 507/35 | 0-4-0DE | YE | 2803 1960 |

Distington Foundry & Engineering Works, Workington.
Gauge : 4'8½".   (NX 985291)

	433/16			
No.203		0-6-0DE	YE	2603 1955	OOU
	433/13	0-4-0DE	YE	2685 1958	OOU
		0-4-0DE	YE	2857 1961	
	(149/30)	4wDH	CE	5846 1971	
		4wBE	GB	2720 1956	

Workington Works, Moss Bay, Workington.
Gauge : 4'8½".   (NX 985291, 990278)

No.201			0-4-0DE	YE	2657 1957	OOU
			0-4-0DE	YE	2656 1957	Dsm
No.272			0-4-0DE	YE	2707 1958	
No.273			0-4-0DE	YE	2711 1960	OOU
No.274			0-4-0DE	YE	2859 1961	OOU
No.302			0-4-0DH	YE	2949 1965	OOU
No.303			0-4-0DH	YE	2950 1965	
No.305			0-4-0DH	YE	2952 1965	
No.306			0-6-0DH	YE	2829 1961	
No.307		153/1510	0-6-0DH	YE	2834 1961	OOU
No.308			0-6-0DH	YE	2826 1961	OOU
No.309		21	0-6-0DH	YE	2825 1961	OOU
No.310			0-6-0DH	YE	2828 1961	
311			0-6-0DH	YE	2827 1961	
312	27	153/1507	0-6-0DH	YE	2831 1962	
No.314		153/1508	0-6-0DH	YE	2832 1962	
	26	153/506	0-6-0DE	YE	2830 1961	Dsm
No.401			0-6-0DH	YE	2691 1959	OOU
No.402			0-6-0DH	HE	7409 1976	
No.403			0-6-0DH	HE	7543 1978	
No.404			0-6-0DH	HE	8978 1979	
			4wWE	HL	3856 1936	
		Rebuilt		Workington	1973	OOU

BRITISH STEEL CORPORATION, SCOTTISH DIVISION, SHAPFELL LIMESTONE QUARRIES,
SHAP, PENRITH.
Gauge : 4'8½".   (NY 571134)

316		0-4-0DE	RH	423659 1958
No.10	72/21/44	0-6-0DH	S	10186 1964
		0-4-0DH	AB	601 1975

BURLINGTON SLATE QUARRIES LTD, KIRKBY-IN-FURNESS.
Gauge : 3'2¼".   (SD 245837)

	4wBE		GB	2051 1946	OOU
	4wDM		RH	320573 1951	OOU

DAVID CAIRD LTD, FURNESS FOUNDRY, BARROW-IN-FURNESS.
Gauge : 4'8½".   (SD 194692)

DAVID CAIRD	4wDM		FH	3685 1954	
	4wDM		RH	435490 1959	Dsm
	4wDM		RH	441935 1960	OOU

CARROCK FELL MINING CO LTD, CARROCK FELL MINE, near MUNGRISDALE.
Gauge : 2'0".   (NY 323329)

105/207	4wBE		WR	F6909 1966	
	0-4-OBE		WR	J7291 1969	
	0-4-OBE		WR		
		Rebuilt	Whiteheaps		
	0-4-OBE		WR		
		Rebuilt	Whiteheaps		

CENTRAL ELECTRICITY GENERATING BOARD, ROOSECOTE POWER STATION, BARROW-IN-FURNESS.
Gauge : 4'8½".   (SD 223680)

	0-4-ODM	JF	4210056 1951	
No.69	0-4-ODH	AB	477 1961	

CUMBERLAND MOSS LITTER INDUSTRY LTD & T.HOWLETT & CO LTD, KIRKBRIDE, WIGTON.
Gauge : 2'0".   (NY 238539)

	4wDM	MR	7191 1937	Dsm
	4wDM	MR	7463 1939	
	4wDM	RH	192887 1939	Dsm
	4wDM	MR	8586 1941	Dsm
	4wDM	MR	8627 1941	Dsm
	4wDM	RH	217993 1943	Dsm
LOD/758032	4wDM	MR	8860 1944	Dsm
LOD/758038	4wDM	MR	8884 1944	Dsm
	4wDM	MR	8905 1944	
	4wDM	RH	223744 1944	Dsm
	4wDM	MR	9231 1946	Dsm
	4wDM	MR	21282 1964	

CUMBRIA COUNTY COUNCIL, THE PORT OF WORKINGTON.
Gauge : 4'8½".   (NX 993294)

No.211	0-4-ODE	YE	2628 1956
No.212	0-4-ODE	YE	2684 1958
No.213	0-4-ODH	JF	4220012 1961
No.214	0-4-ODH	JF	4220013 1961
No.215	0-4-ODH	JF	4220014 1961

CUMBRIA EDUCATION COMMITTEE, STONE CROSS SPECIAL SCHOOL, ULVERSTON.
Gauge : 4'8½".   (SD 283783)

BARROW STEEL CO. No.17	0-4-OST IC	SS	1585 1865

DEPARTMENT OF THE ENVIRONMENT, BROUGHTON MOOR, near MARYPORT.
Gauge : 2'6".   (NY   )

			4wDM		MR	7485	1940
			4wDM		MR	8774	1942
			4wDM		HE	6007	1963

EGREMONT MINING CO LTD, FLORENCE IRON ORE MINE, EGREMONT.
Gauge : 2'6".   (NY 018103)   (Underground)

| | | | BE | | WR | | |

THE ESKDALE (CUMBRIA) TRUST, RAILWAY MUSEUM, RAVENGLASS.
Gauge : 1'3½".   (NY 086967)

| | SYNOLDA | | 4-4-2 | OC | BL | 30 | 1912 |

Also here are the side frames of ELLA 4-6-2 OC DB 2 189x.

FORCE CRAG MINE (U.K.) LTD, BARYTES MINE, BRAITHWAITE, near KESWICK.
Gauge : 1'10".   (NY 200216)

| | | | 4wBE | | WR | 2489 | 1943 |

GLAXO OPERATIONS U.K. LTD, ULVERSTON.
Gauge : 4'8½".   (SD 306777)   RSW.

| 2022 | GLAXO | | 0-4-0F | OC | AB | 2268 | 1949 |

R.J.HARRISON, c/o BORDER POULTRY FARM, HARRABY GREEN ROAD, CARLISLE.
Gauge : 2'0".   (NY 413543)

| | | | 4wDM | | Dtz | 181229 | 1932 |

INTERNATIONAL MILL SERVICE LYCRETE LTD, WORKINGTON WORKS, MOSS BAY, WORKINGTON.
Gauge : 4'8½".   (   )   RTC.

| 5153 | | | 4wDM | R/R | Unimog | | 1977 |
| 5154 | | | 4wDM | R/R | Unimog | | 1977 |

LAKESIDE & HAVERTHWAITE RAILWAY CO LTD, HAVERTHWAITE.
Gauge : 4'8½".   (SD 349843)

2073	(42073)		2-6-4T	OC	Bton		1950
2085	(42085)		2-6-4T	OC	Bton		1951
L.H.R.No.8	(D2117)		0-6-0DM		Sdn		1959
03072			0-6-0DM		Don		1959
12							
	ALEXANDRA		0-4-0ST	OC	AB	929	1902
	ASKHAM HALL		0-4-0ST	OC	AE	1772	1917
	RENISHAW IRONWORKS No.6	0-6-0ST	OC	HC	1366	1919
65			0-6-0ST	OC	HC	1631	1929
No.1							
	CALIBAN		0-4-0ST	OC	P	1925	1937
	PRINCESS		0-6-0ST	OC	WB	2682	1942
	M.P.M. No.2		0-4-0ST	OC	P	2087	1948
	REPULSE		0-6-0ST	IC	HE	3698	1950
1	DAVID		0-4-0ST	OC	AB	2333	1953
94	CUMBRIA		0-6-0ST	IC	HE	3794	1953
	RACHEL		4wPM		MR	2098	1924
No.2	FLUFF		0-4-0DM		JF	21999	1937
	ARMY 248		0-4-0DM		AB	344	1941
	ARMY 601 D1		0-6-0DE		Derby		1945

	L.H.R. P.T.1.	0-4-0DM	JF	4200022 1948	
		0-4-0DH	TH	108C 1961	
		Rebuild of 0-4-0DM	JF	22919 1940	
		2w-2PMR	Wkm	691 1932	
		2w-2PMR	Wkm		DsmT

MINISTRY OF DEFENCE, AIR FORCE DEPARTMENT, No.14 MAINTENANCE UNIT, BRUNTHILL SIDINGS, CARLISLE.
Gauge : 4'8½".   (NY 385599)

| | A.M.W. No.152 | 0-4-0DM | JF | 22602 1939 |
| | A.M.W. No.211 | 0-4-0DM | JF | 22958 1941 |

MINISTRY OF DEFENCE, ARMY DEPARTMENT.
Eskmeals Establishment.
See Section Five for full details.

Longtown Depot, near Carlisle.
See Section Five for full details.

MINISTRY OF DEFENCE, NAVY DEPARTMENT, ROYAL NAVAL ARMAMENT DEPOT, BROUGHTON MOOR, near MARYPORT.
Gauge : 4'8½".   (NY   )

	YARD No.54	0-4-0DM	HE	2642 1941
	YARD No.DP35	0-4-0DM	RH	313390 1953
	YARD No.9970	4wDM	FH	3995 1963

Gauge : 2'6".   (NY 055320)

	YARD No.3	0-4-0DM	HE	2011 1939
	YARD No.24	0-4-0DM	HE	2017 1939
	YARD No.25	0-4-0DM	HE	2018 1939
	YARD No.26	0-4-0DM	HE	2019 1939
	YARD No.B1	0-4-0DM	HE	2021 1939
R4	YARD No.70	4wDM	RH	221623 1942
R6	YARD No.72	4wDM	RH	221626 1942
R7	YARD No.73	4wDM	RH	221625 1942
R9	YARD No.83	4wDM	RH	235727 1943
R10	YARD No.84	4wDM	RH	235728 1943
R12	YARD No.86	4wDM	RH	235730 1943
R8	YARD No.103	4wDM	RH	242918 1947

S.MORGAN & SONS LTD, SCRAP MERCHANT, HINDPOOL ROAD, BARROW-IN-FURNESS.
Gauge : 2'6".   (SD 191692)

2		4wBE	WR	c1922	OOU
		4wBE	WR		OOU
4		4wBE	WR		OOU

NATIONAL COAL BOARD.
See Section Four for full details.

RAVENGLASS & ESKDALE RAILWAY CO LTD, RAVENGLASS.
Gauge : 1'3¼".   (NY 086967)

	RIVER IRT	0-8-2	OC		DB	3 1894
				Rebuilt	Ravenglass	1927
15	BONNIE DUNDEE	0-4-2WT	OC		KS	720 1901
	RIVER ESK	2-8-2	OC		DP	21104 1923
	RIVER MITE	2-8-2	OC	Clarkson	4669 1966
No.10	NORTHERN ROCK	2-6-2	OC	Ravenglass	10 1976
"I.C.L.No.1"		4w-4wDM	Rebuilt	Ravenglass	1925 +
	QUARRYMAN	4wPM		MH	2 1926
I.C.L.No.4	"PRETENDER"	4w-2DM	S/0	Ravenglass	1933
		Rebuild	of 4wPM	MH	NG39A 1929
	SHELAGH OF ESKDALE	4-6-4DH		SL	1969
126	"SILVER JUBILEE"	4w-4DHR		Ravenglass	1976
	LADY WAKEFIELD	4w-4wDH		Ravenglass	1980
		2w-2PMR		Ravenglass	1970

          + In use as a Tool Wagon.

RICHARDSONS MOSS LITTER CO LTD, SOLWAY MOSS WORKS, LONGTOWN, near GRETNA.
Gauge : 2'6".   (NY 338682)

	4wDM	MR	5879 1935	
	4wDM	FH	1985 1936	Dsm
	4wDM	MR	7137 1936	Dsm
	4wDM	MR	7190 1937	
	4wDM	FH	2408 1941	OOU
	4wDM	MR	9709 1952	
	4wDM	MR	9710 1952	
FANNY	4wDM	MR	21619 1957	
	4wDM	MR	26014 1966	
	4wDM	MR	40S371 1970	
	4wDM	Alan Keef	2 1976	
	4wDM	Alan Keef	3 1978	
	4wDM	Alan Keef	6 1981	

Gauge : 2'0".   (Locos here for repairs only.)

4wDM	L	8023 1936	OOU
4wDM	RH	217973 1941	Dsm
4wDM	RH	213853 1942	Dsm

JOHN T. SCOTT, DEALER, CLIBURN.
Gauge : 1'8".   (NY 585256)

4wDM	HE	3595 1948	OOU

SCOTTISH AGRICULTURAL INDUSTRIES LTD, BOLTON FELL MILL, near HETHERSGILL.
Gauge : 2'0".   (NY 487699)   (Subsidiary of I.C.I. Ltd.)

33					
		4wDM	MR	7037 1936	
		4wDM	MR	7188 1937	
		4wDM	MR	8638 1941	Dsm
		4wDM	MR	8655 1941	
3013	LOD 758027	4wDM	MR	8825 1943	
		4wDM	RH	277273 1949	
		4wDM	L	37366 1951	
		4wDM	LB	52726 1961	
		4wDM	LB	55730 1968	

SHAP GRANITE CO LTD, SHAP WORKS.
Gauge : 4'8½".   (NY 569112)   (Subsidiary of Thos W. Ward Ltd.)

		0-4-0DM		JF	4110003 1949

F.G.SHEPHERD, FLOW EDGE COLLIERY, MIDDLE FELL, ALSTON.
Gauge : 1'10".   (NY 738442)

	4wBE		F.G.Shepherd		Dsm
	0-4-0BE		WR	5655 1956	

SOUTH TYNEDALE RAILWAY PRESERVATION SOCIETY, ALSTON STATION.
Gauge : 2'0".   (   )

SAO DOMINGOS	0-6-0WT	OC	OK	11784c1925	
	4wDM		MR	5880 1935	+ Dsm
	4wDM		FH	2325 1941	
	4wDM		HE	2607 1942	
	0-6-0DM		HC	DM819 1952	

        + Converted to a brake-van.

VICKERS SHIPBUILDING GROUP LTD, BARROW SHIPBUILDING WORKS & BARROW ENGINEERING
WORKS, BARROW-IN-FURNESS.
Gauge : 4'8½".   (SD 200685)   (Member Company of British Shipbuilders.)

6331	PLUTO	0-4-0DM	HC	D1089 1958	
6614	MERCURY	0-4-0DM	HC	D1217 1960	OOU

WHITEHAVEN HARBOUR COMMISSIONERS, WHITEHAVEN.
Gauge : 4'8½".   (NX 972186)

	0-4-0DM		RH	408298 1957

DERBYSHIRE

ADAMSON-BUTTERLEY LTD, BUTTERLEY WORKS, RIPLEY.
Gauge : 4'8½".   (SK 405518)

TEUCER	0-4-0DM	(VF	D294 1955
		(DC	2567 1955

ARMYTAGE & SONS LTD, WHITTINGTON MOOR, CHESTERFIELD. (Closed)
Gauge : 4'8½".   (SK 383742)

950019	2w-2PMR	Wkm		OOU
	2w-2PMR	Wkm		OOU

BLUE CIRCLE INDUSTRIES LTD, HOPE CEMENT WORKS, HOPE.
Gauge : 4'8½".   (SK 167823)

PEVERIL	0-6-0DH		S	10087 1963
DERWENT	0-6-0DH		S	10156 1963
	4wDM	R/R	Unilok	1803

W.BUSH & SON LTD, BIRCHWOOD SIDINGS, SOMERCOTES, ALFRETON.
Gauge : 4'8½".   (SK 432545)

WILLIAM	0-4-0DM	JF	22918	1940	OOU
A.M.W. No.270	0-4-0DM	JF	23011	1945	OOU

Gauge : 2'0".

	4wDM	RH	223742	1946	Dsm

CENTRAL ELECTRICITY GENERATING BOARD, MIDLANDS REGION.
Drakelow Power Station, near Burton-On-Trent.
Gauge : 4'8½".   (SK 235200, 241198)

DRAKELOW	No.1	0-6-0DM	RH	326668	1955	Dsm
DRA	No.2	0-6-0DM	RH	326669	1956	OOU
	No.3	0-6-0DM	RH	326670	1957	
	No.4	0-6-0DM	RH	421421	1960	
	No.5	0-6-0DH	AB	484	1963	
	No.6	0-6-0DH	AB	485	1963	
		0-6-0DH	HE	7541	1976	

Spondon Power Station, Thulston Road, Borrowash, Derby.
Gauge : 4'8½".   (SK 405344)

No.1	4wBE/WE	EE	905	1935	
No.2	4wBE/WE	EE	1130	1939	
No.3	4wBE/WE	(EE	1379	1946	
		(Bg	3226	1946	OOU

Willington Power Station, near Burton-On-Trent.
Gauge : 4'8½".   (SK 306292)

1	0-6-0DM	(RSHN	7859	1956
		(DC	2573	1956
2	0-6-0DM	(RSHN	7860	1956
		(DC	2574	1956
No.3	0-4-0DM	HE	5388	1960
No.4	0-4-0DM	HE	5389	1960

CLAY CROSS (IRON & FOUNDRIES) LTD.
Clay Cross Foundry, near Chesterfield.
Gauge : 2'10½".   (SK 398644)

	4wWE	Clay Cross		c1969
26004	4wWE	Clay Cross		c1973
	4wWE	Clay Cross		c1971
	2w-2CE	Clay Cross		c1975
27408	2w-2CE	Clay Cross		1977

Spun Pipe Works, Clay Cross.
Gauge : 2'0".   (SK 401644)

14006	4wDM	L	35811	1950
	4wDM	L	37911	1952
14001	4wDM	L	41803	1955
	4wDM	LB	54684	1965
	4wDM	Clay Cross		1961 +
	4wDM	Clay Cross		1973 +

+ Constructed from parts supplied by Listers.

COURTAULDS ACETATE LTD.
Spondon Sports Ground.
Gauge : 4'8½".   (SK 410350)

	GEORGE	0-4-0ST	OC		RSHN	7214 1945	Pvd

Spondon Works.
Gauge : 4'8½".   (SK 405347)

DL 2	4wDH		S	10177 1964	
DH 1	0-4-0DH		RH	518190 1965	
	4wDH		RR	10280 1968	

W.H.DAVIS & SONS LTD, LANGWITH JUNCTION.
Gauge : 4'8½".   (SK 529683)

23373	ARMY 808	4wDM		RH	224347 1945
		4wDM		RH	321732 1952

DERBYSHIRE COALITE CO LTD, BUTTERMILK LANE, BOLSOVER.
Gauge : 4'8½".   (SK 457715)

No.5	0-4-0DH		NB	27548 1956	OOU
No.6	0-4-0DH		NB	27736 1958	Dsm
	0-6-0DH		RR	10279 1968	
No.8	0-6-0DH		RR	10291 1970	
9	0-6-0DH		TH	237V 1971	

DINTING RAILWAY CENTRE LTD, DINTING, near GLOSSOP.
Gauge : 4'8½".   (SK 021946)

5596	(45596)	BAHAMAS	4-6-0	3C		NBQ	24154 1935
					Rebuilt	HE	5596 1968
6115	(46115)	SCOTS GUARDSMAN	4-6-0	3C		NBQ	23610 1927
1054	(58926)		0-6-2T	IC		Crewe	2979 1888
19	(60019)	BITTERN	4-6-2	3C		Don	1866 1937
No.532	(60532)	BLUE PETER	4-6-2	3C		Don	2023 1948
63601			2-8-0	OC		Gorton	1912
			0-4-0VBTram			BP	2734 1886
			0-6-0ST	OC		AE	1883 1922
No.1704		NUNLOW	0-6-0T	OC		HC	1704 1938
		SOUTHWICK	0-4-0CT	OC		RSHN	7069 1942
W.D.150		WARRINGTON	0-6-0ST	IC		RSH	7136 1944
					Rebuilt	HE	3892 1969
No.2258		TINY	0-4-0ST	OC		AB	2258 1949
		JACOB	0-4-0PM			Bg	680 1916
RS 8			0-4-0DH			SCW	1960
			Rebuild of 0-4-0ST OC AE				1913 1923
950021			2w-2PMR			Wkm	590 1932
1307			2w-2PMR			Wkm	(1307 1933?) Dsm

D.S.F. REFRACTORIES LTD, FRIDEN BRICKWORKS, near HARTINGTON.
Gauge : 2'0".   (SK 166607, 170607)   RTC.

No.10	4wDM		RH	191658 1938	OOU
34204	4wDM		RH	222101 1943	OOU
	4wDM		RH	237914 1946	OOU

GARDNER MACHINERY & METALS, SCRAP MERCHANTS, DOVE HOLES STATION, near BUXTON.
Gauge : 2'0".   (SK 074780)

	4wDM	RH	244487	1946 OOU

Also other locos for scrap or resale occasionally present.

R.D.GEESON (DERBY) LTD, THE OLD QUARRY, PENTRICH, RIPLEY.
Gauge : 4'8½".   (SK 394520)

	4wDM	RH	262994	1948 OOU

Gauge : 2'0".

	4wDM	MR	9381	1948 OOU
	4wDM	MR	22221	1964 OOU

Locos hired out & also other locos for scrap or resale occasionally present.

GLASS BULBS & TUBES LTD, SHEFFIELD ROAD GLASSWORKS, CHESTERFIELD.
Gauge : 4'8½".   (SK 386732)

161 301	4wDM	RH	458959	1961

M.HILL, NEWBRIDGE LANE, OLD WHITTINGTON, CHESTERFIELD.
Gauge : 2'0".   (SK 390739)

	4wDM	RH	170369	1934
LOD/758097	4wDM	RH	202967	1940

IMPERIAL CHEMICAL INDUSTRIES LTD, MOND DIVISION,
Hindlow Limeworks,
Gauge : 4'8½".   (SK 097679)

RS 140	J.B.GANDY	4wDM		RH	299103 1950
			Rebuilt	SCW	OOU
RS/140		4wDM		FH	3892 1958
RS 206		4wDH		FH	3952 1961

Tunstead Limeworks & Quarries (including Lime Group Workshops & Stores), Tunstead,
Gauge : 4'8½".   (SK 101743, 097755)

RS 153	WALLACE AKERS	0-4-0DE		YE	2609 1956 OOU
RS 170		0-6-0DH		RR	10284 1969
			Rebuilt	TH	1974
RS 212	NIDDERDALE 13	4wDH		S	10051 1960
RS 233		4wDH		TH	284V 1979
RS 244	HARRY TOWNLEY	4wDH		TH	289V 1980

INNES LEE INDUSTRIES, CAMPBELL BRICKWORKS, BARROW HILL, STAVELEY.
Gauge : 2'0".   (SK 413756)   RTC.

	4wDM	MR	60S364	1968 OOU

LAPORTE INDUSTRIES LTD.
Ladywash Mine, near Eyam.   (Closed)
Gauge : 1'6".   (SK 219775)   (Underground)

	4wBE	LMM	1028	1949	Dsm
4	4wBE	GB	1326	1933	Dsm
5	4wBE	GB		c1933	
6	4wBE	GB	2493	1946	
7	4wBE	CE	5214	1966	
8	4wBE	CE	5370	1967	
9	4wBE	CE	5818	1970	
10	4wBE	CE	5885	1971	
11	4wBE	CE	B0140	1973	

Sallet Hole Mine, Stoney Middleton.
Gauge : 2'0".   (SK 223743)   (Mainly Underground)

3	4wBE	CE	5180	1966	
4	4wBE	WR	3492	1946	
6	4wBE	CE	B0123	1973	
7	4wBE	CE	B1560	1977	
8	4wBE	CE	5797	1970	
9	4wBE	WR	H7205	1968	OOU
10	4wBE	CE	5950	1972	
	4wBE	CE	B1559	1977	

LEYS MALLEABLE CASTINGS CO LTD, COLOMBO STREET, DERBY.
Gauge : 4'8½".   (SK 359340)

RS36/1	BAGNALL	LEY'S	4wDH	WB	3207 1961

LISTER PLANT SALES, SHEFFIELD ROAD, SHEEPBRIDGE, near CHESTERFIELD.
Gauge : 4'8½".   (SK 376759)

0-4-0DM	JF	4110011 1950	OOU

ALBERT LOOMS LTD, SCRAP DEALERS, SPONDON.
Yard (SK 390354) with locos for scrap or resale occasionally present.

MARKHAM & CO LTD, BROADOAKS WORKS, CHESTERFIELD.
Gauge : 4'8½".   (SK 389710)

4wDM	RH	476141 1963

MIDLAND RAILWAY CENTRE, (MIDLAND RAILWAY TRUST LTD), BUTTERLEY STATION SITE, RIPLEY.
Gauge : 4'8½".   (SK 403520, 412519)

158A			2-4-0	IC		Derby	1866
1708	(41708)		0-6-0T	IC		Derby	1880
4027	(44027)		0-6-0	IC		Derby	1924
6203	(46203)	PRINCESS					
		MARGARET ROSE	4-6-2	4C		Crewe	253 1935
	(47327)		0-6-0T	IC		NB	23406 1926
16440	(47357)		0-6-0T	IC		NB	23436 1926
					Rebuilt	Derby	1973
	(47445)		0-6-0T	IC		HE	1529 1927
	(47564)		0-6-0T	IC		HE	1580 1928
13809	(53809)		2-8-0	OC		RS	3895 1925
73129			4-6-0	OC		Derby	1956

D4	GREAT GABLE	2-6w+6w-2DE			Derby		1959
2271		0-6-0DM			(RSH	7913	1957
					(DC	2615	1957
12077		0-6-0DE			Derby		1950
1311-G		2-2wDM			Mercury	5337	1927
No.4		0-4-0ST	OC		NW	454	1894
	GLADYS	0-4-0ST	OC		Mkm	109	1894
	VICTORY	0-4-0ST	OC		P	1547	1919
	STANTON No.24	0-4-0CT	OC		AB	1875	1925
68012		0-6-0ST	IC		HE	3193	1944
				Rebuilt	HE	3887	1964
	N1	4wVBT	VCG		S	9370	1947
		4wVBT	VCG		S	9632	1957
RS 12		4wDM			MR	460	1917
		4wPM			MR	1930	1919
RS 9		4wDM			MR	2024	1921
77	ANDY	0-4-0DM			JF	16038	1923

Gauge : 3'0".

	HANDYMAN	0-4-0ST	OC	HC	573	1900
No.2		2-4-0DM		JF	20685	1935

<u>MILLER & BAIRD LTD, PLANT DEPOT, TOWN STREET, SANDIACRE.</u>
Gauge : 2'0".   (SK 480368)

	4wDM		L	52031 c1960 +

+ Plate reads 25031.
Loco occasionally used on contract work.

<u>NATIONAL COAL BOARD.</u>

For full details see Section Four.

<u>THE NATIONAL TRAMWAY MUSEUM.</u>
<u>Clay Cross Store, Hepthorne Lane, Clay Cross.</u>
Gauge : Metre.   (SK 402652)

4wDM		RH	373363 1954

<u>Cliff Quarry, Crich.</u>
Gauge : 4'8½".   (SK 345549)

	0-4-0VBTram		BP	2464 1885
	4wWE		EE	717 1927
RUPERT	4wDM		RH	223741 1944 +
G.M.J.	4wDM		RH	326058 1952 @

+ Rebuilt T.M.S. 1964 from 60cm gauge.
@ Rebuilt T.M.S. 1969 from 3'3" gauge.

<u>N.E.I. INTERNATIONAL COMBUSTION LTD.</u>
<u>Clayton Works, Hatton.</u>
(SK 212298)

New CE locos under construction, and locos for repair, usually present.

<u>Sinfin Lane Works, Derby.</u>
Gauge : 4'8½".   (SK 353327)

| | I.C.LTD, DERBY | 0-4-0DM | | HC | D790 1953 |

<u>NORTH WEST WATER AUTHORITY, EASTERN DIVISION, LONGDENDALE HEADWORKS,</u>
<u>BOTTOMS OFFICES AND WORKSHOPS, TINTWISTLE.</u>
Gauge : 2'0".   (SK 026973)

| P 396 | 81 A 03 | 4wDM | | RH | 497542 1963 |

<u>PEAK RAILWAY.</u>
<u>Matlock Site.</u>
Gauge : 4'8½".   (   )

| 4936 | KINLET HALL | 4-6-0 | OC | Sdn | 1929 |
| 80080 | | 2-6-4T | OC | Bton | 1954 |

<u>Midland Railway Station, Buxton.</u>
Gauge : 4'8½".   (SK 060738)

| 48624 | | 2-8-0 | OC | Afd | 1943 |
| 92214 | | 2-10-0 | OC | Sdn | 1959 |

<u>c/o Shaws Metals Ltd, Haydock Park Road, Ascot Drive, Derby.</u>
Gauge : 4'8½".   (SK 373336)

| 34101 | HARTLAND | 4-6-2 | 3C | Elh | 1950 |

<u>PEAKSTONE LTD.</u>
<u>Hindlow Limeworks.</u>
Gauge : 4'8½".   (SK 087689)

| | | 4wDM | | MR | 5765 1959 | OOU |

<u>Holderness Limeworks, Peak Dale.</u>
Gauge : 4'8½".   (SK 088773)

| (07001) | | 0-6-0DE | | RH | 480686 1962 |
| | | 4wDM | | FH | 4006 1963 |

<u>PIMBROOK LTD, TUTBURY.</u>
Gauge : 4'8½".   (SK 214296)

| | | 0-6-0DE | | YE | 2895 1964 |

<u>RAYNESWAY PLANT LTD, PLANT DEPOT, RAYNESWAY, DERBY.</u>
Gauge : 4'8½".   (SK 384352)

| | PLANT No.7926 | 4wDM | | RH | 263000 1949 |
| | BALFOUR BEATTY No.8841 | 4wDH | | Permaquip | 1979 |

Gauge : 2'0".

	PLANT 4950	4wBE		WR	6503 1962
	PLANT 4951	4wBE		WR	6504 1962
	PLANT 4952	4wBE		WR	6505 1962
	PLANT 5495	4wBE		WR	G7125 1967
	PLANT 5666	4wBE		WR	5070 1953
	PLANT 5667	4wBE		WR	5072 1953
	PLANT 5676	4wBE		WR	H7198 1968

```
            PLANT 5677              4wBE            WR    H7199 1968
            PLANT 8194              4wBE            WR    N7787 1974
            PLANT 8203             4wBE            WR     7296 1975
            PLANT 8204             4wBE            WR     7297 1975
            PLANT 8210             4wBE            WR    N7788 1974
        Locos present in yard between contracts.
```

RIBER CASTLE WILDLIFE PARK, MATLOCK.
Gauge : 4'8½".   (SK 308590)

```
                            0-4-0ST   OC        P     1555 1920
```

S.P.O. COAL COMPANY LTD, DOE LEA COLLIERY, DOE LEA, near CHESTERFIELD.
(Subsidiary of S.P.O. Minerals Ltd.)
Gauge : 2'0".   (SK 454667)   (Underground)

```
                            4wDM            RH   252809 1948  OOU
                            4wDM            RH   296047 1950  OOU
```

STANTON & STAVELEY LTD. (Subsidiary Of British Steel Corporation.)
Dale Plant Foundry, Ilkeston.
Gauge : 3'0".   (SK 460389)

```
    1       1031/65            4wBE            GB
    2                          4wBE            GB
```

Stanton Works, Ilkeston.
Gauge : 4'8½".   (SK 471390, 476388)

```
    52                         0-6-0DH         S    10139 1962
    53                         0-6-0DH         S    10140 1962
    54      9111/164           0-6-0DH         RR   10184 1964
    55                         0-6-0DH         RR   10185 1964
    57                         0-6-0DH         RR   10214 1964
    60                         0-4-0DH         RR   10253 1967
    61      GRAHAM             0-4-0DH         RR   10207 1965
    62                         0-4-0DH         TH    227V 1970
    63      9111/173           0-4-0DH         TH    228V 1970
```

Staveley Works, near Chesterfield.
Gauge : 4'8½".   (SK 414750)

```
No.2                           4wDH            S    10013 1959
No.3                           4wDH            S    10045 1960  Dsm
No.4                           4wDH            S    10046 1960
No.5                           4wDH            S    10047 1960
    9611/62   STANTON No.41    0-4-0DE         YE    2597 1955  OOU
        Also locos on hire from B.R.
```

STEETLEY (MFG) LTD, DOWLOW WORKS, DOWLOW, near BUXTON.
Gauge : 4'8½".   (SK 102678)

```
        CYNTHIA                4wDM            RH   412431 1957
                               0-4-0DH         RH   418793 1957
        HEATHCOTE              0-4-0DE         RH   461959 1961
```

TARMAC ROADSTONE HOLDINGS LTD, BUXTON QUARRY.
Gauge : 4'8½".   (SK 083694)

550/6/0357	4wDM		RH	252842 1948	
	4wDM		RH	305306 1952	OOU

TARMAC ROADSTONE (NORTHERN) LTD.
Middle Peak Quarry, Wirksworth.
Gauge : 4'8½".   (SK 287548)

DERBYSHIRE STONE	4wDM		FH	1891 1934	OOU
525 29 441	0-4-0DM		Bg	3227 1951	OOU
	0-4-0DM		Bg	3357 1952	
	6wDH		RR	10274 1968	

Topley Pike Quarry, Buxton.
Gauge : 4'8½".   (SK 102724)

	0-4-0DM		RH	319285 1953	OOU
	0-4-0DE		RH	424839 1959	

T.I.CHESTERFIELD LTD, CHESTERFIELD.
Gauge : 4'8½".   (SK 383702)

850	4wDH		TH	166V 1966

TRIANGLE ALLOYS LTD, BRIDGE STREET, CLAY CROSS.
Gauge : 2'0".   (   )

	4wDM	LB	50191 1957	OOU

WAGGON REPAIRS LTD.
Long Eaton Foundry, Long Eaton.   (Closed)
Gauge : 4'8½".   (SK 496331)

6		4wDM		FH	3044 1945
4	45	4wDM		FH	3909 1959

Sheepbridge Works, Whittington Hill, Chesterfield.
Gauge : 4'8½".   (SK 380746)

2	4wDM		FH	3601 1953

WALKER & PARTNERS LTD, INKERSALL ROAD ESTATE, STAVELEY, CHESTERFIELD.
Dealers yard (SK 435744) with locos for resale occasionally present.

THOS. W. WARD (RAILWAY ENGINEERS) LTD, MIDLAND FOUNDRY, SANDIACRE.
Gauge : 4'8½".   (SK 482362)

4wDM		FH	3884 1958

DEVON

BETHELL GWYN & CO LTD, SHIPOWNERS, VICTORIA WHARF, PLYMOUTH.
Gauge : 4'8½".   (SX 490537)   (Subsidiary of the P. & O. Group.)

		4wDM		FH	3281	1948	

BICTON WOODLAND RAILWAY, EAST BUDLEIGH.
Gauge : 1'6".   (SY 074862)

No.								
No.1	WOOLWICH		0-4-0T	OC	AE	1748	1916	
No.2	BICTON		4wDM		RH	213839	1942	
No.3	CARNEGIE		0-4-4-0DM		HE	4524	1954	
No.4	BUDLEY	LOD/758235	4wDM		RH	235624	1945	OOU

BLUE CIRCLE INDUSTRIES LTD, PLYMSTOCK WORKS.
Gauge : 4'8½".   (SX 506542)

		4wDH.		TH	125V	1963	

DART VALLEY LIGHT RAILWAY LTD.
Gauge : 4'8½".   Locos are kept at :-

Buckfastleigh.   (SX 747663)
Staverton Road.   (SX 785638)

1369			0-6-0PT	OC		Sdn		1934	
1420	BULLIVER		0-4-2T	IC		Sdn		1933	
1450	ASHBURTON		0-4-2T	IC		Sdn		1935	
1638			0-6-0PT	IC		Sdn		1951	
4920	DUMBLETON HALL		4-6-0	OC		Sdn		1929	
6430			0-6-0PT	IC		Sdn		1937	
6435			0-6-0PT	IC		Sdn		1937	
3298	(30587)		2-4-0WT	OC		BP	1412	1874	
80064			2-6-4T	OC		Bton		1953	
	LADY ANGELA		0-4-0ST	OC		P	1690	1926	
1	ASHLEY		0-4-0ST	OC		P	2031	1942	
	MAUREEN		0-6-0ST	IC		HE	2890	1943	
					Rebuilt	HE	3882	1962	
			0-6-0ST	IC		WB	2766	1944	
	GLENDOWER		0-6-0ST	IC		HE	3810	1954	
	M.F.P.No.4		0-4-0DM			JF	4210141	1958	
			2w-2PMR			Wkm	946	1933	DsmT
			2w-2PMR			Wkm	4146	1947	
			2w-2PMR			Wkm	4149	1947	DsmT
PWM 3290			2w-2PMR			Wkm	4840	1948	DsmT
PWM 3944			2w-2PMR			Wkm	6929	1955	
			2w-2PMR			Wkm	8198	1958	

Gauge : 7'0¼".

151	TINY	0-4-0VBWT	VCG		Sara	1868	Pvd

DARTINGTON AMENITY TRUST, MORWELLHAM QUAY, near TAVISTOCK.
Gauge : 2'0".   (SX 448699)

PLANT 4862	4wBE	WR	6298	1960	
PLANT 5494	4wBE	WR	G7124	1967	
S259	4wBE	WR	H7197	1968	

DEPARTMENT OF THE ENVIRONMENT, OKEHAMPTON GUN RANGES.
Gauge : 2'6".   (SX 593910)

WD 767138		2w-2PM		Wkm	3284 1943
WD 767139		2w-2PM		Wkm	3282 1943

ENGLISH CHINA CLAYS LTD, MARSH MILLS DRYING WORKS, PLYMPTON.
Gauge : 4'8½".   (SX 521574)

	0-4-0DH		EEV	3987 1970

G.W.GLOVER, UPTON PYNE, near EXETER.
Gauge : 3'0".   (SX 906984)

	4wDM		JF	3930048 1951

G.D.MASSEY, 57, SILVER STREET, THORVERTON, EXETER.
Gauge : 2'0".   (SS 932018)

A.R.40		4wBER		Massey 1967

MINISTRY OF DEFENCE, NAVY DEPARTMENT.
Devonport Dockyard.
Gauge : 4'8½".   (SX 449558)

YARD No.4857	4wDM		FH	3741 1955
YARD No.4858	4wDM		FH	3744 1955 +
YARD No.4860	4wDM		FH	3747 1955 +
YARD No.5197	4wDM		FH	3773 1955
YARD No.5198	4wDM		FH	3774 1955
YARD No.5199	4wDM		FH	3775 1955 +
YARD No.5200	4wDM		FH	3776 1956

+ These locos usually work South Yard, the remainder usually working the
    North Yard.

Royal Naval Armament Depot, Ernesettle.
Gauge : 4'8½".   (SX 45x60x)

" No.11 "		4wDM		RH	224349 1945
V3	YARD No.13	0-4-0DM		HE	3133 1949
52 RN 33		4wDM	R/R	Unimog	
52 RN 35		4wDM	R/R	Unimog	

NATIONAL TRUST, SALTRAM HOUSE, PLYMPTON.
Gauge : 4'6".   (SX 521556)

LEE MOOR No.2	0-4-0ST	OC		P	784 1899

NORTH DEVON CLAY CO LTD, PETERS MARLAND.
Gauge : 4'8½".   (SS 503125)

PETER	0-4-0DM		JF	22928 1940	OOU
PROGRESS	0-4-0DH		JF	4000001 1945	
49	4wDM		RH	443642 1960	

PLYM VALLEY RAILWAY ASSOCIATION, MARSH MILLS, PLYMPTON.
Gauge : 4'8½".   (SX 472542)(SX 517564)

4160		2-6-2T	OC	Sdn	1948
34007	WADEBRIDGE	4-6-2	3C	Bton	1945

W.L.A.PRYOR, LYNTON RAILWAY STATION, STATION HILL, LYNTON.
Gauge : 2'0".   (   )

No.3	BRUNEL	4wDM		RH	179880 1936

SEATON AND DISTRICT ELECTRIC TRAMWAY CO, COLYTON.
Gauge : 2'9".   (SY 247901)

		4wDM		RH	435398 1959

TIVERTON MUSEUM SOCIETY, TIVERTON MUSEUM, TIVERTON.
Gauge : 4'8½".   (SS 955124)

1442		0-4-2T	IC	Sdn	1935

TORBAY AND DARTMOUTH RAILWAY, PARK SIDING, PAIGNTON.
Gauge : 4'8½".   (SX 889606)

4555		2-6-2T	OC	Sdn	1924	
4588	WARRIOR	2-6-2T	OC	Sdn	1927	
5239	GOLIATH	2-8-0T	OC	Sdn	1924	
7827	LYDHAM MANOR	4-6-0	OC	Sdn	1950	
D1023	WESTERN FUSILIER	6w-6wDH		Sdn	1963	
(D2192)	No.2	0-6-0DM		Sdn	1961	
	ENTERPRISE	0-6-0DM		HC	D810 1953	
(900391)		2w-2PMR		Wkm	671 1932	Dsm
PWM 2210		2w-2PMR		Wkm	4127 1946	DsmT
PWM 2802		2w-2PMR		Wkm	4980 1948	DsmT
PWM 3773		2w-2PMR		Wkm	6652 1953	
PWM 3957	DART VALLEY No.4	2w-2PMR		Wkm	6942 1955	

Locos are returned to D.V.R. Buckfastleigh for overhaul as required.

WATTS, BLAKE, BEARNE & CO LTD, No.10 ADIT, WEST GOLDS MINE, near NEWTON ABBOT.
Gauge : 2'6".   (SX 857727)   (Usually Underground)

		4wBE		CE	5382 1968

WESTERN FUEL CO LTD, EXMOUTH JUNCTION COAL CONCENTRATION DEPOT.
Gauge : 4'8½".   (SX 934938)

		0-4-0DM		(VF	D98 1949
				(DC	2269 1949

DORSET

BEDFORD & JESTY LTD, WATERCRESS GROWERS, DODDINGS FARM, BERE REGIS.
Gauge : 1'6".   (SY 847947)

		4wPM		B.J.Fry	1948

MINISTRY OF DEFENCE, NAVY DEPARTMENT, ROYAL NAVAL UNDERWATER WEAPONS ESTABLISHMENT,
BINCLEAVES, WEYMOUTH.
Gauge : 2'0".   (SY 684780)

		4wBE		GB	2345 1950

P.D.FUELS LTD. BALLAST WHARF, HAMWORTHY.
Gauge : 4'8½".   (SZ 010901)

1135		4wDM		RH	242867	1946

SWANAGE RAILWAY SOCIETY, SWANAGE STATION.
Gauge : 4'8½".   (SZ 026789)

6695		0-6-2T	IC	AW	983	1928
80078		2-6-4T	OC	Bton		1954
	CUNARDER	0-6-0ST	OC	HE	1690	1931
ED 10	RICHARD TREVITHICK	0-4-0ST	OC	AB	2354	1954
	MARDY No.1	0-6-0ST	OC	P	2150	1954
	BERYL	4wPM		FH	2054	1938
	MAY	0-4-0DM		JF	4210132	1957
		4wDM		RH	402812	1957

WEYMOUTH MINIATURE RAILWAY, RADIPOLE LAKE, WEYMOUTH.
Gauge : 1'3".   (SY 675794)

		4w-4wPM		G.& S.	1953

WINCHESTER AND ALTON RAILWAY LTD. TUCKTON LEISURE PARK, STOUR ROAD, CHRISTCHURCH.
Gauge : 4'8½".   (SZ 148925)

2	HAMPSHIRE	0-4-0ST	OC	WB	2842	1946	Pvd

DURHAM

J.BARTLETT, NEWFIELD DRIFT MINE, BISHOP AUCKLAND.
Gauge : 2'6".   (NZ 209332)

6/44	0-4-0BE		WR	6595 1962
6/53	0-4-0BE		WR	6704 1962
6/44	0-4-0BE		WR	C6710 1963

BLUE CIRCLE INDUSTRIES LTD. WEARDALE CEMENT WORKS, near EASTGATE.
Gauge : 4'8½".   (NY 949384)

	4wDH		TH	148V 1965
ELIZABETH	4wDH		RR	10232 1965

BRITISH STEEL CORPORATION, SCUNTHORPE DIVISION.
Blackdene Fluor Mines, Ireshopeburn, near Wearhead,   (Closed)
Gauge : 2'0".   (NY 868389)

3	4wBE		CE	B01502 1977

Blanchland Fluor Mines, Grove Rake Mine, near Rookhope.
Gauge : 2'0".   (NY 895441)   (Underground)

			0-4-0BE		WR			
			0-4-0BE		WR			
		Rebuilt			Whiteheaps	1975	OOU
8			4wBE		WR	4812	1952	
			4wBE		GB	6068	1962	OOU
			0-4-0BE		WR	7481	1972	
			0-4-0BE		WR	7549	1973	
			0-4-0BE		WR	7728	1976	
			0-4-0BE		WR	R7735	1977	
			0-4-0BE		WR	P7846	1975	
			0-4-0BE		WR	T8007	1979	
7			0-4-0BE		WR	T8012	1979	
			0-4-0BE		WR		1979	
			0-4-0BE		WR		1979	

Blanchland Fluor Mines, Whiteheaps Mine, Ramshaw.
Gauge : 2'0".   (NY 948466)   (Underground)

	0-4-0BE	WR	T8005	1979	
	0-4-0BE	WR			
	0-4-0BE	Whiteheaps			Dsm

BRITISH STEEL CORPORATION, TEESSIDE DIVISION, CONSETT & JARROW WORKS,
CONSETT WORKS, CONSETT. (Closed)
Gauge : 4'8½".   (NZ 099503, 107505, 110500)

31		0-6-0DH	RR	10285	1969
41		0-6-0DH	S	10079	1961
43		0-6-0DH	S	10088	1962
		0-4-0WE	Consett		1972
		Rebuild of 4wWE	GB	2368	1952

CHEMICAL & INSULATING CO LTD, FAVERDALE, DARLINGTON.
Gauge : 4'8½".   (NZ 272165)   RTC.

	4wWE	GEC	(1928?)	

Gauge : 2'0".   (NZ 272166)

	4wBE	GB	2848	1957

Gauge : 1'8".   (NZ 272166)

No.1		4wDM	RH	375360	1954	
		4wDM	RH	402428	1956	Dsm
No.3		4wDM	RH	476124	1962	
No.4	MOSELEY	4wDM	RH	354013	1953	

CLEVELAND BRIDGE & ENGINEERING CO LTD, SMITHFIELD ROAD, DARLINGTON.
Gauge : 4'8½".   (NZ 295135)

185	DAVID PAYNE	0-4-0DM	JF	4110006	1950	OOU
		4wDH	TH/S	111C	1961	

DARLINGTON (NORTH ROAD STATION) RAILWAY MUSEUM SOCIETY, DARLINGTON.
Gauge : 4'8½".   (NZ 289157)

	LOCOMOTION	0-4-0	VC	RS	1	1825
25	DERWENT	0-6-0	OC	Kitching		1845
17		0-4-0VBT	OC	HW	33	1873
No.1275		0-6-0	IC	D	708	1874
No.1463		2-4-0	IC	Dar		1885
	MET	0-4-0ST	OC	HL	2800	1909
	PATONS	0-4-0F	OC	WB	2898	1948

DARLINGTON RAILWAY PRESERVATION SOCIETY, MONTROSE STREET, DARLINGTON.
Gauge : 4'8½".   (   )

78018	2-6-0	OC	Dar	1954 +

   + Currently under renovation at Henry Boot Engineering Ltd.

DARLINGTON & SIMPSON ROLLING MILLS LTD, RISE CARR ROLLING MILLS, DARLINGTON.
Gauge : 4'8½".   (NZ 284172, 286164)

No.1	4wDH	TH	129V	1963
No.2	4wDH	TH	131V	1963

DURHAM COUNTY COUNCIL EDUCATION DEPARTMENT, DINSDALE PARK RESIDENTIAL SCHOOL,
near DARLINGTON.
Gauge : 4'8½".   (NZ 341121)

No.4	0-4-0DM	JF	4210087	1953

FORDAMIN COMPANY (SALES) LTD, CLOSEHOUSE BARYTES MINE, near MIDDLETON-IN-TEESDALE.
Gauge : 1'7½".   (NY 850227)

4wDM	HE	4569	1956	OOU
0-4-0BE	WR	D6754	1964	
0-4-0BE	WR	P7664	1975	
0-4-0BE	WR	P7731	1975	

HANRATTY BROS, SCRAP IRON & STEEL, WHESSOE ROAD, DARLINGTON.
Gauge : 4'8½".   (NZ 289158)

0-4-0DM	HE	2839	1943

ALF LISTER, SCRAP MERCHANT, BOYD STREET, CONSETT.
Yard (NZ 113503) with locos for scrap & resale occasionally present.

J. & J. MADDISON, YEW TREE MINE, BOLLIHOPE.
Gauge : 2'0".   (NY 991351)

4wBE	WR	6299	1960
0-4-0BE	WR	K7377	1970
4wBE	GB	420288	1971
4wBE	WR		

NATIONAL COAL BOARD.

          For full details see Section Four.

NORTH EASTERN IRON REFINING CO LTD, STILLINGTON.
Gauge : 4'8½".   (NZ 372237)

| | | 4wDM | | RH | 312427 1951 OOU |
| | | 0-4-0DH | | JF | 4220027 1964 |

NORTH OF ENGLAND OPEN AIR MUSEUM, BEAMISH HALL.
Gauge : 4'8½".   (NZ 217547)

876	(65033)	0-6-0	IC	Ghd	1889
		0-4-0	VC	Geo Stephenson	1822
18		0-4-0T	OC	Lewin	1863
		0-4-0VBT	VC	HW	1871
E No.1		2-4-0VBCT	OC	BH	897 1887
No.3 2320/69	TWIZELL	0-6-0T	IC	RS	2730 1891
	SOUTH DURHAM				
	MALLEABLE No.5	0-4-0ST	OC	Stockton	1900
No.14		0-4-0ST	OC	HL	3056 1914
	LOCOMOTION	0-4-0	VC	Loco.Ent	No.1 1975
		4wDM		RH	476140 1963

Gauge : 2'0".   (NZ 212548)

| | | 4wDM | | RH | | Dsm |

RAISBY QUARRIES LTD, GARMONDSWAY QUARRY, near COXHOE.
Gauge : 4'8½".   (NZ 338351)

| | | 4wDH | | S | 10077 1961 |

R.H.P.BEARINGS LTD, AUTOMOTIVE BEARINGS DIVISION, ANNFIELD PLAIN. (Closed)
Gauge : 4'8½".   (NZ 159508)

| | | 4wDM | | RH | 275881 1949 OOU |

SEAHAM HARBOUR DOCK CO LTD, SEAHAM HARBOUR.
Gauge : 4'8½".   (NZ 432493)

D1		0-6-0DH	EEV	D1191 1967
D2		0-6-0DH	EEV	D1192 1967
D3		0-6-0DH	EEV	D1193 1967
D4		0-6-0DH	EEV	D1194 1967
D5		0-6-0DH	EEV	D1195 1967

SEDGEFIELD DISTRICT COUNCIL, TIMOTHY HACKWORTH MUSEUM, SHILDON.
Gauge : 4'8½".   (   )

| | | 0-6-0 | OC | Hackworth | 1835 |
| | SANS PAREIL | 0-4-0 | VC | BREL | 1980 |

SWISS ALUMINIUM MINING (U.K.) LTD,
Burtree Pasture Mine, Cowshill. (Closed)
Gauge : 2'0".   (NY 859412)

| | | 4wBE | | WR | |
| | | 4wBE | | WR | |

Cambokeels Mine, Eastgate, Stanhope.
Gauge : 60cm.   (NY 934383)

1		4wBE	WR	S7968	1978
		4wBE	WR	7888R	1977
		4wBE	WR	R7965	1977
4		4wBE	WR	R7964	1977
5	MBS 248	4wBE	WR	6907	1965
6	MBS 249	4wBE	WR	E6908	1965
No.7	M 204	4wBE	WR	P7789	1975
No.8	M 206	4wBE	WR	Q7796	1976

Hope Level Mine, Stanhope.
Gauge : 2'0".   (NY 991397)

	0-4-0BE	WR	8079	1980

Redburn Mine, Rookhope, Stanhope. (Closed)
Gauge : 2'0".   (NY 927431)   (Underground)

	4wBE	WR		(c1936?)
	4wBE	WR	1192	1938
	4wBE	WR	5299	1955
	4wBE	WR	5601	1956
	4wBE	WR	D6805	1964
	4wBE	CE	5889	1971
2-6	4wBE	WR		

Stanhopeburn Mine, Shield Hurst Level, Stanhope.
Gauge : 2'0".   (NY 987413)

1	4wBE	WR	S7966	1978
2	4wBE	WR	S7967	1978
3	0-4-0BE	WR	N7644	1973

TARMAC ROADSTONE HOLDINGS LTD, WEST CORNFORTH LIMESTONE QUARRY.
Gauge : 4'8½".   (NZ 317347)   RTC.

6/1698	R.F.SPALDING	4wDM	RH	262996	1949	00U
6/1701		4wDM	RH	326071	1954	00U

TYNE-TEES TUNNELLING GREAT BRITAIN, PLANT DEPOT, WARD'S SAWMILLS, WOLSINGHAM.
Gauge : 60cm.   (NZ 084372)

4wBE	SIG	
4wBE	SIG	
4wBE	SIG	
4wBE	SIG	

WHESSOE HEAVY ENGINEERING LTD, BRINKBURN ROAD, DARLINGTON.
Gauge : 4'8½".   (NZ 284160)

316	DERWENT II	0-4-0DE	RH	312988	1952

WHORLTON LIDO, near BARNARD CASTLE.
Gauge : 1'3".   (NZ 106146)

	KING GEORGE	4-4-2	OC	BL	21	1912
	WENDY	4-4wDM		CoSi		1972
278		2-8-0PH	S/o	SL	R9	1976

<u>WOLSINGHAM STEEL CO LTD, WOLSINGHAM STEELWORKS.</u>
Gauge : 4'8½".   (NZ 083368)

4wDM		RH	432480 1959

EAST SUSSEX

<u>BRITISH GYPSUM LTD, MOUNTFIELD.</u>
Gauge : 4'8½".   (TQ 730199)

1		4wDH		TH	183V 1967
2		4wDH		TH	184V 1967

<u>DEPARTMENT OF EDUCATION & SCIENCE, PRESTON PARK, BRIGHTON.</u>
Gauge : 4'8½".   (TQ 302061)

2092 (S 10656, S 12123)	4w–4w+4w–4wRER		1937

<u>DRUSILLA'S ZOO PARK, BERWICK, near EASTBOURNE.</u>
Gauge : 2'0".   (TQ 524050)

EMILY	4wDM		RH	226294 1943	
	4wPM		FH	3116 1946	Pvd
	4wDM		MR	9409 1948	

<u>THE GREAT BUSH RAILWAY, TINKERS PARK, HADLOW DOWN, near UCKFIELD.</u>
Gauge : 2'0".   (TQ 538241)

1	AMINAL	4wDM		MR		
		Rebuilt Ludlay Brick			c1931	
2	SEZELA No.2	0-4-0T	OC	AE	1720 1915	
4	MILD	4wDM		MR	8687 1941	
5	ALPHA	4wDM		RH	183744 1937	
6	SEZELA No.6	0-4-0T	OC	AE	1928 1923	
8	FIDO	4wDM		FH	2586 1941	Dsm
12		4wDM		FH	2535 1942	
14	ALBANY	4wDM		RH	213840 1941	
19		4wDM		FH		
No.22	LANA	4wBE		WR	5033 1953	
		4wBE		WR	5035 1954	
No.24	TITCH	0-4-0BE		WR	M7535 1972	

<u>REDLAND BRICKS LTD, CROWBOROUGH BRICKWORKS, JARVIS BROOK, CROWBOROUGH.</u>
Gauge : 2'0".   (TQ 532296)

No.1		0-4-0BE		WR	M7534 1972
No.4		0-4-0BE		WR	4634 1951
		0-4-0BE		WR	T8033 1979
		4wBE		WR	

<u>RUGBY PORTLAND CEMENT CO LTD, LEWES WORKS, SOUTHERHAM, LEWES.</u>
Gauge : 4'8½".   (TQ 426094)

0-4-0DM		(RSHN	7924 1959
		(DC	2591 1959

ESSEX

ALLIED BREWERIES LTD, MANNINGTREE.
Gauge : 4'8½".   (   )

  3053 (CAR No.92, CAR No.93)   4w-4w+4w-4wRER

B.P.OIL LTD, PURFLEET TERMINAL.
Gauge : 4'8½".   (TQ567775)

No.25   0-6-0DE   YE   2641 1957

ALEXANDER BRUCE (GRAYS) LTD, STEVEDORES, BRUCES WHARF, GRAYS.
Gauge : 4'8½".   (TQ 612776)

  41   4wDM   FH   3885 1958

BUTLINS LTD, CLACTON HOLIDAY CAMP.
Gauge : 2'0".   (TM 164136)

  148   C.P.HUNTINGDON   4-2-4DH S/O   Chance 76.50-148-24 1976

CARLESS SOLVENTS LTD, HARWICH REFINERY, REFINERY ROAD, PARKESTON, HARWICH,
Gauge : 4'8½".   (TM 232323)

                                    0-4-0F   OC   RSHN   7803 1954 Pvd
  BWC 687F   FP 41 CO   4wDM   R/R   S&H/Whc   4001
                                    4wDH   R/R   NNM   73511 1979

COLNE VALLEY RAILWAY, near CASTLE HEDINGHAM.
Gauge : 4'8½".   (TL 774362)

No.1   (D2041)   0-6-0DM   Sdn   1959
  1875   BARRINGTON   0-4-0ST OC   AE   1875 1921
No.1   0-4-0ST OC   HL   3715 1928
  2199   VICTORY   0-4-0ST OC   AB   2199 1945
  72   2235/72   0-6-0ST IC   VF   5309 1945
  WD 190   CASTLE HEDINGHAM   0-6-0ST IC   HE   3790 1952
No.40   0-6-0T   OC   RSH   7765 1954
                                    4wPM   Lake & Elliot   c1924
              YD. No.43   4wDM   RH 221639 1943
                                    2w-2PMR   Wkm   DsmT

ESSO PETROLEUM CO LTD.
Bitumen Terminal, Harrison's Wharf, Purfleet,
Gauge : 4'8½".   (TQ 552782)

  2   0-6-0DH   HC   D1373 1965

Purfleet Tank Farm,
Gauge : 4'8½".   (TQ 563777)

                                    0-4-0DM   Bg/DC   2161 1941
                                    0-4-0DE   YE   2686 1958

G.A. & R.J. FELDWICK, WICKFORD.
Gauge : 2'0".   (   )

No.29   AYALA   4wDM   MR   7374 1939

FISONS LTD, STANFORD-LE-HOPE.
Gauge : 4'8½".   (TQ 696815)

			0-4-0DH	JF	4220001 1959

J.GEVERTZ LTD, DEALERS, THE OLD MILL CORNER, ATTHORPE RODING.
Yard (TL 593143) with locos for scrap or resale occasionally present.

HARLOW DEVELOPMENT CORPORATION, LOWER MEADOW PLAY CENTRE, off PARINGDON ROAD,
SOUTHERN WAY, HARLOW.
Gauge : 4'8½".   (TL 452078)

		4wDM	FH	3596 1953

LAFARGE ALUMINOUS CEMENT CO LTD, LONDON ROAD, WEST THURROCK.
Gauge : 4'8½".   (TQ 573779)

1		4wDH	RR	10247 1966
2		4wDH	RR	10248 1966
3		4wDH	RR	10249 1966

LILLEY/WADDINGTON LTD, PLANT DEPOT, HARVEY ROAD, BURNT MILLS INDUSTRIAL AREA,
BASILDON.
Gauge : 2'0".   (TQ 736904)

EL 1		0-4-0BE	WR	G6304 1960
EL 3		4wBE	WR	C6575 1963
EL 5		4wBE	WR	E6808 1965
EL 6		4wBE	WR	E6809 1965
EL 11		4wBE	CE	5852B 1970
EL 12		4wBE	CE	5852/1 1970
EL 18 W		0-4-0BE	WR	D6879 1964
EL 20 W	66/4/1	0-4-0BE	WR	4475 1950
EL 21 W		0-4-0BE	WR	4476 1950
EL 24 W		0-4-0BE	WR	3219 1945
EL 25 W		4wBE	WR	4213 1949
EL 26 W		4wBE	WR	4355 1950
EL 27 W		4wBE	WR	4653 1951
EL 28 W		4wBE	WR	4654 1951
EL 29 W		4wBE	WR	4352 1950
EL 30 W		4wBE	WR	4897 1952
EL 31 W		4wBE	WR	4898 1952
EL 32 W		4wBE	WR	5005 1952
EL 33 W		4wBE	WR	5007 1952
EL 34 W		4wBE	WR	
EL 35 W		4wBE	WR	
EL 38 W		4wDM	MR	22031 1959
EL 41 W		4wDM	MR	22032 1959
EL 42 W	66/4/7	0-4-0BE	WR	D6878 1964

Gauge : 1'6".

EL 4		0-4-0BE	WR	G7216 1967
EL 7		4wBE	CE	5373/1 1967
EL 8		4wBE	CE	5740 1970
EL 9		4wBE	CE	5740/2 1970
EL 10		4wBE	CE	5464 1968
EL 11		4wBE	CE	5373/2 1967
EL 14		4wBE	CE	5953A 1972
EL 15		4wBE	CE	5953B 1972
EL 16		4wBE	CE	B0110A 1973
EL 17		4wBE	CE	B0110B 1973

```
EL 19 W    66/4/4        0-4-OBE           WR      4580 1950
EL 22 W    66/4/6        0-4-OBE           WR      3788 1948
EL 23 W    66/4/3        0-4-OBE           WR      4579 1950
EL 36 W    PN66/8/1       4wBE             CE      5920 1972
EL 37 W    66/6/1         4wBE             CE      5827 1970
```

Locos present in yard between contracts.
See also entries under Greater Manchester, Warwickshire & Strathclyde.

PETER LIND LTD, CONTRACTORS, AVELEY PLANT CENTRE.
Yard (TQ 557787) with locos occasionally present between contracts.

MILTON HALL (SOUTHEND) BRICK CO LTD.
Gauge : 2'0".   Locos are kept at :-

    C = Cherry Orchard Lane Works, Hawkwell, Rochford.   (TQ 859899)
    S = Star Lane Works, Great Wakering.   (TQ 934873)

```
                       4wDM           MR      8614 1941 C
                       4wDM           MR     21520 1955 C
                       4wDM           RH    441951 1960 S
```

MINISTRY OF DEFENCE, ARMY DEPARTMENT, SHOEBURYNESS ESTABLISHMENT.
    For full details see Section Five.

MOBIL OIL CO LTD, CORYTON BULK TERMINAL.
Gauge : 4'8½".   (TQ 746828)

```
1                      0-4-ODH           AB     506/1 1969 +
2                      0-4-ODH           AB     506/2 1969 +
No.24                  0-4-ODH           TH      239V 1971
                       0-6-ODH           TH      291V 1980
```
    + Rebuild of AB 506 1965.

NATIONAL COAL BOARD, SOUTHERN DEPOT LTD, SOUTHEND COAL CONCENTRATION DEPOT.
Gauge : 4'8½".   (TQ 881865)

```
D2184                  0-6-ODM           Sdn           1962
                       0-4-ODM           JF   4200035 1949   OOU
```

PROCTER & GAMBLE LTD, SOAP MANUFACTURERS, WEST THURROCK.
Gauge : 4'8½".   (TQ 595773)

```
                       4wDH              TH       144V 1964
```

PURFLEET DEEP WHARF & STORAGE CO LTD, PURFLEET.
Gauge : 4'8½".   (TQ 565776)

```
1                      0-4-ODH           RH    437362 1960
2                      0-4-ODH           RH    457303 1963
3                      0-4-ODH           RH    512463 1965
4                      0-4-ODH           RH    512464 1965
                       0-4-ODM          (RSH     7922 1957
                                        (DC      2589 1957
6                      0-4-ODM          (VF      D297 1956
                                        (DC      2583 1956
```

ROM RIVER CO LTD, WITHAM.
Gauge : 4'8½".   (TQ 826146)

		6wDM		KS	4421	1929	
	ROM RIVER	4wDM		FH	3491	1951	

SEALINK LTD, HARWICH HARBOUR.
Gauge : 4'8½".   (   )

		4wDH	R/R	NNM	80504	1980	
		4wDH	R/R	NNM	80505	1980	
		4wDH	R/R	NNM	80508	1980	

SHARPE BROS, GABLES SERVICE STATION, RAYLEIGH.
Gauge : 4'8½".   (TQ 784920)

139	BEATTY	0-4-0ST	OC	HL	3240	1917	Pvd

SHELL U.K., SHELL HAVEN REFINERIES, STANFORD-LE-HOPE.
Gauge : 4'8½".   (TQ 720816, 729815, 740817)

No.25		4wDH		TH	279V	1978	
No.26		4wDH		TH	280V	1978	
No.27		4wDH		TH	281V	1978	
No.28		4wDH		TH	282V	1979	

L.J.SMITH, THE BUNGALOW, RECTORY LANE, BATTLESBRIDGE.
Gauge : 2'0".   (TQ 783965)

	SMUDGE	4wDM		MR	8729	1941	
		4wDM		RH	218016	1943	
02	HAYLEY	0-4-0DM	S/O	Bg	3232	1947	
03		4wBE		CE	5667	1969	

SOUTHEND PIER RAILWAY, SOUTHEND-ON-SEA.
Gauge : 3'6".   (TQ 884850)   RTC.

		2w-2PMR		Wkm	10943	1976	OOU

STOUR VALLEY RAILWAY PRESERVATION SOCIETY, CHAPPEL & WAKES COLNE STATION.
Gauge : 4'8½".   (TL 898289)

(69621)	No.999	0-6-2T	IC	Str		1924	
80151		2-6-4T	OC	Bton		1957	
(2279)	No.2	0-6-0DM		(RSH	8097	1960	
				(DC	2656	1960	
3051	CAR No.88	4w-4wRER					
No.11	STOREFIELD	0-4-0ST	OC	AB	1047	1905	
	JUBILEE	0-4-0ST	OC	WB	2542	1936	
68067	GUNBY	0-6-0ST	IC	HE	2413	1941	
54	PENN GREEN	0-6-0ST	IC	RSH	7031	1941	
		0-4-0ST	OC	P	2039	1943	
		0-6-0T	OC	RSHN	7597	1949	
	JUPITER	0-6-0ST	IC	RSH	7671	1950	
8410/45	BELVOIR	0-6-0ST	OC	AB	2350	1954	
No.13		0-6-0T	OC	RSH	7846	1955	
		4wPM		MR	2029	1920	
A.M.W.No.144	PAXMAN	0-4-0DM		AB	333	1938	
		4wWE		MV		1941	
		4wDM		FH	3294	1948	
No.23		0-4-0DH		JF	4220039	1965	
		2w-2PMR		Wkm	1583	1934	DsmT
RT 960232		2w-2PMR		Wkm			Dsm

THAMES MATEX LTD, OLIVER ROAD, WEST THURROCK.
Gauge : 4'8½".   (TQ 576766)

(D2953)	0-4-0DM	AB	395 1955	OOU
	0-6-0DH	RSHD/WB	8343 1962	

TUNNEL INDUSTRIAL SERVICES LTD, TUNNEL INDUSTRIAL ESTATE, WEST THURROCK.
Gauge : 4'8½".   (TQ 576777)

No.8	0-4-0DE	YE	2856 1961

WALTON-ON-THE-NAZE PIER RAILWAY.
Gauge : 2'0".   (TM 255214)

	0-4-0DM S/O	Bg	3024 1939

THOS. W. WARD LTD, COLUMBIA WHARF, GRAYS.
Gauge : 4'8½".   (TQ 611776)

7	0-4-0DM	AB	419 1957

Also other locos for scrap occasionally present.

GLOUCESTERSHIRE

BRITISH WATERWAYS BOARD, SHARPNESS DOCKS.
Gauge : 4'8½".   (SO 667023)

REGD. No.DL1	4wDM	RH	463150 1961
DL2	0-6-0DM	WB	3151 1962

CAWOODS SOLID FUELS LTD, CHELTENHAM COAL CONCENTRATION DEPOT.
Gauge : 4'8½".   (SO 936234)

No.6	0-4-0DH	YE	2676 1959	
	0-4-0DH	TH	106C 1961	
	Rebuild of 0-4-0DM	JF	4200011 1947	OOU

DEAN FOREST RAILWAY SOCIETY LTD, NORCHARD STEAM CENTRE, LYDNEY, FOREST OF DEAN.
Gauge : 4'8½".   (SO 629044)

4121		2-6-2T	OC	Sdn	1937
5521		2-6-2T	OC	Sdn	1927
5532		2-6-2T	OC	Sdn	1928
5541		2-6-2T	OC	Sdn	1928
9681		0-6-0PT	IC	Sdn	1949
18	JESSIE	0-6-0ST	IC	HE	1873 1937
	USKMOUTH I	0-4-0ST	OC	P	2147 1952
No.4	G.B.KEELING	0-6-0ST	IC	HE	3806 1953
2145		0-4-0DM		HE	2145 1940
55	4210101	0-4-0DM		JF	4210101 1955
DS 3057		4wPMR		Wkm	4254 1947
9045		2w-2PMR		Wkm	8774 1960

DOW-MAC (CONCRETE) LTD, NAAS LANE, QUEDGELEY.
Gauge : 4'8½".   (SO 821128)

	0-4-0DE		RH	418602 1958

DOWTY GROUP LTD, ASHCHURCH.
Gauge : 4'8½".   (SO 925335)

| 80 AB 61 | 4wPM | | Dowty | 1959 + |
| | 4wPM | | Dowty | + |

+ Converted lorry.

DOWTY RAILWAY PRESERVATION SOCIETY, ASHCHURCH.
Gauge : 4'8½".   (SO 925335)

No.1		0-4-0T	OC	AE	1977 1925
	DRAKE	0-4-0ST	OC	AB	2086 1940
		0-4-0F	OC	AB	2126 1942
2		0-4-0ST	OC	AB	2221 1946
21		0-4-0DM		JF	4210130 1957
19		0-6-0DH		JF	4240016 1964

Gauge : 1'10¾".

| 1 | GEORGE B. | 0-4-0ST | OC | HE | 680 1898 |

Gauge : 2'0".

2		4wDM		L	34523 1949
3	SPITFIRE	4wPM		MR	7053 1937
4		4wDM		RH	354028 1953
5		4wDM		RH	181820 1936
6		4wDM		RH	166010 1932
7	JUSTINE	0-4-0WT	OC	Jung	939 1906
8		4wDM		L	28039 1945
	CHAKA	0-4-2T	OC	HE	2075 1940
	PETER PAN	0-4-0ST	OC	KS	4256 1922

GLOUCESTER RAILWAY CARRIAGE & WAGON CO LTD, BRISTOL ROAD, GLOUCESTER.
Gauge : 4'8½".   (   )

| 4184 | 4w-4RER | | GRC | 1923 Pvd |

GLOUCESTERSHIRE WARWICKSHIRE RAILWAY SOCIETY, TODDINGTON GOODS YARD.
Gauge : 4'8½".   (SP 050322)

2807		2-8-0	OC	Sdn	1905
5952	COGAN HALL	4-6-0	OC	Sdn	1935
7821	DITCHEAT MANOR	4-6-0	OC	Sdn	1950
7828	ODNEY MANOR	4-6-0	OC	Sdn	1950
		0-4-0ST	OC	P	1976 1939
		0-6-0ST	IC	HE	2409 1942
D1	STEAMTOWN	0-6-0DM		HC	D615 1938

Gauge : 2'0".

| | 4wDM | | HE | |

MINISTRY OF DEFENCE, ARMY DEPARTMENT, ASHCHURCH DEPOT.
For full details see Section Five.

SOUTHERN DEPOT CO LTD, STONEHOUSE COAL CONCENTRATION DEPOT.
Gauge : 4'8½".   (SO 799055)

DOUGAL		0-4-0DM	(VF	D77	1947
			(DC	2251	1947

WAGON REPAIRS LTD, REPARCO WORKS, INDIA ROAD, GLOUCESTER.
Gauge : 4'8½".   (SO 845177)

L 13					
		4wDM	RH	279591	1949
	YARD No.4856	4wDM	FH	3737	1955
	YARD No.5332	4wDM	FH	3816	1956	OOU

R.J.WASHINGTON, 404, GLOUCESTER ROAD, CHELTENHAM.
Gauge : 2'0".   (SO 926217)

3				
	4wDM	L	(8022	1936?)
	4wPM	R.Thomas		1941
	4wDM	RH	213834	1942
	4wDM	MR	26007	1964
	2w-2PM	R.J.Washington		1980

G.F.WILLIAMS, TANHOUSE FARM, FRAMPTON ON SEVERN.
Gauge : 2'6".   (SO 745069)

	4wDM	RH	170374	1934	Dsm

GREATER LONDON

BRITISH GYPSUM LTD, ERITH.
Gauge : 4'8½".   (TQ 507789)   RTC.

20287	PRINCESS MARGARET	0-4-0DM	AB	376	1948	OOU
	CRABTREE	4wDM	RH	338416	1953	OOU
		4wDM	RH	512572	1965	OOU

BRITISH RAILWAYS BOARD, EASTERN REGION, STRATFORD DIESEL REPAIR SHOP.
Gauge : 4'8½".   (TQ 383849)

Industrial locos occasionally present for overhaul.

CENTRAL ELECTRICITY GENERATING BOARD.
Croydon 'B' Power Station, Waddon Marsh.
Gauge : 4'8½".   (TQ 305665)

	HENGIST	0-4-0DH	RSHD/WB	8367	1962
2		0-4-0DH	EEV	D1122	1966

COAL MECHANISATION (TOLWORTH) LTD, TOLWORTH COAL CONCENTRATION DEPOT.
Gauge : 4'8½".   (TQ 198656)

D2310		0-6-0DM	(RSH	8169	1960
			(DC	2691	1960
20188		0-4-0DM	AB	375	1948

GEORGE COHEN, SONS & CO LTD.
Bidder Street Scrapyard, Canning Town.
Gauge : 4'8½". (TQ 391818)

| No.3 | ELIZABETH II | | 4wDM | | FH | 3736 1955 | |
| | | | 0-4-0DM | | Bg | 3589 1962 | |

Also other locos for scrap occasionally present.

Park Royal Scrapyard.
Gauge : 4'8½". (TQ 200824)

| | | | 4wDM | | FH | 1977 1936 | OOU |

Also other locos for scrap occasionally present.

DAY & SON, BRENTFORD TOWN GOODS DEPOT.
Gauge : 4'8½". (TQ 166778)

| (12049) | | | 0-6-0DE | | Derby | 1948 | |

P.ELMS, 73, CROW LANE, ROMFORD.
Gauge : 3'6½". (TQ 500878)

| | WOTO | | 0-4-0ST OC | | WB | 2133 1924 | Pvd |
| | SIR TOM | | 0-4-0ST OC | | WB | 2135 1925 | Pvd |

600 FERROUS FRAGMENTISERS LTD, SCRUBS LANE, WILLESDEN.
Gauge : 4'8½". (   )

| (03018) | 600 No.2 | | 0-6-0DM | | Sdn | 1958 | |

FORD MOTOR CO LTD, DAGENHAM.
Gauge : 4'8½". (TQ 496825, 499827)

1	(D2267)		0-6-0DM		(RSH	7897 1958	
					(DC	2611 1958	
2	(D2280)	P 1381 C	0-6-0DM		(RSH	8089 1960	
					(DC	2657 1960	
3	(D2333)	P 1062 C	0-6-0DM		(RSH	8192 1961	
					(DC	2714 1961	
4	(2051)		0-6-0DM		Don	1959	
5			0-6-0DH		EEV	D1229 1967	
				Rebuilt	HE	8900 1977	
6		P 1215 C	0-6-0DH		HC	D1396 1967	
	GT/PL/1	P 260 C	4wDM		Robel 21 11 RK1		

M.J.GLEESONS (CONTRACTORS) LTD, PLANT DEPOT, DRAKE ROAD, MITCHAM.
Gauge : 2'0". (TQ 281673)

| | | | 0-4-0BE | | WR | N7639 1973 | |

Loco present in yard between contracts.

GREENWICH NARROW GAUGE RAILWAY SOCIETY, off NATHAN WAY, WOOLWICH.
Gauge : 2'0". (   )

| | | | 4wDM | | RH | 172892 1934 | |

ARTHUR GUINNESS, SON & CO (PARK ROYAL) LTD, PARK ROYAL BREWERY.
Gauge : 4'8½".   (TQ 195828)

	CARPENTER	0-4-ODM		FH	3270	1948
	WALRUS	0-4-ODM		FH	3271	1949

G.W.R. PRESERVATION GROUP, BRIDGE ROAD DEPOT, SOUTHALL.
Gauge : 4'8½".   (   )

2885		2-8-0	OC	Sdn		1938
4110		2-6-2T	OC	Sdn		1936
	BIRKENHEAD	0-4-OST	OC	RSHN	7386	1948
	A.E.C. No.1	4wDM		AEC		1938

HARINGEY LIBRARIES, MUSEUM & ARTS SERVICE, BRUCE CASTLE PARK, TOTTENHAM.
Gauge : 2'0".   (TQ 335907)

	4wAtmospheric Car		c1865	Pvd

KINGS COLLEGE, STRAND.
Gauge : 1'3".   (TQ 308808)

PEARL	2-2-2WT	IC	Brotherhood	1860	Pvd

LONDON TRANSPORT EXECUTIVE.
Gauge : 4'8½".   Locos are kept at :-

Acton Works, Bollo Lane.	(TQ 196791)
Amersham Station.	(SP 964982)
Cockfosters Depot.	(TQ 288962)
Ealing Common Depot, Uxbridge Road.	(TQ 189802)
Golders Green Depot, Finchley Road.	(TQ 253875)
Hainault Depot.	(TQ 450918)
Highgate Depot.	(TQ 279886)
Lillie Bridge Depot.	(TQ 250782)
Morden Depot.	(TQ 255680)
Neasden Depot.	(TQ 206858)
Northfields Depot, Northfields Avenue.	(TQ 167789)
Rickmansworth Station.	(TQ 057946)
Stonebridge Park.	(TQ 192845)
Upminster Depot.	(TQ 570871)
Wembley Park Signal Engineers Depot.	(TQ 194864)
West Ruislip Depot.	(TQ 094862)

L 11	4w-4wRE		MC	1931
		Rebuilt	Acton	1964
L 13A/L 13B	4w-4w+4w-4wRE		MC	1938
		Rebuilt	Acton	1974
L 15	4w-4wBE/RE		MC	1970
L 16	4w-4wBE/RE		MC	1970
L 17	4w-4wBE/RE		MC	1971
L 18	4w-4wBE/RE		MC	1971
L 19	4w-4wBE/RE		MC	1971
L 20	4w-4wBE/RE		MC	1964
L 21	4w-4wBE/RE		MC	1964
L 22	4w-4wBE/RE		MC	1965
L 23	4w-4wBE/RE		MC	1965
L 24	4w-4wBE/RE		MC	1965
L 25	4w-4wBE/RE		MC	1965
L 26	4w-4wBE/RE		MC	1965
L 27	4w-4wBE/RE		MC	1965
L 28	4w-4wBE/RE		MC	1965
L 29	4w-4wBE/RE		MC	1965

L 30	4w–4wBE/RE		MC	1965
L 31	4w–4wBE/RE		MC	1965
L 32	4w–4wBE/RE		MC	1965
L 33	4w–4wBE/RE		Acton	1962
L 35	4w–4wBE/RE		GRC	1938
L 36	4w–4wBE/RE		GRC	1938
L 37	4w–4wBE/RE		GRC	1938
L 38	4w–4wBE/RE		GRC	1938
L 39	4w–4wBE		GRC	1938
L 40	4w–4wBE/RE		GRC	1938
L 44	4w–4wBE/RE		Don	1973
L 45	4w–4wBE/RE		Don	1974
L 46	4w–4wBE/RE		Don	1974
L 47	4w–4wBE/RE		Don	1974
L 48	4w–4wBE/RE		Don	1974
L 49	4w–4wBE/RE		Don	1974
L 50	4w–4wBE/RE		Don	1974
L 51	4w–4wBE/RE		Don	1974
L 52	4w–4wBE/RE		Don	1974
L 53	4w–4wBE/RE		Don	1974
L 54	4w–4wBE/RE		Don	1974
L 55	4w–4wBE/RE		RYP	1951
L 56	4w–4wBE/RE		RYP	1951
L 57	4w–4wBE/RE		RYP	1951
L 58	4w–4wBE/RE		RYP	1951
L 59	4w–4wBE/RE		RYP	1951
L 60	4w–4wBE/RE		RYP	1951
L 61	4w–4wBE/RE		RYP	1952
L 126	4w–4wRE		GRC	1938
		Rebuilt	Acton	1971
L 127	4w–4wRE		GRC	1938
		Rebuilt	Acton	1971
L 128	4w–4wRE		GRC	1938
		Rebuilt	Acton	1971
L 129	4w–4wRE		GRC	1938
		Rebuilt	Acton	1971
L 130	4w–4RE		MC	1934
		Rebuilt	Acton	1967
L 131	4w–4RE		MC	1934
		Rebuilt	Acton	1967
L 134	4w–4RE		M	1927
		Rebuilt	Acton	1967	OOU
L 135	4w–4RE		MC	1934
		Rebuilt	Acton	1967	OOU
L 140	4w–4wRE		MC	1938
		Rebuilt	Acton	1980
L 141	4w–4wRE		MC	1938
		Rebuilt	Acton	1973
L 142	4w–4wRE		MC	1938
		Rebuilt	Acton	1973
L 143	4w–4wRE		MC	1938
		Rebuilt	Acton	1973
L 144	4w–4wRE		MC	1938
		Rebuilt	Acton	1975
L 145	4w–4wRE		MC	1938
		Rebuilt	Acton	1975
L 146	4w–4wRE		MC	1938
		Rebuilt	Acton	1976
L 147	4w–4wRE		MC	1938
		Rebuilt	Acton	1976
L 148	4w–4wRE		MC	1938
		Rebuilt	Acton	1977
L 149	4w–4wRE		MC	1938
		Rebuilt	Acton	1977

L 150		4w-4wRE	MC	1938	
	Rebuilt		Acton	1978	
L 151		4w-4wRE	MC	1938	
	Rebuilt		Acton	1978	
L 152		4w-4wRE	MC	1938	
	Rebuilt		Acton	1978	
L 153		4w-4wRE	MC	1938	
	Rebuilt		Acton	1978	
L 154		4w-4wRE	MC	1938	
	Rebuilt		Acton	1978	
L 155		4w-4wRE	MC	1938	
	Rebuilt		Acton	1978	
L 156		4w-4wRE	MC	1938	
	Rebuilt		Acton	1981	
L 157		4w-4wRE	MC	1938	
	Rebuilt		Acton	1981	
L 158		4w-4wRE	MC	1938	
	Rebuilt		Acton	1981	
ESL 100		4w-4-4-4wRE	BRC	1903	
	Rebuilt		Acton	1938	OOU
ESL 101		4w-4-4-4wRE	BRC/Met.Amal	1903	
	Rebuilt		Acton	1939	
ESL 102		4w-4-4-4wRE	BRC	1903	
	Rebuilt		Acton	1939	
ESL 104		4w-4-4-4wRE	BRC	1903	
	Rebuilt		Acton	1939	
ESL 105		4w-4-4-4wRE	BRC/Met.Amal	1903	
	Rebuilt		Acton	1939	OOU
ESL 106		4w-4-4-4wRE	BRC	1903	
	Rebuilt		Acton	1939	
ESL 107		4w-4-4-4wRE	BRC/Met.Amal	1903	
	Rebuilt		Acton	1939	
ESL 108		4w-4-4-4wRE	BRC	1903	
	Rebuilt		Acton	1939	
ESL 111		4w-4-4-4wRE	Met.Amal	1903	
	Rebuilt		Acton	1940	OOU
ESL 114		4w-4-4-4wRE	BRC	1903	
	Rebuilt		Acton	1940	OOU
ESL 116		4w-4-4-4wRE	BRC/Met.Amal	1903	
	Rebuilt		Acton	1940	OOU
ESL 117		4w-4-4-4wRE	BRC/Met.Amal	1903	
	Rebuilt		Acton	1942	
ESL 118A/ESL 118B		4w-4-4-4wRE	BRC	1932	
	Rebuilt		Acton	1961	
TCC 1		4w-4wRE	MC	1938	
	Rebuilt		Acton	1978	
TCC 5		4w-4wRE	MC	1938	
	Rebuilt		Acton	1978	
TRC 910		4w-4wRE	Cravens	1960	
	Rebuilt		Acton	1978	
TRC 911		4w-4wRE	Cravens	1960	
	Rebuilt		Acton	1978	
53003		4w-4wRER	BRC	1937	
53028		4w-4wRER	BRC	1937	
53210		4w-4wRER	GRC	1939	
53223		4w-4wRER	GRC	1939	
53262		4w-4wRER	BRC	1939	
54003		4w-4wRER	BRC	1937	
54035		4w-4wRER	GRC	1937	
54210		4w-4wRER	GRC	1939	
54211		4w-4wRER	GRC	1939	
54235		4w-4wRER	GRC	1939	
12	SARAH SIDDONS	4w-4wRE	VL	1922	

DL 81		0-6-0DH	RR	10278 1968	
DL 82		0-6-0DH	RR	10272 1967	
DL 83		0-6-0DH	RR	10271 1967	
40/173		2w-2PMR	Wkm	7819 1957	
		2w-2PMR	Wkm	8822 1961	Dsm
		2w-2PMR	Wkm	9522 1963	Dsm
N		2w-2PMR	Wkm	9523 1963	
		2w-2PMR	Wkm	9813 1965	
		2w-2PMR	Wkm	9814 1965	
40/443		4wBER	RM	13502 c1968	

LONDON TRANSPORT EXECUTIVE, COVENT GARDEN MUSEUM.
Gauge : 4'8½".   (TQ 303809)

23		4-4-0T OC	BP	710 1866	
		4wWT G	AP	807 1872	
5	JOHN HAMPDEN	4w-4wRE	VL	1922	
4248		4w-4wRER	GRC	1923	
11182		4w-4wRER	MC	1939	

M.E.ENGINEERING LTD, PLANT DEALERS, NEASDEN LANE, NEASDEN.
Gauge : 2'0".   (TQ 215853)

		4wDM	RH	195846 1939	Dsm
47		4wDM	RH	221610 1943	Dsm
		4wDM	RH	223700 1943	Dsm
		4wDM	MR	9543 1950	OOU
		4wDM	FH	3582 1954	OOU
		4wDM	MR	11111 1959	OOU

Gauge : 3'0".

		4wDM	MR	40S323 1968	OOU

Gauge : 2'4".

		4wBE	Bilsthorpe c1966	OOU

Also other locos occasionally present for hire & resale.

J.MURPHY & SONS LTD, PLANT DEPOT, KENTISH TOWN.
Yard (TQ 287855) with locos present between contracts.

See entry under Staffordshire for loco fleet details.

MUSEUM OF LONDON, LONDON WALL, E.C.2.
Gauge : 2'0".   (TQ 322816)

		4wAtmospheric Car	c1865 Pvd

PLASSER RAILWAY MACHINERY (G.B.) LTD, DRAYTON GREEN ROAD, WEST EALING.
Gauge : 4'8½".   (TQ 161809)

		4wDM Plasser

POST OFFICE RAILWAY.
Gauge : 2'0".   Locos are kept at :-

	King Edward Building, St. Pauls.		(TQ)
	Mount Pleasant Parcels Office, Clerkenwell.		(TQ 311823)
	New Western District Office, Rathbone Place.		(TQ 296814)

1		4wBE	EE	702	1926	
2		4wBE	EE	703	1926	
3		4wBE	EE	704	1926	
		4wRE	EE	601	1925	Pvd
		4wRE	EE	652	1925	+
	BREAKDOWN No.1	4wRE	EE			a
	BREAKDOWN No.2	4wRE	EE			a
		2w-2-2-2wRE				b
		2w-2-2-2wRE				b
752		2w-2-2-2wRE	EE	752	1930	
754		2w-2-2-2wRE	EE	754	1930	
755		2w-2-2-2wRE	EE	755	1930	
756		2w-2-2-2wRE	EE	756	1930	
758		2w-2-2-2wRE	EE	758	1930	OOU
759		2w-2-2-2wRE	EE	759	1930	OOU
760		2w-2-2-2wRE	EE	760	1930	
761		2w-2-2-2wRE	EE	761	1930	
762		2w-2-2-2wRE	EE	762	1930	
763		2w-2-2-2wRE	EE	763	1930	OOU
793		2w-2-2-2wRE	EE	793	1930	
794		2w-2-2-2wRE	EE	794	1930	
795		2w-2-2-2wRE	EE	795	1930	
796		2w-2-2-2wRE	EE	796	1930	
797		2w-2-2-2wRE	EE	797	1930	
799		2w-2-2-2wRE	EE	799	1930	OOU
801		2w-2-2-2wRE	EE	801	1930	
802		2w-2-2-2wRE	EE	802	1930	OOU
803		2w-2-2-2wRE	EE	803	1930	
804		2w-2-2-2wRE	EE	804	1930	OOU
805		2w-2-2-2wRE	EE	805	1930	OOU
806		2w-2-2-2wRE	EE	806	1930	
807		2w-2-2-2wRE	EE	807	1930	OOU
808		2w-2-2-2wRE	EE	808	1930	
810		2w-2-2-2wRE	EE	810	1930	
811		2w-2-2-2wRE	EE	811	1930	
812		2w-2-2-2wRE	EE	812	1930	
813		2w-2-2-2wRE	EE	813	1930	OOU
814		2w-2-2-2wRE	EE	914	1930	
815		2w-2-2-2wRE	EE	815	1930	
816		2w-2-2-2wRE	EE	816	1930	
817		2w-2-2-2wRE	EE	817	1930	OOU
818		2w-2-2-2wRE	EE	818	1930	OOU
819		2w-2-2-2wRE	EE	819	1930	
820		2w-2-2-2wRE	EE	820	1931	OOU
822		2w-2-2-2wRE	EE	822	1931	OOU
823		2w-2-2-2wRE	EE	823	1931	
824		2w-2-2-2wRE	EE	824	1931	OOU
826		2w-2-2-2wRE	EE	826	1931	
827		2w-2-2-2wRE	EE	827	1931	
828		2w-2-2-2wRE	EE	828	1931	
830		2w-2-2-2wRE	EE	830	1931	OOU
925		2w-2-2-2wRE	EE	925	1936	OOU
926		2w-2-2-2wRE	EE	926	1936	
928		2w-2-2-2wRE	EE	928	1936	
929		2w-2-2-2wRE	EE	929	1936	
930		2w-2-2-2wRE	EE	930	1936	
931		2w-2-2-2wRE	EE	931	1936	OOU
932		2w-2-2-2wRE	EE	932	1936	OOU

1		2w-2-2-2wRE		EE	3334 1962	Dsm
2		2w-2-2-2wRE		EE	3335 1962	
501		2w-2-2-2wRE		GB	420461/1 1980	
502		2w-2-2-2wRE		GB	420461/2 1980	
503		2w-2-2-2wRE		GB	420461/3 1980	
504		2w-2-2-2wRE		GB	420461/4 1980	
505		2w-2-2-2wRE		GB	420461/5 1980	
506		2w-2-2-2wRE		GB	420461/6 1980	
507		2w-2-2-2wRE		GB	420461/7 1980	
508		2w-2-2-2wRE		GB	420461/8 1980	
509		2w-2-2-2wRE		GB	420461/9 1981	
510		2w-2-2-2wRE		GB	420461/10 1981	
511		2w-2-2-2wRE		GB	420461/11 1981	
512		2w-2-2-2wRE		GB	420461/12 1981	
513		2w-2-2-2wRE		GB	420461/13 1981	
514		2w-2-2-2wRE		GB	420461/14 1981	
515		2w-2-2-2wRE		GB	420461/15 1981	
516		2w-2-2-2wRE		GB	420461/16 1981	
517		2w-2-2-2wRE		GB	420461/17 1981	
518		2w-2-2-2wRE		GB	420461/18 1981	
519		2w-2-2-2wRE		GB	420461/19 1981	
520		2w-2-2-2wRE		GB	420461/20 1981	
521		2w-2-2-2wRE		GB	420461/21 1981	
522		2w-2-2-2wRE		GB	420461/22 1981	

+ Converted to a battery carrier.
a Converted to a wagon.
b Converted to a passenger car.

PURFLEET DEEP WHARF & STORAGE CO LTD, ERITH WHARF.
Gauge : 4'8½".   (TQ 517779)

1	THETIS	0-4-0DM	(RSHN	7815	1954
			(DC	2502	1954
2	TAURUS	0-4-0DM	(VF	D139	1951
			(DC	2273	1951
	PRIAM	0-4-0DM	(VF	D293	1955
			(DC	2566	1955

RESCO (RAILWAYS) LTD.
c/o Fielding & Bacon Ltd, Erith.
Gauge : 4'8½".   (TQ 525774)

63.000.428		0-6-0ST	IC		RSH	7135 1944
63.000.440		0-6-0ST	IC		HE	3696 1950
No.4	(W4W)	4w-4wDMR			AEC	1934
		4wDM			FH	3658 1953
21 90 03	J.D.THOMSON	0-6-0DH			EEV	3990 1970

Locos for repair and/or resale usually present.

Nathan Way Industrial Estate, Woolwich.
Gauge : 4'8½".   (TQ 45x79x)

No.15	HASTINGS	0-6-0ST	IC		HE	469 1888
No.14	CHARWELTON	0-6-0ST	IC		MW	1955 1917
No.10	GERVASE	0-4-0VBT	VCG		S	6807 1928
		Rebuild of 0-4-0ST			MW	1472 1900 +
		0-6-0DE			EE	1554 1948
		4wDM			HE	5308 1960

+ Carries plate S 6710.
Locos for repair and/or resale usually present.

S.C.ROBINSON, 47, WAVERLEY GARDENS.
Gauge : 2'0".   (TQ 187831)

| | | 4wDM | | RH | 209429 | 1942 |

ROM RIVER CO LTD, STEEL REINFORCEMENT ENGINEERS, STEWARTS ROAD, BATTERSEA.
Gauge : 4'8½".   (TQ 293765)

| 33009 | B308 | 0-4-0DH | | NB | 27814 | 1958 |

SCIENCE MUSEUM, SOUTH KENSINGTON.
Gauge : 5'0".   (TQ 268793)

| | "PUFFING BILLY" | 4w | VCG | Wm.Hedley | 1813 |

Gauge : 4'8½".

	ROCKET	0-2-2	OC	RS	19	1829
	"SANS PAREIL"	0-4-0	VC	Hackworth		1829
	BAUXITE No.2	0-4-0ST	OC	BH	305	1874
4073	CAERPHILLY CASTLE	4-6-0	4C	Sdn		1923
	ROCKET	0-2-2	OC	RS	4089	1934 +
No.1		4wRE		M. & Platt/BP		1890
	DELTIC	6w-6wDE		EE	2007	1955
3327		4w-4wRER		MC		1927

        + Replica of loco in original condition as built in 1829.

Gauge : 1'11½".

| | | 0-4-0DM | | HE | 4369 | 1951 + |

        + Not on current display.

SHELMERDINE & MULLEY LTD, EDGEWARE ROAD, CRICKLEWOOD, N.W.2.
Gauge : 1'3".   (TQ 233864)

| | | 2-6-2 | OC | J.Lemon-Burton | 1967 |

TARMAC ROADSTONE LTD, HAYES.
Gauge : 4'8½".   (TQ 105795)

| 655/29/38 | | 4wDM | | RH | 518494 | 1967 |

TAYLOR WOODROW PLANT LTD, PLANT DEPOT, GREENFORD.
Gauge : 2'0".   (TQ 126826)   (Subsidiary of Taylor Woodrow Ltd)

		4wBE	WR	6865	1964
EL 2		4wBE	WR	6867	1964
		4wBE	WR	6868	1964	Dsm
		4wBE	WR		

        Locos present in yard between contracts.

THAMES METAL CO LTD, SCRAP MERCHANTS, ANGERSTEIN WHARF, GREENWICH.
Gauge : 4'8½".   (TQ 403792)

| 48 | | 4wDM | | FH | 3912 | 1959 |

THAMES WATER AUTHORITY, METROPOLITAN PUBLIC HEALTH DIVISION, HOGSMILL VALLEY
SEWAGE WORKS, BERRYLANDS.
Gauge : 2'0".   (TQ 195685)   RTC.

| 1 | | 4wDM | | HE | 4848 | 1957 | OOU |
| 5 | | 4wDM | | HE | 7120 | 1969 | OOU |

THOS W.WARD LTD, SILVERTOWN MACHINERY WORKS, THAMES ROAD.
Gauge : 4'8½".   (TQ 417801)

S 75		4wDM	RH	305314	1951	OOU
		0-4-0DM	JF	4210076	1952	OOU
		0-4-0DM	HC	D1009	1956	
		4wDM	FH	3900	1959	

SAMUEL WILLIAMS (DAGENHAM DOCK) LTD, DAGENHAM DOCK.
Gauge : 4'8½".   (TQ 489817, 490820)   RTC.

23	4wDM	FH	3722	1955	OOU
24	4wDM	FH	3768	1955	OOU
25	4wDM	FH	3799	1956	OOU
26	4wDM	FH	3813	1956	OOU
27	4wDM	FH	3945	1960	OOU
28	4wDM	FH	3949	1960	OOU
	4wDH	FH	3997	1963	OOU

GREATER MANCHESTER

ALLIED STEEL & WIRE LTD, GRIDWELD DIVISION, WOODHOUSE LANE, WIGAN.
Gauge : 4'8½".   (SD 565066)

397	0-6-0DE	YE	2757	1959

ARNOTT YOUNG & CO LTD, AGECROFT.
Gauge : 4'8½".   (SJ 811998)

D3	0-4-0DE	RH	398119	1957	OOU

ASHTON CANAL CARRIERS, HANOVER STREET NORTH, GUIDE BRIDGE.
Gauge : 2'0".   (SJ 920978)

4wDM	RH	200761	1941
4wDM	HE	2820	1943
4wDM	HE	6012	1960

ASSOCIATED TUNNELLING CO LTD, PLANT DEPOT, LOWTON ST MARYS, near WARRINGTON.
Gauge : 2'0".   (SJ 632974)

No.3	6	4wBE	WR	6502	1962
		4wBE	GB	420253	1970

Locos present in yard between contracts.

P.BLACKHAM, ? STOCKPORT.
Gauge : 1'3".   (   )

BLUE PACIFIC	4-6-2	OC	K.L.Guinness c1935

BRITISH FUEL CO, BROADWAY DEPOT, CHADDERTON.
Gauge : 4'8½".   (SD 903054)

35	0-6-0DM	HC	D1036	1958
	4wDH	TH/S	107C	1961

BRITISH RAIL ENGINEERING LTD. HORWICH WORKS.
Gauge : 4'8½".   (   )

| D1048 | WESTERN LADY | 6w-6wDH | | Crewe | | 1962 | OOU |

BRITISH STEEL CORPORATION, PROFIT CENTRES, B.S.C.FORGES, FOUNDRIES AND ENGINEERING, RAILWAY & RING ROLLED PRODUCTS, TRAFFORD PARK WORKS, ASHBURTON ROAD, MANCHESTER.
Gauge : 4'8½".   (SJ 779971)

10		0-4-0DM		(RSH	7810	1954	
				(DC	2513	1954	
14		0-4-0DM		(RSH	8089	1959	
				(DC	2654	1959	

BRITISH STEEL CORPORATION, REDPATH ENGINEERING LTD, MANUFACTURING DIVISION, WARRINGTON BRANCH, MANCHESTER WORKS, WESTINGHOUSE ROAD; TRAFFORD PARK, MANCHESTER.
Gauge : 4'8½".   (SJ 793964)

| | | 4wDM | | FH | 3886 | 1958 | |

CARREX METAL GROUP LTD, SCRAP MERCHANTS, REGENT STREET, ROCHDALE.
Gauge : 4'8½".   (SD 901142)

| | | 0-6-0F | OC | HL | 3805 | 1932 | Pvd |

CENTRAL ELECTRICITY GENERATING BOARD.
Agecroft Power Station.
Gauge : 4'8½".   (SD 802016)

AG-1	AGECROFT No.1	0-4-0ST	OC	RSHN	7416	1948	OOU
AG-2	AGECROFT No.2	0-4-0ST	OC	RSHN	7485	1948	OOU
AG-3	AGECROFT No.3	0-4-0ST	OC	RSHN	7681	1951	OOU

Carrington Power Station, Partington.
Gauge : 4'8½".   (SJ 727933, 731934)

		0-4-0DM		JF	4210059	1951	OOU
		0-6-0DH		HE	6973	1969	
		0-6-0DH		HE	8977	1980	

Chadderton Power Station, Oldham.
Gauge : 4'8½".   (SD 890038)

1	(D2593)	0-6-0DM		HE	7179	1969	
			Rebuild of	HE	5642	1959	
2	(D2587)	0-6-0DM		HE	7180	1969	
			Rebuild of	HE	5636	1959	
3		0-6-0DH		HE	8976	1979	

Kearsley Power Station, Stoneclough, Radcliffe.
Gauge : 4'8½".   (SD 761049)

2		4w-4wWE		HL	3872	1936	
3		4w-4wWE		RSHN	7078	1944	OOU
4		4w-4wWE		RSHN	7284	1945	OOU
		0-4-0DM		JF	4210078	1952	

CLAYTON ANILINE CO LTD, DYESTUFF WORKS, CLAYTON.
Gauge : 4'8½".   (SJ 877984)   RTC.

	0-4-0DM		JF	4210074	1952	OOU

GEORGE COHEN, SONS & CO LTD, FREDERICK ROAD SCRAPYARD, SALFORD.
Gauge : 4'8½".   (SJ 815994)

L/15	0-4-0DM		JF	22895	1940	
	0-4-0DM		JF	23009	1944	OOU

CROXDEN GRAVEL LTD, TWELVE YARDS ROAD, IRLAM.
Gauge : 2'0".   (SJ 715966)

4wDM	LB	51651	1960
4wDM	Alan Keef	No.5	1979

J.F.DONELAN & CO LTD, PLANT DEPOT.
Gauge : 1'6".   (   )

4wBE	CE	B0151A	1973
4wBE	CE	B0171	1974
4wBE	CE	B0171C	1974
4wBE	CE	B0171D	1974
4wBE	CE	B0171A	1974
0-4-0BE	WR	N7608	1974
0-4-0BE	WR	N7609	1974

Locos present in yard between contracts.

EAST LANCASHIRE RAILWAY PRESERVATION SOCIETY, BURY TRANSPORT MUSEUM,
CASTLECROFT ROAD, BURY.
Gauge : 4'8½".   (SD 803109)

D832		4w-4wDH		Sdn		1960
D1041	WESTERN PRINCE	6w-6wDH		Crewe		1962
32	GOTHENBURG	0-6-0T	IC	HC	680	1903
No.945		0-4-0ST	OC	AB	945	1904
No.70		0-6-0T	IC	HC	1464	1921
No.1		0-4-0ST	OC	AB	1927	1927
	M.E.A. No.1	0-6-0T	OC	RSHN	7683	1951
	WINFIELD	4wDM		MR	9009	1948
		4wDM		FH	3438	1950

ESSO PETROLEUM CO LTD, MODE WHEEL REFINERY, TRAFFORD PARK.
Gauge : 4'8½".   (SJ 795976)

4wDH	RR	10198	1965

G.E.C. TURBINE GENERATORS LTD, TRAFFORD PARK.
Gauge : 4'8½".   (SJ 793957)

84002	4w-4wWE	NB	27794	1960 +
84010	4w-4wWE	NB	27802	1960 +
	0-4-0DE	RR	10254	1967

+ In use as a stationary generator.

GREATER MANCHESTER SCIENCE & RAILWAY MUSEUM, LIVERPOOL ROAD, MANCHESTER.
Gauge : 4'8½".   (   )

	"NOVELTY"	2-2-0VBWT	VC	Science Mus	1929	+	
		0-4-0ST	OC	AB	1964	1929	
1		4w-4wWE		HL	3682	1927	a
		4wBE		EE	1378	1944	

+ Replica of original loco, built 1829, by Braithwaite & Ericsson.
a Currently at Greater Manchester Museum of Transport,
  Boyle Street, Chetham, Manchester. (SD 846006)

Gauge : 3'0".

No.3	PENDER	2-4-0T	OC	BP	1255 1873

ALF HALL, DELPH STATION.
Gauge : 4'8½".   (SD 987074)

	BROOKES No.1	0-6-0ST	IC	HE	2387 1941	
(DB 965100)		2w-2PMR		Wkm	7615 1957	Dsm

JOSE K. HOLT GORDON LTD, SCRAP MERCHANTS, CHEQUERBENT, WESTHOUGHTON.
Gauge : 4'8½".   (SD 674062)

		0-6-0ST	OC	AE	1600 1912	OOU
63.000.431	HARRY	0-6-0ST	IC	HC	1776 1944	Pvd

T.KILROE & SONS LTD, PLANT DEPOT, LOMAX STREET, RADCLIFFE.
Gauge : 2'0".   (SD 784068)

	4wBE	WR

Gauge : 1'6".

	0-4-0BE	WR	6600	1962	
	0-4-0BE	WR	C6711	1963	
	4wBE	CE	5858	1971	
	4wBE	CE	B0105A	1973	
	4wBE	CE	B0105B	1973	
	4wBE	CE	B0172	1973	+
	0-4-0BE	WR	N7611	1972	
	0-4-0BE	WR	N7612	1973	
	0-4-0BE	WR	N7613	1973	
	0-4-0BE	WR	N7614	1973	

+ Plate reads CE 0487.
Locos present in yard between contracts.

LILLEY/WADDINGTON LTD, PLANT DEPOT, HORWICH.
Yard (SD 627111) with locos present between contracts.
      See entry under Essex for loco fleet details.

LITTLE MILL INN, ROWATH, STOCKPORT.
Gauge : 4'8½".   (   )

3051	CAR No.89	4w-4wRER

S.LITTLER, WILLIAM TODD & CO (PEMBERTON) LTD, WIGAN.
Gauge : 4'8½".   (SD 558036)

	0-4-0DM	JF	22598 1938	OOU

MANCHESTER SHIP CANAL CO LTD, MODE WHEEL SHED & WORKSHOPS, MODE WHEEL ROAD, off ECCLES NEW ROAD, WEASTE, SALFORD.
Gauge : 4'8½".   (SJ 798977)

4001	ALNWICK CASTLE	0-6-0DE		HC	D1075	1959	
4002	ARUNDEL CASTLE	0-6-0DE		HC	D1076	1959	
D2		0-6-0DM		HC	D1187	1960	OOU
D3		0-6-0DM		HC	D1188	1960	
D5		0-6-0DM		HC	D1190	1960	OOU
D6		0-6-0DM		HC	D1191	1960	
	ENGINEER'S E1	0-6-0DM		HC	D1199	1960	OOU
D7		0-6-0DM		HC	D1251	1962	
D8		0-6-0DM		HC	D1252	1962	OOU
D11		0-6-0DM		HC	D1255	1962	OOU
D12		0-6-0DM		HC	D1256	1962	
D13		0-6-0DM		HC	D1257	1962	
D14		0-6-0DM		HC	D1258	1962	
DH22		4wDH		RR	10196	1964	Dsm
DH23		4wDH		RR	10226	1965	
DH24		4wDH		RR	10227	1965	
DH26		4wDH		RR	10229	1965	
CE 9123		4wDM	Robel	54-12-65 RR1			

See also entry under Cheshire.

MATHER & PLATT LTD, PARK WORKS, MILES PLATTING.
Gauge : 4'8½".   (SJ 872998)   RTC.

	0-4-0DM		HE	2144	1940	OOU
	4wDM		FH	3470	1951	OOU

THOMAS MITCHELL (SCRAP MERCHANTS & PLANT HIRE) LTD, EDGAR STREET, BOLTON.
Gauge : 4'8½".   (SD 713088)

	0-4-0DM		RSH	6977	1939	OOU

Other locos for resale occasionally present.

MOSELEY INDUSTRIAL TRAMWAY MUSEUM, BULKELEY SCHOOL, NORTHDOWNS ROAD, CHEADLE, STOCKPORT.
Gauge : 2'0".   (SJ 864871)

No.8		4wPM		KC		c1926
		4wPM		MR	4565	1928
		4wPM		L	4404	1932
LM 1		4wDM		RH	177639	1936
2	8 1A 126	4wDM		MR	7333	1938
65		4wDM		MR	8663	1941
64	L 55	4wDM		RH	223667	1943
3		4wDM		RH	229647	1943
1	8 1A 136	4wDM		MR	8878	1944
	8 1A 116	4wDM		MR	8934	1944
		4wDM		MR	7522	1948
	ALD HAGUE	4wDM		FH	3465	1954

NATIONAL COAL BOARD.

For full details see Section Four.

NOBELS EXPLOSIVES CO LTD, ROBURITE WORKS, SHEVINGTON, near WIGAN.
Gauge : 2'0".   (SD 543075)   (Subsidiary of I.C.I. Ltd.)

1		4wDM	RH	273500	1949
2		4wDM	RH	381705	1955
3		4wDM	RH	381704	1955
4		4wDM	RH	260716	1949
5		4wDM	RH	280866	1949
6		4wDM	RH	280865	1949
7		4wDM	RH	260719	1948
		4wDM	RH	304439	1950

NORTH WEST WATER AUTHORITY, EASTERN DIVISION, ASHTON WORKS, DUKINFIELD.
Gauge : 2'0".   (SJ 932973)

CHAUMONT	4wDH	HU	LX1002	1968

NORTH WEST WATER AUTHORITY, WATER SUPPLY DIVISION, LOWER RIVINGTON RESERVOIR, HORWICH.
Gauge : 2'0".   (SD 631121)   RTC.

No.2	8105	4wDM	RH	422573	1958	OOU

NORTHERN METALS LTD, TRAFFORD WHARF ROAD, TRAFFORD PARK.
Gauge : 4'8½".   (   )

1	0-6-0DH	NB	28057	1962	OOU
2	0-6-0DH	NB	28056	1962	OOU

PEATCO PRODUCTS LTD, ASTLEY ROAD, IRLAM.
Gauge : 2'0".   (SJ 699954)

4wDM	L	7954	1936

PROCTER & GAMBLE LTD, SOAP MANUFACTURERS, TRAFFORD PARK.
Gauge : 4'8½".   (SJ 786979)

XTB 435H	4wDM R/R	Unimog		1970

SALFORD METROPOLITAN DISTRICT COUNCIL, GEORGE THORNS RECREATION CENTRE,
LIVERPOOL ROAD, IRLAM.
Gauge : 4'8½".   (SJ 722943)

0-4-OF OC	P	2155	1955

STANDARD RAILWAY WAGON CO LTD.
Heywood Works.
Gauge : 4'8½".   (SD 868102)

4wDM	FH	3673	1954	OOU
0-4-0DM	JF	4210111	1956	

Walmsley Sidings, South Reddish, near Stockport.
Gauge : 4'8½".   (SJ 897933)

THE ROLAND BELLE	0-4-0DM	JF	4210017	1950

THOMPSON & CO (INCE) LTD, WAGON WORKS, LOWER INCE, WIGAN.
Gauge : 4'8½".   (SD 597047)

4wDM	RH	349032	1953	OOU

VERNON & ROBERTS LTD, PEEL STREET, STALYBRIDGE.
Gauge : 4'8½".   (SJ 957983)

| | BELLA | 4wDM | | RH | 186309 | 1937 | OOU |

K.WALSH, ASTLEY GREEN COLLIERY.
Gauge : 2'0".   (   )

| 6 | | 4wDM | | MR | 8937 | 1944 | |

THOS W. WARD LTD, SCRAP MERCHANTS, BRINDLE HEATH SCRAPYARD, SALFORD.
Gauge : 4'8½".   (SD 805003)

113		4wDM		RH	201980	1940	OOU
		0-4-0DE		RH	381752	1955	OOU
	BELFAST	0-4-0DM		JF	4210109	1956	
DH3		0-6-0DH		YE	2838	1961	
	F.W.W.	4wDM		MR	9930	1966	OOU

J.P.WHITTER, P.O. BOX 96, WIGAN.
Gauge : 2'6".   (   )

| | | 0-4-0DM | | HE | 2252 | 1940 | OOU |

HAMPSHIRE

B.P. OIL LTD, HAMBLE.
Gauge : 4'8½".   (SU 478065)

| No.21 | | 0-6-0DM | | HC | D707 | 1950 | |
| No.24 | | 0-6-0DH | | HE | 6950 | 1967 | |

Gauge : 1'8".   (SU 477062)

| | | 2w-2-4BE | | GB | 6132 | 1966 | |

BRITISH GAS CORPORATION, SOUTHERN REGION, PORTSMOUTH WORKS.
Gauge : 4'8½".   (SU 663027)

| 2 | | 0-4-0ST OC | | P | 2100 | 1949 | OOU |
| | FLEET No.1139 | 4wDM | | RH | 463153 | 1961 | |

GEORGE COHEN, SONS & CO LTD, POLLOCK BROWN, NORTHAM IRON WORKS, SOUTHAMPTON.
Gauge : 4'8½".   (SU 436127)

PB 1000	29968	0-4-0DM		JF	22968	1942	
PB 1001	22996	0-4-0DM		JF	22996	1943	
		0-4-0DM		JF	4200002	1946	Dsm
PB 56		4wDM		RH	305310	1952	OOU
PB 1002		0-4-0DM		HE	4262	1952	

Also other locos for resale occasionally present.

N.O.COURTNEY, WADES FARM, LONGPARISH, near ANDOVER.
Gauge : 2'0".   (   )

	4wDM	MR	7192	1937
5	4wDM	MR	8724	1941

EASTLEIGH RAILWAY PRESERVATION SOCIETY, B.R.E.L. EASTLEIGH WORKS.
Gauge : 4'8½".   (SU 457185)

30828	4-6-0	OC	Elh	1928

ESSO PETROLEUM CO LTD, FAWLEY REFINERY.
Gauge : 4'8½".   (SU 452046, 453040, 462037)

	0-6-0DH	EES	8423	1963
2034	4wDH	RR	10197	1965
553	0-6-0DH	HE	7542	1978
	0-6-0DH	HE	8999	1981

GENERAL ESTATES CO LTD, HYTHE PIER RAILWAY.
Gauge : 2'0".   (SU 423081)

	4wRE	BE	16302	1917
2	4wRE	BE	16307	1917

HAMPSHIRE NARROW GAUGE RAILWAY SOCIETY, "FOUR WINDS", DURLEY, BISHOPS WALTHAM.
Gauge : 2'0".   (SU 522173)

	CLOISTER	0-4-0ST	OC	HE	542	1891
		0-8-0T	OC	Hano	8310	1918
	WENDY	0-4-0ST	OC	WB	2091	1919
No.2		0-4-2ST	OC	HE	1842	1936
LO 20	BRAMAGE HALL	4wPM		MR	5226	1930
		4wDM		OK	4013	1930
		4wDM		OK	5125	1935
		0-4-0DM		OK	20777	1936
		0-4-0DM		OK	21160	1938
	AGWI PET	4wPM		MR	4724	1939
		4wDM		MR	8998	1946	Dsm
	STINKER	4wDM		RH	392117	1956

INTERNATIONAL SYNTHETIC RUBBER CO LTD, HYTHE.
Gauge : 4'8½".   (SU 442058)

4wDM	RH	416568	1957

MINISTRY OF DEFENCE, ARMY DEPARTMENT.
Bramley Depot.

        For full details see Section Five.

Marchwood Depot.

        For full details see Section Five.

R.A.O.C. Aldershot.

        For Full Details see Section Five.

MINISTRY OF DEFENCE, NAVY DEPARTMENT.
Portsmouth Dockyard.
Gauge : 4'8½".   (SU 642012)   RTC.

YD No.9266		4wDH	R/R	S&H	7503 1967	OOU

Royal Naval Armament Depot, Bedenham, Bridgemary.
Gauge : 4'8½".   (SU 593035)

211		0-4-0DM	(VF	4860 1942	
			(DC	2168 1942	
220		0-4-0DM	HE	3130 1944	OOU
No.218		0-4-0DM	HE	3396 1946	OOU
	YARD No.766	0-4-0DM	RH	414300 1957	
No.1	YD. No.26653	4wDH	BD	3730 1977	
No.2	YD. No.26654	4wDH	BD	3731 1977	
No.3	YD. No.26655	4wDH	BD	3732 1977	
No.4	YD. No.26656	4wDH	BD	3733 1977	
		0-4-0DH	HE	9045 1980	
		0-4-0DH	HE	9046 1980	

Also one loco kept at Priddys Hard, near Gosport (SU 614015) and Frater (SU 592028).

P.D.FUELS LTD, FUEL DISTRIBUTORS, CORRALLS DIBLES WHARF COAL CONCENTRATION DEPOT,
NORTHAM.
Gauge : 4'8½".   (SU 432123)

0-4-0DM	Bg	3568 1961	
0-6-0DM	HC	D1253 1962	

POUNDS, SHIPOWNERS & SHIPBREAKERS LTD, PORTSMOUTH.
Gauge : 4'8½".   (SU 644033)

COLIN	0-4-0DH	HE	7346 1973	OOU

Also other locos for resale or scrap occasionally present.

G.W.SMITH, WAYLAND HOUSE, 31, MANOR ROAD SOUTH, WOOLSTON.
Gauge : 1'6".   (SU 440113)

G.N.R. No.1	2-2-2	IC	c1863

T.W.SMITH, THE BUNGALOW, WHITWORTH CRESCENT, BITTERNE, SOUTHAMPTON.
Gauge : 4'8½".   (SU 439140)

4wPM	MR	5355 1931

Gauge : 1'6".

4-2-2	OC	WB	1425 1893

W.SMITH, DEALER, HILLSIDE COTTAGES, BAUGHURST, near TADLEY.
Gauge : 2'6".   (SU 579597)

YARD No.P 9261	0-4-0DM	HE	2254 1940	OOU
YARD No.P 19774	4wDM	HE	6008 1963	OOU

Yard with locos for resale occasionally present.

CAPTAIN SOMERTON-RAYNER, QUARLEY, near ANDOVER.
Gauge : 2'0".   (   )

4wDM	MR	9774 1952

SOUTHERN COUNTIES DEMOLITION & TRADING CO LTD, PLANT DEALERS, BEDHAMPTON, HAVANT.
Yard (SU 698065) with locos for resale occasionally present.

SOUTHSEA MINIATURE RAILWAY, S.M.R. LTD, SOUTHSEA.
Gauge : 1'5".   (SZ 641982)

		4w-4wRER		Holland? 1976

THOS W. WARD LTD, PLANT DEALERS, RINGWOOD.
Yard (SU 154047) with locos for scrap occasionally present.

WINCHESTER AND ALTON RAILWAY LTD.
Gauge : 4'8½".   Locos are kept at :-

New Alresford Station.   (SU 588325)
Ropley Station.   (SU 629324)

(30120)	120	4-4-0	IC	9E	572	1899
30506		4-6-0	OC	Elh		1920
31625		2-6-0	OC	Afd		1929
31806		2-6-0	OC	Bton		1926
31874	BRIAN FISK	2-6-0	OC	Woolwich		1925
34016	BODMIN	4-6-2	3C	Bton		1945
34067	TANGMERE	4-6-2	3C	Bton		1947
34105	SWANAGE	4-6-2	3C	Bton		1950
35018	BRITISH INDIA LINE	4-6-2	3C	Elh		1945
47324		0-6-0T	IC	NB	23403	1926
76017		2-6-0	OC	Hor		1953
	SLOUGH ESTATES No.3	0-6-0ST	OC	HC	1544	1924
196	ERROL LONSDALE	0-6-0ST	IC	HE	3796	1953
		0-4-0DM		JF	22889	1939
		0-4-0DH		NB	27078	1953
(DS 3317)	DS 3319 "MERCURY"	2w-2PMR		Wkm	6642	1953
DS 9023	RLC/009023	2w-2PMR		Wkm	8087	1958
(TP 57P)	ENGINEER No.1	2w-2PMR		Wkm	8267	1959

HEREFORD & WORCESTER

BEWDLEY WILD LIFE PARK, BEWDLEY.
Gauge : 1'3".   (   )

278	RIO GRANDE	2-8-0DH	S/O	SL	15.2.79	1979

BIRDS COMMERCIAL MOTORS LTD, LONG MARSTON DEPOT.
Gauge : 4'8½".   (SP 154458)

(D2857)		0-4-0DH		YE	2816	1960
		0-4-0F	OC	AB	1772	1922	OOU
		0-4-0DM		AB	325	1937

Also other locos for scrap occasionally present.

BRITISH SUGAR CORPORATION LTD, FOLEY PARK, KIDDERMINSTER.
Gauge : 4'8½".   (SO 825749)

	0-4-0DM		RH	281269	1950
	0-4-0DE		RH	408866	1959

H.P.BULMER LTD, CIDER MANUFACTURERS, MOORFIELDS, HEREFORD.
Gauge : 4'8½".   (SO 505402)

No.		Name	Type		Builder	Works No.	Date	Notes
5786			0-6-0PT	IC	Sdn		1930	+ Pvd
6000		KING GEORGE V	4-6-0	4C	Sdn		1927	Pvd
35028		CLAN LINE	4-6-2	3C	Elh		1948	Pvd
(46201)	6201	PRINCESS ELIZABETH	4-6-2	4C	Crewe	107	1933	Pvd
46512			2-6-0	OC	Sdn		1952	Pvd
1579		PECTIN	0-4-0ST	OC	P	1579	1921	Pvd
47		CARNARVON	0-6-0ST	IC	K	5474	1934	+ Pvd
193		SHROPSHIRE	0-6-0ST	IC	HE	3793	1953	Pvd
No.1		WOODPECKER	0-4-0DM		JF	22871	1939	
(D2578)	2	CIDER QUEEN	0-6-0DM		HE	6999	1968	
			Rebuild of		HE	5460	1958	

          + Property of Worcester Locomotive Society Ltd.

CENTRAL ELECTRICITY GENERATING BOARD, STOURPORT POWER STATION.
Gauge : 2'6".   (SO 815708)   RTC.

No.1	4wBE	EE	688	1925	OOU
No.2	4wBE	EE	689	1925	OOU

DROITWICH CANAL TRUST, SALWARPE.
Gauge : 2'0".   (SO 873618)

4wDM	MR	7471	1940

GARRINGTONS LTD, GARRINGTONS NEWTON WORKS, BROMSGROVE.
Gauge : 4'8½".   (SO 966692)

PLANT No.1301	0-4-0DM	JF	4100013	1948

HEREFORDSHIRE WATERWORKS MUSEUM TRUST, BROOMY HILL, HEREFORD.
Gauge : 2'0".   (SO 497394)

3101				
	4wPM	MR	1381	1918
	4wDM	LB	52886	1962

MINISTRY OF DEFENCE, ARMY DEPARTMENT, MORETON-ON-LUGG DEPOT.

          For full details see Section Five.

W.MORRIS, BROMYARD & LINTON LIGHT RAILWAY, BROADBRIDGE HOUSE, BROMYARD.
Gauge : 2'6".   (SO 657548)

4wDM	Bg	3406	1953

Gauge : 2'0".

MESOZOIC	0-6-0ST	OC	P	1327	1913	
	4wPM		MR	6031	1936	+
	4wDM		RH	187101	1937	
	4wDM		RH	195849	1939	@
L 10	4wDM		RH	198241	1939	

		4wVBT		Jaywick Rly 1939 DsmT
		4wDM	RH	213848 1942
No.3	NELL GWYNNE·	4wDM	RH	229648 1944
No.6	PRINCESS	4wDM	RH	229655 1944
LM 30		4wDM	RH	229656 1944
		4wDM	RH	246793 1947
		4wDM	MR	9382 1948
1		4wDM	MR	9676 1952
2		4wDM	MR	9677 1952
No.7		4wDM	MR	20082 1953
		4wDH	RH	437367 1959
		4wDM	MR	102G038 1972
		2w-2PM	Wkm	3034 1941

+ Currently under renovation elsewhere.
@ In use as a generating unit.

OWENS TRAILERS LTD, ROTHERWAS.
Gauge : 1'3".   (   )

   303   0-6-0PM   J.Taylor

PAINTER BROS LTD, ENGINEERS, MORTIMER ROAD, HEREFORD.
Gauge : 2'0".   (SO 508413)

	4wDM	L	40407 1954
	4wDM	LB	54181 1964

SEVERN VALLEY RAILWAY CO LTD, BEWDLEY STATION & ARLEY STATION.
Gauge : 4'8½".   (SO 793753, 800764)
     For details of locos see under Salop entry.

S.SIMMONS, WILDEN, near STOURPORT.
Gauge : 1'6".   (   )

                        4-4-2   OC   Curwen

D.TURNER, "FAIRHAVEN", WYCHBOLD.
Gauge : 2'0".   (SO 922660)

                        4wDM   MR   8600 1940

UNDERWOOD & CO LTD, DROITWICH COAL CONCENTRATION DEPOT.
Gauge : 4'8½".   (SO 895636)
     THE SHERIFF   4wDM   RH   458961 1962

THE WOOLHOPE LIGHT RAILWAY, P.J.FORTEY, THE HORNETS NEST, CHECKLEY,
MORDIFORD, near HEREFORD.
Gauge : 1'3".   (SO 608378)
   202   0-6-0PM   J.Taylor   c1974

WORCESTER LOCOMOTIVE SOCIETY LTD, MORETON-ON-LUGG SITE.
Gauge : 4'8½".   (SO 508472)

	4wPM	MR	4217 1931
PWM 3767	2w-2PMR	Wkm	6646 1953

HERTFORDSHIRE

HEMEL HEMPSTEAD LIGHTWEIGHT CONCRETE LTD, CUPID GREEN, HEMEL HEMPSTEAD.
Gauge : 4'8½".   (TL 076094)   RTC.

(D2203)		0-6-0DM	(VF	D145	1952	OOU
			(DC	2400	1952	
		4wDM	MR	9921	1959	Dsm

C. & D. LAWSON, 11, OKELEY LANE, HIGHFIELD ESTATE, TRING.
Gauge : 2'6".   (SP 913114)

No.3		4wDM	RH	297066	1950	
No.4		4wDM	RH	402439	1957	
No.5		4wDM	RH	432654	1959	
No.6		4wDM	RH	224315	1944	
No.7	ELLEN	4wDM	RH	200069	1939	
No.8		4wDM	RH	244559	1946	
		4wPM	L	34652	1949	
		2w-2PM	Wkm	3431	1943	Dsm
		2w-2PM	Wkm	3578	1944	Dsm

Gauge : 2'1½".

| No.1 | | 4wDM | RH | 166045 | 1933 | |
| No.2 | | 4wDM | RH | 247178 | 1947 | |

Gauge : 1'8".

| | | 4wDM | RH | 229657 | 1945 | |

Locos are currently stored elsewhere.

MARPLES RIDGWAY LTD, CONTRACTORS, METROPOLITAN STATION APPROACH ROAD, WATFORD.
Yard (TQ 095965) with locos between contracts occasionally present.

JOHN MOWLEM & CO LTD, (WELHAM PLANT), CONTRACTORS & STEELWORKERS, PLANT DEPOT,
WELHAM GREEN.
Gauge : 4'8½".   (TL 230065)

| | LONDON JOHN | 0-4-0DH | HC | D1291 | 1964 | |

Gauge : 2'6".

| JM 92 | | 4wBE | CE | B1547A | 1977 | |
| JM 93 | | 4wBE | CE | B1547B | 1977 | |

Gauge : 2'0"..

JM 75		4wBE	WR	5665	1957	
JM 77		4wBE	WR	D6800	1964	
JM 78		4wBE	WR	6769	1964	
JM 79		4wBE	WR	C6770	1964	
JM 83		4wBE	CE	5942A	1972	
JM 84		4wBE	CE	5942B	1972	
JM 85		4wBE	CE	5942C	1972	
JM 86		4wBE	CE	B0148A	1973	
JM 87		4wBE	CE	B0148B	1973	
JM 88		4wBE	CE	B0402A	1974	
JM 89		4wBE	CE	B0402B	1974	
JM 90		4wBE	CE	B0402C	1974	
JM 91		4wBE	CE	B0402D	1974	
		4wBE	CE	B0445	1975	

Gauge : 1'6".

JM 82   4wBE   CE   5806 1970

    Narrow gauge locos present in yard between contracts.

NABISCO FOODS LTD, WELWYN GARDEN CITY.
Gauge : 4'8½".   (TL 242131)
                                    0-4-0DM   JF   20337 1934

ROBIN PEARMAN, 96, PARK AVENUE, POTTERS BAR.
Gauge : 2'0".   (TL 266007)
                                    4wDM   MR   2059 1920

PLEASURE-RAIL LTD, KNEBWORTH WEST PARK & WINTER GREEN RAILWAY,
KNEBWORTH HOUSE, near STEVENAGE.
Gauge : 1'11½".   (TL 228208)
                                    0-6-0ST   OC   P   1270 1911
                  TRIASSIC   0-4-0T   OC   AE   1738 1915
                  SEZELA No.4   0-4-0ST   OC   HE   1429 1922
    No.1   LADY JOAN   4-4-0T   OC   WB   2820 1945
    No.5   ISIBUTU   4wDM   MR   8717 1941 + Dsm
                                    4wDM   MR   8738 1942
      9   L3   LB4   4wDM   S/O   RH   217967 1942
    No.6   HORATIO   4wDM   MR   8993 1946
    No.2   4wDM   MR   8995 1946 + Dsm
                                    4wDM   RH   373359 1958
      3   4wDM   MR   40S273 1966

           + Converted into a brake van.

Gauge : 1'10¼".
                  LILLA   0-4-0ST   OC   HE   554 1891

Gauge : 4'8½".
    900338   2w-2PMR   Wkm   (626?)

M.SAUL, WENGEO LANE, WARE.
Gauge : 4'8½".   (TL 346147)
                  NEWSTEAD   0-6-0ST   IC   HE   1589 1929

D.WICKHAM & CO LTD, WARE.
(TL 362140)
    New Wickham railcars usually present.

HUMBERSIDE

BLUE CIRCLE INDUSTRIES LTD.
Central Works, Kirton Lindsey.
Gauge : 4'8½".   (SE 950012)

DON ATKINSON	0-4-0DH	RH	525947	1968

Humber Works, Melton, near Hull.
Gauge : 4'8½".   (SE 965258)

THE HERBERT TURNER	0-4-0DE	RH	425478	1959
	0-4-0DH	RH	513139	1967

WILLIAM BLYTH.
Barton Brick & Tile Yard, Barton-On-Humber.
Gauge : 2'0".   (TA 038234)   RTC.

4wDM	RH	247182	1947	OOU

Far Ings Tileries, Barton-On-Humber.
Gauge : 2'0".   (TA 023233)

4wDM	RH	260708	1948

B.P. CHEMICALS INTERNATIONAL LTD, SALT END REFINERY, HULL.
Gauge : 4'8½".   (TA 165275)

801	4wDM	RH	275882	1950
802	0-6-0DH	HE	7041	1971

B.P. MARKETING LTD, SALT END, HULL.
Gauge : 4'8½".   (TA 162277)

No.16	0-4-0DM	Bg/DC	2164	1941

BRITISH STEEL CORPORATION, SCUNTHORPE DIVISION, SCUNTHORPE WORKS.
Appleby-Frodingham Works, Scunthorpe.
Gauge : 4'8½".   (SE 910110, 913109, 915110, 916105)

1	0-6-0DE	YE	2877	1963	
3	0-6-0DE	YE	2595	1956	OOU
6	0-6-0DE	YE	2863	1962	OOU
7	0-6-0DE	YE	2634	1957	
9	0-6-0DE	YE	2864	1962	
10	0-6-0DE	YE	2865	1962	
13	0-6-0DE	YE	2715	1958	
15	0-6-0DE	YE	2901	1963	
16	0-6-0DE	YE	2902	1963	
17	0-6-0DE	YE	2788	1960	OOU
21	0-6-0DE	YE	2900	1963	OOU
25	0-6-0DE	YE	2936	1964	
26	0-6-0DE	YE	2727	1959	OOU
27	0-6-0DE	YE	2937	1964	
29	0-6-0DE	YE	2938	1964	OOU
30	0-6-0DE	YE	2943	1965	
31	0-6-0DE	YE	2903	1963	
34	0-6-0DE	YE	2876	1963	
36	0-6-0DE	YE	2799	1961	OOU

37			0-6-0DE		YE	2738 1959	
39			0-6-0DE		YE	2790 1960	OOU
40			0-6-0DE		YE	2764 1959	
41			0-6-0DE		YE	2765 1959	
42			0-6-0DE		YE	2766 1960	OOU
43			0-6-0DE		YE	2767 1960	OOU
44			0-6-0DE		YE	2768 1960	
45			0-6-0DE		YE	2944 1965	
46			0-6-0DE		YE	2945 1965	OOU
47			0-6-0DE		RR	10236 1967	
49			0-6-0DE		RR	10237 1967	
50			0-6-0DE		RR	10238 1967	
52	DE 2		0-6-0DE		YE	2773 1959	
53			0-6-0DE		YE	2793 1961	
54	DE 4		0-6-0DE		YE	2908 1963	
55			0-6-0DE		YE	2690 1959	
70			4w-4wDE		HE	7281 1972	
71			4w-4wDE		HE	7282 1972	
72			4w-4wDE		HE	7283 1972	
73			4w-4wDE		HE	7284 1972	
74			4w-4wDE		HE	7285 1972	
75			4w-4wDE		HE	7286 1972	
76			4w-4wDE		HE	7287 1973	
77			4w-4wDE		HE	7288 1973	
78			4w-4wDE		HE	7289 1973	
79			4w-4wDE		HE	7290 1973	
80			4w-4wDE		HE	7474 1977	
No.123	SIR DOUGLAS		0-8-0DH		RR	10171 1963	
No.124	HENRY CLIVE		0-8-0DH		RR	10172 1963	
1	0448/73/01		0-4-0DE		(BD	3734 1977	
					(GECT	5434 1977	
2	0448-73-02		0-4-0DE		(BD	3735 1977	
					(GECT	5435 1977	
3	0448/73/03		0-4-0DE		(BD	3736 1977	
					(GECT	5436 1977	
4	0448-73-04		0-4-0DE		(BD	3737 1977	
					(GECT	5437 1977	
5	0448-73-05		0-4-0DE		(BD	3738 1977	
					(GECT	5438 1977	
6	0448-73-06		0-4-0DE		(BD	3739 1977	
					(GECT	5439 1977	
7	0449-73-07		0-4-0DE		(BD	3740 1977	
					(GECT	5440 1977	
No.1		714/37	4wDM	Robel		21.12 RK3 1969	OOU
No.2	0714/78/06	714/24	4wDM	Robel		21.12 RN5 1973	
No.3	0714/78/05	714/22	4wDM	Robel		54.12-56-RT1 1966	
No.4	0714/78/07	714/26	4wDM	Robel		54.12-56-RW3 1974	
5	0714/69/29		4wDM	Robel		54.12-56-AA169 1978	
	0714/69/09		4wDM			Donelli 1980	

Dawes Lane Coke Ovens, Scunthorpe.
Gauge : 4'8½".   (SE 921118)

		4wWE		GB	420383/1 1977	
		4wWE		GB	420383/2 1977	

Frodingham Coke Ovens.
Gauge : 4'8½". (SE 917108)

4		0-4-0WE	WSO	6609/2	1956	OOU
5		4wRE	Schalker		1973	
6		4wRE	Schalker		1973	
7		4wRE	Schalker		1979 +	

+ Built under license by Starco Engineering, Winterton Road,
Scunthorpe, Humberside.

Normanby Park Works, Scunthorpe. (Closed)
Gauge : 4'8½". (SE 885137)

5		0-6-0DE		YE	2909	1963	
No.6		0-6-0DE		YE	2744	1960	OOU
No.8	GEORGE	0-6-0DE		YE	2871	1962	OOU
No.10		0-6-0DH		S	10106	1963	Dsm
11		0-6-0DE		YE	2899	1963	OOU
14		0-6-0DE		YE	2716	1958	OOU
No.18		0-6-0DE		YE	2897	1962	OOU
20		0-6-0DE		YE	2668	1958	OOU
24		0-6-0DE		YE	2728	1959	
No.28		0-6-0DE		YE	2791	1962	OOU
33		0-6-0DE		YE	2737	1959	OOU
35		0-6-0DE		YE	2789	1960	OOU
No.51		0-6-0DE		YE	2709	1959	OOU
56		6wDH		RR	10277	1968	OOU
No.59	CLEM	0-6-0DE		HE	7400	1975	OOU
No.60	RALPH	0-6-0DE		HE	7401	1975	
No.61	SALLY	0-6-0DE		HE	7473	1976	OOU
No.62	PETER	0-6-0DE		HE	7499	1977	
No.113	LIONEL	0-6-0DH		S	10109	1963	OOU
No.114	GEOFFREY	0-6-0DH		S	10110	1963	OOU
No.117	RICHARD	0-6-0DH		S	10113	1963	OOU
No.120	JUDITH	0-6-0DH		S	10116	1963	OOU
No.121	TANIS	0-6-0DH		S	10117	1963	OOU
No.1	7714/70/26	4wDM	Robel	54.12.62-RP9		1971	OOU
	7714/70/11	4wDM	R/R	Unimog		1974	OOU
		0-4-0WE		RSHN	6966	1938	OOU
No.1		0-4-0WE		RSHN	7678	1951	OOU
No.2		4wRE		GB	2472	1953	OOU
		4wRE		GB	2858	1958	OOU

Locos also work Flixborough Wharf. (SE 860144)

BRITISH SUGAR CORPORATION LTD. BRIGG FACTORY.
Gauge : 4'8½". (SE 991061)

	0-4-0DM	RH	281266	1950

BRITISH TRANSPORT DOCKS BOARD.
Alexandra Dock, Hull.
Gauge : 4'8½". (TA 127291)

C 101	2w-2PMR	Wkm	6603	1953

Goole Docks.
Gauge : 4'8½". (SE 740233)

900331	2w-2PMR	Wkm	496	1932

B.T.P. TIOXIDE LTD, PYEWIPE WORKS, GRIMSBY.
Gauge : 4'8½".   (TA 254113)

4		0-4-0DM	RH	375713 1954
(5)		4wDM	RH	412429 1957 OOU
6		0-4-0DM	RH	414303 1957
7		4wDM	RH	421418 1958

CAPPER PASS & SON LTD, FERRIBY, near HULL.
Gauge : 4'8½".   (SE 974255)

	0-4-0DM	JF	22060 1937

CENTRAL ELECTRICITY GENERATING BOARD, KEADBY POWER STATION.
Gauge : 4'8½".   (SE 825117)   RTC.

FARADAY	0-4-0DM	HC	D920 1955
EDISON	0-4-0DH	HC	D1341 1966

P.CLARK, near GRIMSBY.
Gauge : 4'8½".   (   )

FFIONA JANE	0-4-0ST	OC	P	1749 1928
	0-4-0ST	OC	RSHN	7680 1950

CLEETHORPES MINIATURE RAILWAY, MARINE EMBANKMENT, CLEETHORPES.
Gauge : 1'2¼".   (TA 321073)

		2-8-0Gas	S/O	SL	7217 1972
800		2-8-0Gas	S/O	SL	15.5.78 1978
4472	THE FLYING SCOTSMAN	4-6-2Gas	S/O	Cook	+

+ Currently in a Council Store. (TA   )

CLUGSTON CONSTRUCTION, PLANT DEPOT, SCUNTHORPE.
Gauge : 1'0".   (SE   )

2w-2BE	Iso	
2w-2BE	Iso	

Locos present in yard between contracts.

CLUGSTON SLAG LTD, NORTH LINCOLN WORKS, FRODINGHAM. (Closed)
Gauge : 4'8½".   (SE 921106)   RTC.

0-6-0DE	YE	2631 1957 OOU

CONOCO LTD, REFINERY, KILLINGHOLME.
Gauge : 4'8½".   (TA 163168)

M.F.P. No.1	0-4-0DM	JF	4210131 1957
M.O.P. No.8	0-4-0DM	JF	4210145 1958 OOU
	0-4-0DH	HE	6981 1968

COURTAULDS LTD, GREAT COATES, GRIMSBY.
Gauge : 4'8½".   (TA 238124)

"GEORGE"	4wVBT VCG	S	9596 1955 OOU
WILLIAM	4wVBT VCG	S	9599 1956
	4wDM	FH	3817 1956

<u>ALBERT DRAPER & SON LTD, NEPTUNE STREET, HULL.</u>
Yard (TA 086275) with locos for scrap occasionally present.

<u>FISONS LTD, HORTICULTURE DIVISION, BRITISH MOSS PEAT WORKS, SWINEFLEET, near GOOLE.</u>
Gauge : 3'0".   (SE 770169)

L12							
3							
	02-14		4wDM	MR	10160	1950	
			4wDM	MR	10455	1955	
			4wDM	RH	432661	1959	+
	02-05	TANIA	4wDM	RH	432665	1959	
			4wDM	RH	466594	1961	+
	02-03		4wDM	LB	53976	1964	
	02-04		4wDM	LB	53977	1964	
			4wDM	LB	54184	1964	
			4wDM	LB	55471	1967	
			4wDM	Fisons		1976	
			4wDM	SMH	40SD507	1978	

+ One of these is named SIMBA and the other is 02-07 SHEEBA.

<u>FISONS LTD, IMMINGHAM WORKS.</u>
Gauge : 4'8½".   (TA 200157)

	4wDH	S	10037	1960
	0-4-0DH	JF	4220029	1964

<u>GOXHILL TILERIES LTD, BARROW HAVEN WORKS.</u>
Gauge : 2'0".   (TA 064238)   RTC.

HI 6186	4wDM	RH	175418	1936	OOU
	4wDM	RH	223692	1943	OOU
	4wDM	RH	235654	1946	OOU

<u>GRANT LYON EAGRE LTD, CIVIL ENGINEERS, PLANT DEPOT, SCOTTER ROAD, SCUNTHORPE.</u>
Gauge : 4'8½".   (SE 871114)

No.1		4wDM	RH	200793	1940
	511 83 04	4wDM	RH	294269	1951
	ALFRED HENSHALL	0-4-0DM	RH	313392	1952
		4wDH R/R	S&H	7510	1967
		4wDM R/R	S&H	7512	1972

Locos present in yard between contracts.

<u>HUMBERSIDE LOCOMOTIVE PRESERVATION GROUP, B.R. DAIRYCOATES DEPOT, HULL.</u>
Gauge : 4'8½".   (   )

30777	SIR LAMIEL	4-6-0	OC	NB	23223	1925
(45305)	5305	4-6-0	OC	AW	1360	1937

<u>KINGSTON-UPON-HULL CORPORATION, TRANSPORT MUSEUM, 36, HIGH STREET, HULL.</u>
Gauge : 3'0".   (TA 102284)

1	0-4-0Tram OC	K	T56	1882

<u>KIRTON LINDSEY WINDMILL AND MUSEUM, KIRTON LINDSEY.</u>
Gauge : 4'8½".   (   )

	0-6-0ST OC	HC	1604	1928

LINCOLNSHIRE COAST LIGHT RAILWAY CO LTD, HUMBERSTON, CLEETHORPES.
Gauge : 2'0".   (TA 326059)

1	PAUL	4wDM		MR	3995 1927
2	JURASSIC	0-6-0ST	OC	P	1008 1903
3	ELIN	0-4-0ST	OC	HE	705 1899
4	WILTON	4wDM		MR	7481 1940
7	NOCTON	4wDM		MR	1935 1920
	"GRICER"	4wDM		MR	8622 1941
	MAJOR	4wDM		MR	8874 1944

LINDSEY OIL REFINERY LTD, KILLINGHOLME REFINERY.
Gauge : 4'8½".   (TA 160176)

	SPRINGBOK	4wDH	TH	212V 1969
	BEAVER	0-6-0DH	AB	630 1978
	BADGER	0-6-0DH	AB	658 1980

MINISTRY OF DEFENCE, ARMY DEPARTMENT, ARMY SCHOOL OF MECHANICAL TRANSPORT, R.C.T. MUSEUM, LECONFIELD.

For full details see Section Five.

NYPRO (U.K.) LTD, FLIXBOROUGH.
Gauge : 4'8½".   (SE 858148)   (Subsidiary of Fisons Ltd.)

	4wDM	RH	476139 1963	OOU

JONATHAN POTTS LTD, SCRAP METAL DEALERS, ESTATE ROAD No.1, GRIMSBY.
Gauge : 4'8¼".   (TA 255110)

	4wDM	RH	382823 1955

SCUNTHORPE CORPORATION, JUBILEE PLAYING FIELDS, ASHBY, SCUNTHORPE.
Gauge : 4'8½".   (SE 893088)

	LYSAGHT'S	0-6-0ST	OC	RSH	7035 1940

SCUNTHORPE SLAG LTD, SANTON WORKS, SCUNTHORPE.
Gauge : 4'8½".   (SE 923123)

E/1/3		4wDM	RH	210478 1941	OOU
E/1/4	ARNOLD MACHIN	0-6-0DE	YE	2661 1958	

SEVERN TRENT WATER AUTHORITY, TRENT RIVER MANAGEMENT DIVISION, OWSTON FERRY PLANT DEPOT.
Gauge : 2'0".   (SK 814994)

20	U 194	4wDM	RH	7002/0967/5 1967	OOU
21	U 477	4wDM	RH	7002/0967/6 1967	OOU

Locos used on river bank work, etc, as required.

J.W.STAMP & SON, HOLYDYKE, BARTON-ON-HUMBER.
Gauge : 2'0".   (TA 029220)

	4wDM	RH	235628 1945	Dsm
	4wDM	RH		Dsm

Gauge : 1'0".   (   )

L2		2w-2BE	Iso	
		2w-2BE	(Iso?)	

Locos present in yard between contracts.

ISLE OF MAN

CIVIL AVIATION AUTHORITY, LAXEY.
Gauge : 3'6".   (SC 432847)

	4wDMR	Wkm	7642 1958
	4wDMR	Wkm	10956 1976

J.EDWARDS, BALLAKILLINGAN HOUSE, CHURCHTOWN, near RAMSEY.
Gauge : 3'0".   (SC 425945)

No.14	THORNHILL	2-4-0T	OC	BP	2028	1880	Pvd

ISLE OF MAN HARBOUR COMMISSIONERS, QUEENS PIER, RAMSEY.
Gauge : 3'0".   (SC 449947, 456941)

	4wPM	S/o	FH	2027	1937

ISLE OF MAN RAILWAYS.
Isle Of Man Steam Railway.
Gauge : 3'0".   Locos are kept at :-

Douglas.   (SC 374754, 375755)
Port Erin.   (SC 198689)

No.4	LOCH	2-4-0T	OC	BP	1416	1874	
No.5	MONA	2-4-0T	OC	BP	1417	1874	OOU
No.6	PEVERIL	2-4-0T	OC	BP	1524	1875	OOU
No.7	(TYNWALD)	2-4-0T	OC	BP	2038	1880	Dsm
No.8	FENELLA	2-4-0T	OC	BP	3610	1894	Pvd
No.9	DOUGLAS	2-4-0T	OC	BP	3815	1896	OOU
No.10	G.H.WOOD	2-4-0T	OC	BP	4662	1905	
No.11	MAITLAND	2-4-0T	OC	BP	4663	1905	
No.12	HUTCHINSON	2-4-0T	OC	BP	5126	1908	
No.13	KISSACK	2-4-0T	OC	BP	5382	1910	
(19)		0-4-0+4DMR		Wkb/Dundalk		1950	
(20)		0-4-0+4DMR		Wkb/Dundalk		1951	
		2w-2PMR		Wkm	5763	1950	
		2w-2PMR		Wkm	7442	1956	
		2w-2PMR		Wkm	8849	1961	

Manx Electric Railway Board, Manx Electric Railway, Ramsey Tram Museum, Ramsey Station.
Gauge : 3'0".   (SC 454943)

No.23		4w-4wWE	c1900	Pvd

Manx Electric Railway Board, Snaefell Mountain Railway, Laxey.
Gauge : 3'6".   (SC 432847)

			4wPMR		Wkm	5864	1951	OOU

PORT ERIN RAILWAY MUSEUM, STRAND ROAD, PORT ERIN.
Gauge : 3'0".   (SC 198689)

No.1		SUTHERLAND	2-4-0T	OC	BP	1253	1873
No.4	15	CALEDONIA	0-6-0T	OC	D	2178	1885
No.16		MANNIN	2-4-0T	OC	BP	6296	1926

ISLE OF WIGHT

MEDINA VALLEY LIGHT RAILWAY, SHANKLIN.
Gauge : 1'3".   (   )

	PRINCESS	4-6-2DM	S/O	H.N.Barlow
		4-6-2DM	S/O	H.N.Barlow

    Currently stored at a secret location.

F.H.REEVE, THE OLD MILL, ST. HELENS.
Gauge : 1'3".   (SZ   )

		4-6-2	OC	Longfleet	1968

WIGHT LOCOMOTIVE SOCIETY, HAVEN STREET STATION.
Gauge : 4'8½".   (SZ 556898)

W8		FRESHWATER	0-6-0T	IC	Bton		1876	
W11	(32640)	NEWPORT	0-6-0T	IC	Bton		1878	
W24		CALBOURNE	0-4-4T	IC	9E	341	1891	
37		VECTIS	0-4-0ST	OC	HL	3135	1915	
38		AJAX	0-6-0T	OC	AB	1605	1918	
39		SPITFIRE	4wDM		RH	242868	1946	
			0-4-0DH		NB	27415	1954	
			4wDMR		Bg/DC	1647	1927	Dsm
PWM 3766	(DS 3320)		2w-2PMR		Wkm	6645	1953	

KENT

ACE SAND & GRAVEL CO LTD, MARSH HOUSE QUARRIES, OARE, near FAVERSHAM.
Gauge : 2'0".   (TR 013625)   RTC.

4wDM	MR	5877	1935	Dsm
4wDM	MR	7469	1940	OOU
4wDM	MR	8606	1941	OOU
4wDM	MR	8704	1942	OOU

BERRY WIGGINS LTD, KINGSNORTH ON THE MEDWAY, HOO, ROCHESTER.
Gauge : 4'8½".   (TQ 807732, 808730)

TITAN	0-4-0DM	(VF	D140	1951
		(DC	2274	1951

D.BEST, ULCOMBE, near HEADCORN.
Gauge : 2'0".   (   )

	0-4-0BE	WR	F7116	1966
	0-4-0BE	WR	F7117	1966
E 1	4wBE	Riordan	T6664	1967

D.,W. & S.BEST, "THE WARREN", SWANTON STREET, BREDGAR, near SITTINGBOURNE.
Gauge : 2'0".   (TQ 873585)

BRONHILDE	0-4-0WT	OC	Sch	9124	1927
	0-6-0WT	OC	Jung	3872	1931
No.15	4wDM		RH	177604	1936
OLDE	4wDM		HE	2176	1940

BLUE CIRCLE INDUSTRIES LTD.
Holborough Works, Snodland.
Gauge : 4'8½".   (TQ 706624)

4wDH	S	10033	1960

Swanscombe Works.
Gauge : 4'8½".   (TQ 601753)

No.1		0-4-0DM	RH	408301	1957	OOU
No.3		4wDH	S	10006	1959	
No.4		4wDH	S	10007	1959	
No.5	SIMON	4wDH	S	10020	1959	

BOWATERS UNITED KINGDOM PAPER CO LTD, SITTINGBOURNE WORKS.
Gauge : 4'8½".   (TQ 920667)

08596	0-6-0DE	Derby	1959

B.P. OIL KENT REFINERY LTD, GRAIN REFINERY.
Gauge : 4'8½".   (TQ 864758, 865753)

	0006	0-4-0DM		AB	412	1957
WIN 4820-0007	0-6-0DH		HE	6971	1968
	0-6-0DH		HE	6972	1968
MAN OF KENT	0-6-0DH		TH	294V	1981
KENTISH MAID	0-6-0DH		TH	295V	1981
WIN 4820-0005	4wDM	R/R	S&H	7508	1967

ROBERT BRETT & SONS LTD, STURRY BALLAST PITS.
Gauge : 2'0".   (TR 184603)   RTC.

		4wDM	MR	8730 1941	OOU
5		4wDM	RH	283871 1950	OOU
6		4wDM	RH	349061 1953	OOU
		4wDM	RH	444193 1960	OOU

W.R.BRETT, GOOSE FARM, CULVERSTONE, MEOPHAM.
Gauge : 2'6".   (TQ 628626)

83		4wDM	RH	213838 1943	OOU

WILLIAM CORY & SON LTD, SOLID FUEL DIVISION.
Rochester Coal Wharf.
Gauge : 4'8½".   (TQ 747686)

R39	THALIA	0-4-0DM	(RSHN	7816 1954
			(DC	2503 1954

Strood Coal Wharf.
Gauge : 4'8½".   (TQ 741694)

R40	TELEMON	0-4-0DM	(VF	D295 1955
			(DC	2568 1955
R41	P.B.A.31	0-6-0DM	HC	D1172 1959

DEPARTMENT OF THE ENVIRONMENT.
Hoo Ness Island.
Gauge : 2'0".   (TQ 783703)

1		4wDM	FH	3982 1962
2		4wDM	FH	3983 1962

Lydd Gun Ranges, Lydd, Romney Marsh.
Gauge : 60cm.   (TR 033198)

L.R.No.1	LOD 758263	AD41	4wDM	RH	191646 1938	
L.R.No.2	LOD 758366	AD40	4wDM	RH	202000 1940	
			4wDM	RH	201999 1940	Dsm
	RTT/767149		2w-2PM	Wkm	3151 1942	
	RTT/767150		2w-2PM	Wkm	3152c1943	
	RTT/767151		2w-2PM	Wkm	3153c1943	OOU
	RTT/767152		2w-2PM	Wkm	3154c1943	Dsm
	RTT/767153		2w-2PM	Wkm	3155c1943	OOU
	RTT/767154		2w-2PM	Wkm	3156c1943	OOU
	RTT/767155		2w-2PM	Wkm	3157c1943	Dsm
	RTT/767156		2w-2PM	Wkm	3158c1943	
	RTT/767157		2w-2PM	Wkm	3159c1943	Dsm
	RTT/767158		2w-2PM	Wkm	3160c1943	OOU
	RTT/767159		2w-2PM	Wkm	3161c1943	
	RTT/767160		2w-2PM	Wkm	3162c1943	
	RTT/767161		2w-2PM	Wkm	3234 1943	Dsm
	RTT/767162		2w-2PM	Wkm	3235 1943	
	RTT/767163		2w-2PM	Wkm	3236 1943	
	RTT/767164		2w-2PM	Wkm	3237 1943	Dsm
	RTT/767165		2w-2PM	Wkm	3238 1943	
	RTT/767166		2w-2PM	Wkm	3239 1943	Dsm
	RTT/767167		2w-2PM	Wkm	3240 1943	OOU
	RTT/767168		2w-2PM	Wkm	3241 1943	OOU
	RTT/767169		2w-2PM	Wkm	3242 1943	
	RTT/767170		2w-2PM	Wkm	3243 1943	Dsm

RTT/767171	2w-2PM	Wkm	3163c1943	OOU
RTT/767172	2w-2PM	Wkm	3164c1943	
RTT/767173	2w-2PM	Wkm	3165c1943	OOU
RTT/767174	2w-2PM	Wkm	3166c1943	OOU
RTT/767175	2w-2PM	Wkm	3167c1943	
RTT/767176	2w-2PM	Wkm	3168c1943	OOU
RTT/767177	2w-2PM	Wkm	3169c1943	Dsm
RTT/767178	2w-2PM	Wkm	3170c1943	
RTT/767179	2w-2PM	Wkm	3171c1943	Dsm
RTT/767180	2w-2PM	Wkm	3149 1942	
RTT/767181	2w-2PM	Wkm	3150 1942	Dsm
RTT/767182	2w-2PM	Wkm	2522 1938	OOU
RTT/767189	2w-2PM	Wkm	2561 1939	Dsm
RTT/767190	2w-2PM	Wkm	2562 1939	Dsm

Also uses M.O.D.,A.D. locos; for full details see Section Five.

F.J.HAM & SON LTD, SCRAP MERCHANTS, MAYS RAILWAY SIDINGS, WESTED LANE, SWANLEY.
Yard (TQ 524677) with locos for scrap occasionally present.

INDEPENDANT SEA TERMINALS, RIDHAM DOCK.
Gauge : 4'8½".   (TQ 918684)

08157		0-6-0DE		Dar	1955

KENT COUNTY COUNCIL, POOR PRIESTS HOSPITAL, STOUR STREET, CANTERBURY.
Gauge : 4'8½".   (   )

(INVICTA)	0-4-0	OC	RS	24 1830

MINISTRY OF DEFENCE, ARMY DEPARTMENT, ASHFORD DEPOT.

For full details see Section Five.

MINISTRY OF DEFENCE, NAVY DEPARTMENT, CHATHAM DOCKYARD.
Gauge : 4'8½".   (TQ 764698)

YARD No.361	AJAX	0-4-0ST	OC	RSHN	7042 1941	OOU
YARD No.562	ROCHESTER CASTLE	4wDM		FH	3738 1955	
YARD No.570	UPNOR CASTLE	4wDM		FH	3742 1955	
YARD No.5219	LEEDS CASTLE	4wDM		FH	3745 1955	
YARD No.18	DOVER CASTLE	4wDM		FH	3770 1955	
YARD No.533	COOLING CASTLE	4wDM		FH	3771 1955	
YARD No.218	DEAL CASTLE	4wDM		FH	3772 1955	
YARD No.12228	ALLINGTON CASTLE	0-4-0DH		HE	6975 1968	

NATIONAL COAL BOARD,

For full details see Section Four.

NORTH DOWNS STEAM RAILWAY, HIGHAM STATION, near GRAVESEND.
Gauge : 4'8½".   (TQ 716726)

	4wDM	RH	412427 1957
	4wDM	MR	9019 1950

NORTHFLEET TERMINAL LTD, CRETE HALL ROAD, GRAVESEND.
Gauge : 4'8½".   (TQ 628745)   RTC.

No.3	0-4-0DE	RH	416209 1957	OOU

REDLAND BRICKS LTD, FUNTON WORKS, near SITTINGBOURNE.
Gauge : 2'0".   (TQ 875677)

		2w-2BE		Redland		
		4wBE		Redland	1979	

REED PAPER & BOARD (U.K.) LTD.
Aylesford Paper Mills, New Hythe.
Gauge : 4'8½".   (TQ 714591, 716593)

	HORNBLOWER	0-4-ODE		RH	416211	1957
	BOUNTY	4wDM		RH	476142	1963

Empire Paper Mills, Greenhithe.
Gauge : 4'8½".   (TQ 595753)

	AIRMAN	0-4-ODE		RH	412713	1957
	PAXMAN	0-4-ODE		RH	412718	1958
	BATMAN	0-4-ODE		RH	512842	1965

ROMNEY HYTHE & DYMCHURCH RAILWAY, NEW ROMNEY.
Gauge : 1'3".   (TR 074249)

1	GREEN GODDESS	4-6-2	OC	DP	21499	1925
2	NORTHERN CHIEF	4-6-2	OC	DP	21500	1925
3	SOUTHERN MAID	4-6-2	OC	DP	22070	1926
4	THE BUG	0-4-OT	OC	Krauss	8378	1926
5	HERCULES	4-8-2	OC	DP	22071	1926
6	SAMSON	4-8-2	OC	DP	22072	1926
7	TYPHOON	4-6-2	OC	DP	22073	1926
8	HURRICANE	4-6-2	OC	DP	22074	1926
9	WINSTON CHURCHILL	4-6-2	OC	YE	2294	1931
10	DOCTOR SYN	4-6-2	OC	YE	2295	1931
11	BLACK PRINCE	4-6-2	OC	Krupp	1664	1937
4		4wDM		MR	7059	1938
PW2		4wPM		RHDR		1949
PW3	REDGAUNTLET	4wPM		Alan Keef	1977	

RUGBY PORTLAND CEMENT CO LTD, ROCHESTER WORKS, HALLING.
Gauge : 4'8½".   (TQ 704650)

		4wDM		RH	421417	1958	OOU
R13		4wDH		S	10040	1960	
R14		4wDH		S	10035	1960	

SHEERNESS IRON & STEEL CO LTD, SHEERNESS.
Gauge : 4'8½".   (TQ 912747)

		0-6-ODE		YE	2759	1959	
		0-4-ODH		YE	2804	1960	+
		0-4-ODH		YE	2806	1960	+

+ These locos can work in tandem.

<u>SHIPBREAKERS (QUEENBOROUGH) LTD. QUEENBOROUGH WHARF SCRAPYARD.</u>
Gauge : 4'8½".   (TQ 896716, 911716, 912719)

(D2070)		0-6-0DM	Don		1959
(2294)	01	0-6-0DM	(RSH	8127	1960
			(DC	2674	1960
03027		0-6-0DM	Sdn		1958
		4wDM	FH	3051	1945	OOU

Also other locos occasionally present for resale, and used in
yards as required.

<u>SITTINGBOURNE & KEMSLEY LIGHT RAILWAY LTD. SITTINGBOURNE & KEMSLEY.</u>
Gauge : 4'8½".   (TQ 905643, 920662)

	BEAR	0-4-0ST	OC	P	614	1896	Pvd
No.1		0-4-0F	OC	AB	1876	1925	Pvd
No.4		0-4-0ST	OC	HL	3718	1928	Pvd

Gauge : 2'6".

	PREMIER	0-4-2ST	OC	KS	886	1905
	LEADER	0-4-2ST	OC	KS	926	1905
	UNIQUE	2-4-0F	OC	WB	2216	1923
	MELIOR	0-4-2ST	OC	KS	4219	1924
	ALPHA	0-6-2T	OC	WB	2472	1932
	TRIUMPH	0-6-2T	OC	WB	2511	1934
	SUPERB	0-6-2T	OC	WB	2624	1940
	VICTOR	4wDM		HE	4182	1953
	EDWARD LLOYD	0-4-0DM		RH	435403	1961

<u>SOUTH EASTERN STEAM CENTRE, ASHFORD M.P.D. (Closed)</u>
Gauge : 4'8½".   (TR 022416)

(31065)	65		0-6-0	IC	Afd		1896
4902	13003, 13004	4w-4w+4w-4wRER			York		1971
	NORTHMET		0-6-0F	OC	RSHN	7056	1942
		4wBE			GB	1210	1930

<u>TENTERDEN RAILWAY CO LTD. (KENT & EAST SUSSEX RAILWAY).</u>
Gauge : 4'8½".   Locos are kept at :-

	Bodiam Station.	(TQ 782250)
	Northiam Station.	(TQ 834266)
	Rolvenden Station.	(TQ 865328)
	Tenterden Station.	(TQ 882336)

(30065)	22	MAUNSELL	0-6-0T	OC	VIW	4441	1943
(30070)	21	WAINWRIGHT	0-6-0T	OC	VIW	4433	1943
(31556)	11	PRIDE OF SUSSEX	0-6-0T	IC	Afd		1909
(32650)	No.10	SUTTON	0-6-0T	IC	Bton		1876
(32670)	No.3	BODIAM	0-6-0T	IC	Bton		1872
(W20W)	20		4w-4wDMR		AEC		1940
(W79978)			4wDMR		A.C.Cars		1958
		MINNIE	0-6-0ST	OC	FW	358	1878
	17	ARTHUR	0-6-0ST	IC	MW	1601	1903
	38	DOLOBRAN	0-6-0ST	IC	MW	1762	1910
5	18	WESTMINSTER	0-6-0ST	OC	P	1378	1914
376	No.19		2-6-0	OC	Nohab	1163	1919
8310/41		RHYL	0-6-0ST	IC	MW	2009	1921
	No.12	MARCIA	0-4-0T	OC	P	1631	1923
7086		ROLVENDEN	0-6-0ST	IC	RSH	7086	1943
5	No.29		0-6-0ST	IC	RSHN	7667	1950
		LINDA	0-6-0ST	IC	HE	3781	1952

	No.23	HOLMAN F.STEPHENS	0-6-0ST	IC		HE	3791	1952
	25	NORTHIAN	0-6-0ST	IC		HE	3797	1953
	No.24	WILLIAM.H.AUSTEN	0-6-0ST	IC		HE	3800	1953
16			4w-4wDE			BTH		1932
			4wDM			RH	252823	1947
	No.42		0-6-0DM			HE	4208	1948
	No.41	BAGLAN	0-4-0DH			RSHD/WB	8377	1962
	No.43		0-4-0DH			JF	4220031	1964
			4wDM			RH		
			4wPM	North London Polytechnic				
(900312)	1		2w-2PMR			Wkm		1931
(900393)			2w-2PMR			Wkm	400	1931
			2w-2PMR			Wkm	473	1931	DsmT
(900393)			2w-2PMR			Wkm	673	1932	DsmT
9043			2w-2PMR			Wkm	6965	1955
7438			2w-2PMR			Wkm	7438	1956

LANCASHIRE

WILLIAM AINSCOUGH & SONS LTD, MOSSY LEA ROAD, WRIGTHINGTON, near WIGAN.
Gauge : 2'0½".   (SJ 535125)

No.6		4wDM	MR	11102	1959	OOU
16		4wDM	MR	11165	1960	OOU
8		4wDM	MR	11223	1963	OOU
15		4wDM	MR	11246	1963	OOU
12		4wDM	MR	11258	1964	OOU
4		4wDM	MR	60S333	1968	OOU
9		4wDM	MR	60S363	1968	OOU

BRITISH FUEL CO, BLACKBURN COAL CONCENTRATION DEPOT.
Gauge : 4'8½".   (SD 677275)

D2272	ALFIE	0-6-0DM	(RSH	7914	1958
			(DC	2616	1958
2588		0-4-0DM	(RSH	7921	1957
			(DC	2588	1957

BRITISH LEYLAND MOTOR CORPORATION, LEYLAND.
Gauge : 4'8½".   (SD 544237)

5	0-4-0DM	JF	4210108	1955

BRITISH NUCLEAR FUELS LTD, URANIUM FUEL CENTRE, SPRINGFIELDS FACTORY,
SALWICK, near PRESTON.
Gauge : 4'8½".   (SD 468317)

4301 A	0-4-0DM	HC	D628	1943
4301 A	0-4-0DM	HC	D629	1945

BROOK VICTOR ELECTRIC VEHICLES LTD, VICTORIA ROAD, BURSCOUGH BRIDGE, near ORMSKIRK.
(SD 441123)

New BV locos under construction occasionally present.

CENTRAL ELECTRICITY GENERATING BOARD.
Fleetwood Power Station. (Closed)
Gauge : 4'8½". (SD 335468)

F/W 6	SIR JAMES	0-6-0F	OC		AB	1550 1917	OOU
F/W/7	LORD ASHFIELD	0-6-0F	OC		AB	1989 1930	OOU
		0-4-0DM			JF	4210068 1952	OOU

Heysham Power Station.
Gauge : 4'8½". (   )

	LANCASTER No.1	0-6-0F	OC		AB	1572 1917	OOU
	No.2	0-4-0F	OC		AB	1950 1928	OOU

Huncoat Power Station, Accrington.
Gauge : 4'8½". (SD 779314)   RSW.

HU 11	HUNCOAT No.1	0-4-0F	OC		WB	2989 1951	
		0-4-0F	OC		WB	3022 1951	OOU
HU 12	HUNCOAT No.3	0-6-0F	OC		HL	3746 1929	
HU 7		0-4-0DM			JF	4210011 1949	

Padiham Power Station, Burnley.
Gauge : 4'8½". (SD 783334)

No.1		0-4-0DH		AB	473 1961	
No.2		0-4-0DH		AB	474 1961	

FLEETWOOD MINIATURE RAILWAY, THE PROMENADE, FLEETWOOD.
Gauge : 1'3". (SD 319481, 331482)

278		2-8-0PH	S/O	SL	7218 1972

J.M.HITCHEN LTD, SCRAP DEALERS, HARPERS LANE, CHORLEY.
Gauge : 4'8½". (SD 587186)

No.8		0-4-0DM		JF	22936 1941 OOU

R.O.HODGSON LTD, WARTON ROAD, CARNFORTH.
Gauge : 4'8½". (SD 499709)

23	MERLIN	0-6-0DM		HC	D761 1951

R.B.A.HOLDEN, 7, BARNMEADOW CRESCENT, RISHTON, BLACKBURN.
Gauge : 2'0". (   )

	4wPM		Lancs Tanning	1958

IMPERIAL CHEMICAL INDUSTRIES LTD, AGRICULTURAL DIVISION, HEYSHAM WORKS,
MORECAMBE & HEYSHAM.
Gauge : 4'8½". (SD 423592)

	LION	0-6-0DH		TH/S	150C 1965
	KING	0-6-0DH		TH	167V 1966

IMPERIAL CHEMICAL INDUSTRIES LTD, MOND DIVISION, HILLHOUSE WORKS, THORNTON.
Gauge : 4'8½". (SD 345435)

No.2		0-4-0DM	JF	4210010 1949
No.3		0-4-0DM	JF	4210058 1952

LANCASHIRE TAR DISTILLERS LTD, PRESTON WORKS, DOCK ESTATE, PRESTON.
Gauge : 4'8½".   (SD 508298)

		4wDM		FH	3906	1959	

LEYLAND METAL CO LTD, LEYLAND.
Gauge : 4'8½".   (SD 546228)

		4wDM		RH	466626	1962	OOU

LYTHAM MOTIVE POWER MUSEUM, DOCK ROAD, LYTHAM.
Gauge : 4'8½".   (SD 381276)

No.42	68095	0-4-0ST	OC	Cowlairs		1887
SNIPEY	HODBARROW No.6	0-4-0CT	IC	N	4004	1890
	VULCAN	0-4-0ST	OC	VF	1828	1902
	GARTSHERRIE No.20	0-4-0ST	OC	NB	18386	1908
	PENICUIK	0-4-0ST	OC	HL	3799	1935
		0-4-0ST	OC	HC	1661	1936
	LYTHAM No.1	0-4-0ST	OC	P	2111	1949
7	SUSAN	4wVBT	VCG	S	9537	1952
		4wPM		H	965	1930

Gauge : 2'0".

		4wDM		HE	2198	1940

DONALD MARTINDALE LTD, CHORLEY.
Gauge : 2'6".   (SD 596172)

		4wDM		FH	3545	1952	OOU
		4wDM		FH	3753	1955	OOU

MINISTRY OF DEFENCE, ROYAL ORDNANCE FACTORY, CHORLEY.
Gauge : 4'8½".   (SD 562203)

18238	R.O.F.CHORLEY No.3	0-4-0DH	JF	4220021	1962
18242	R.O.F.CHORLEY No.4	0-4-0DH	JF	4220022	1962
21811	No.5	0-4-0DH	HE	7161	1970

MORECAMBE MINIATURE RAILWAY, PLEASURE PARK, WEST END PROMENADE, MORECAMBE. (Closed)
Gauge : 1'8".   (SD 428639)

4472	ROBIN HOOD	4-6-4DM	S/O	HC	D570	1932	OOU
	FLYING SCOTSMAN						
	MAY THOMPSON	4-6-2DM	S/O	HC	D582	1933	OOU

NORTH WEST WATER AUTHORITY, PRESTON & DISTRICT WATER SUPPLY UNIT,
SPADE MILL No.2 RESERVOIR, LONGRIDGE.
Gauge : 2'0".   (SD 622376)

		4wDM		MR	8669	1941

NORTH WEST WATER AUTHORITY, RIVERS DIVISION, HAVELOCK ROAD, BAMBER BRIDGE,
near PRESTON.
Gauge : 2'0".   (SD 558255)

81 A 138		4wDM		RH	462365	1960	OOU

PLEASURE BEACH RAILWAY, SOUTH SHORE, BLACKPOOL.
Gauge : 1'9".   (SD 305332)

4472	MARY LOUISE	4-6-2DM	S/O	HC	D578	1933
4473	CAROL JEAN	4-6-4DM	S/O	HC	D579	1933
6200	THE PRINCESS ROYAL	4-6-2DM	S/O	HC	D586	1935

PORT OF PRESTON AUTHORITY, PRESTON DOCKS.
Gauge : 4'8½".   (SD 525294)

ENERGY	4wDH	RR	10281	1968
ENTERPRISE	4wDH	RR	10282	1968
PROGRESS	4wDH	RR	10283	1968

RED ROSE LIVE STEAM GROUP,
Private Farm, Newton-Le-Willows.
Gauge : 4'8½".   (SD 53x13x)

	0-4-0ST	OC	AE	1563	1908

The Willows Childrens Home, Skelmersdale.
Gauge : 4'8½".   SD 497049)

4wDM	RH	244580	1946

RIBBLESDALE CEMENT LTD, CLITHEROE.
Gauge : 4'8½".   (SD 748438)

"No.8"	(D8568)	4w-4wDE	CE	4365U/69	1963	
		0-4-0DM	RH	327970	1954	Dsm
		0-4-0DM	RH	327971	1954	
No.6		0-6-0DH	JF	4240010	1960	

ROSSENDALE FOREST RAILWAY, EDENFIELD, ROSSENDALE.
Gauge : 2'0".   (SD 805188)

4wPM	Bg	2095	1936

SKELMERSDALE NEW TOWN DEVELOPMENT CORPORATION.
Digmoor Shopping Parade Play Area.
Gauge : 4'8½".   (SD 497048)

26	0-6-0ST	OC	AE	1810	1918

New Church Farm Play Area.
Gauge : 4'8½".   (SD 479061)

DAPHNE	0-4-0ST	OC	P	737	1899

STEAMTOWN RAILWAY MUSEUM, WARTON ROAD, CARNFORTH.
Gauge : 4'8½".   (SD 496708)

No.	Old No.	Name	Wheel	Cyl	Builder	Works No.	Date	Notes
5643			0-6-2T	IC	Sdn		1925	
9629			0-6-0PT	IC	Sdn		1946	
(30850)	850	LORD NELSON	4-6-0	4C	Elh		1926	
34092		CITY OF WELLS	4-6-2	3C	Bton		1949	
35005		CANADIAN PACIFIC	4-6-2	3C	Elh		1941	
44871		SOVEREIGN	4-6-0	OC	Crewe		1945	
44932			4-6-0	OC	Hor		1945	
(45407)	5407		4-6-0	OC	AW	1462	1937	
45699		GALATEA	4-6-0	3C	Crewe		1936	
(46441)	6441		2-6-0	OC	Crewe		1950	
(52322)	1122		0-6-0	IC	Hor	420	1896	
(60007)	No.4498	SIR NIGEL GRESLEY	4-6-2	3C	Don	1863	1937	
(60103)	No.4472	FLYING SCOTSMAN	4-6-2	3C	Don	1564	1923	
D2381			0-6-0DM		Sdn		1961	
231 K 22		LA FRANCE	4-6-2	4CC	Cail		1912	
						Rebuilt	1936	
012 104-6			4-6-2	3C	Sch	11360	1940	
		LINDSAY	0-6-0ST	IC	WCI		1887	
No.1		GLENFIELD	0-4-0CT	OC	AB	880	1902	
No.11		SIRAPITE	4wWT	G	AP	6158	1906	
		JOHN HOWE	0-4-0ST	OC	AB	1147	1908	
			0-4-0ST	OC	P	1370	1915	
No.1969		JANE DARBYSHIRE	0-4-0ST	OC	AB	1969	1929	
7			4wVBT	VCG	S	8024	1929	
No.39		GREAT CENTRAL	0-6-0T	OC	RSH	6947	1938	
2134		CORONATION	0-4-0ST	OC	AB	2134	1942	
8310/11		CRANFORD No.2	0-4-0ST	OC	WB	2668	1942	
No.1			0-4-0ST	OC	AB	2230	1947	
			0-6-0F	OC	WB	3019	1952	
1		FIREFLY	0-6-0T	OC	HC	1864	1952	
4		BRITISH GYPSUM	0-4-0ST	OC	AB	2343	1953	
6			0-6-0ST	IC	HE	3855	1954	
		"NOVELTY"	2-2-0VBWT	VC	Loco.Ent	No.3	1980	
7049		TOM ROLT	0-6-0DM		HE	2697	1944	
		NEW JERSEY	0-4-4-0DE		GEU	30483	1949	
D3		THE FLYING FLEA	4wDM		RH	294266	1951	
EYU 338C			4wPM	Rebuilt	McAlpine			
A 144		PWM 2176	2w-2PMR		Wkm	4153		
			2w-2PMR		Wkm			DsmT

Gauge : 1'3".

No.	Name	Wheel	Cyl	Builder	Works No.	Date
	GEORGE THE FIFTH	4-4-2	OC	BL	18	1911
22	PRINCESS ELIZABETH	4-4-2	OC	BL	22	1914
	ROYAL ANCHOR	4w-4wDH		Lane		1956

WALSH & MIDGELEY LTD, DARWEN.
Gauge : 4'8½".   (SD 705194)

Wheel	Builder	Works No.	Date	Notes
0-4-0DM	HC	D1031	1956	OOU

WEST LANCASHIRE LIGHT RAILWAY, STATION ROAD, HESKETH BANK, near PRESTON.
Gauge : 2'0".   (SD 448229)

No.1		CLWYD		4wDM		RH	264251	1951	
No.2		TAWD		4wDM		RH	222074	1943	
No.3		IRISH MAIL		0-4-0ST	OC	HE	823	1903	
No.4				4wPM		FH	1777	1931	
No.5				4wDM		RH	200478	1940	
No.7				4wDM		MR	8992	1946	
No.8				4wDM		HE	4478	1953	+
No.9				0-6-0T	OC	KS	2405	1915	
No.10				4wDM		FH	2555	1946	
11				4wDM		MR	5906	1934	
No.12				4wDM		MR	7955	1945	Dsm
No.16	L22	20		4wDM		RH	202036	1941	
No.17	13			4wDM		MR	11142	1960	
No.18	U192	T.R.A.No.13	TRENT	4wDM		RH	283507	1949	
No.19				4wPM		L	10805	1939	
No.20				4wPM		Bg	3002	1937	
No.21				4wDM		HE	1963	1939	
No.22				4wDM		RH	408430	1957	
No.25				4wDM		RH	297054	1950	

                    + Plate reads 4480 1953.

Gauge : 1'8".

| No.24 | 2 | | 4wDM | | RH | 187057 | 1937 | Dsm |

THE WEST LANCS BLACK 5 FUND, c/o I.C.I. Ltd, Hillhouse Works, Thornton.
Gauge : 4'8½".   (   )

| 45491 | | 4-6-0 | OC | Derby | | 1943 | |

THE WORLD OF ANIMALS ZOOLOGICAL GARDENS, STANLEY PARK, GREAT MARTON, BLACKPOOL.
Gauge : 1'3".   (SD 335362)

| 279 | | 2-8-0PH | S/O | SL | 7219 | 1972 |

LEICESTERSHIRE

AMEY ROADSTONE CORPORATION LTD, EASTERN REGION, LOUGHBOROUGH.
Gauge : 4'8½".   (SK 542205)

| 3101 | | 6wDE | | Derby | | 1955 |

BARDON HILL QUARRIES (ELLIS & EVERARD) LTD, BARDON HILL GRANITE QUARRY, COALVILLE.
Gauge : 4'8½".   (SK 446129)

48	2		0-6-0DM		RSH	7697	1953	OOU
			4wDM		RH	393304	1956	OOU
No.28		DUKE OF EDINBURGH	0-4-0DM		AB	400	1956	

FRANK BERRY LTD, WESTERN BOULEVARD, LEICESTER.
Yard (SK 580035) with locos for scrap occasionally present.

REV. E.R.BOSTON, BOSTONS LODGE, CADEBY RECTORY, MARKET BOSWORTH.
Gauge : 2'0".   (SK 426024)

	MARGARET	0-4-0ST	OC	HE	605	1894
No.2		0-4-0WT	OC	OK	7529	1914
	PIXIE	0-4-0ST	OC	WB	2090	1919
87004		4wDM		MR	2197	1922
		0-4-0PM	S/O	Bg	1695	1928
		4wDM		MR	4572	1929
		4wDM.		HC	D558	1930
1		4wDM		MR	5609	1931
	NEW STAR	4wPM		L	4088	1931
24		4wDM		MR	5853	1934
		4wDM		RH	179870	1936
42		4wDM		MR	7710	1939
20		4wDM		MR	8748	1942

BRUSH ELECTRICAL MACHINES LTD, TRACTION DIVISION, LOUGHBOROUGH.
(SK 543207)

New BT locos under construction occasionally present.

CENTRAL ELECTRICITY GENERATING BOARD, CASTLE DONINGTON POWER STATION.
Gauge : 4'8½".   (SK 433284)   RSW.

	C.E.G.B. 1	0-4-0ST	OC	RSHN	7817	1954
	C.E.G.B. 2	0-4-0ST	OC	RSHN	7818	1954
No.1		0-4-0DH		AB	415	1957
	CASTLE DONINGTON					
	POWER STATION No.2	0-4-0DM		AB	416	1957
No.2		0-4-0DE		RH	420142	1958

CLASSIC CARS OF COVENTRY, SOUTHFIELD ROAD, HINCKLEY.
Gauge : 4'8½".   (   )

	4wPM	FH	2895	1944

E.C.C. QUARRIES LTD, CROFT GRANITE DIVISION, CROFT.
Gauge : 4'8½".   (SP 517960)

	EDWIN	0-4-0DH	RH	437363	1960
Q407T	EX 256	0-4-0DH	JF	4220016	1962

GREAT CENTRAL RAILWAY (1976) LTD, LOUGHBOROUGH (G.C.) STATION.
Gauge : 4'8½".   (SK 543194)

No.1			4-2-2	OC	Don	50	1870
5224			2-8-0T	OC	Sdn		1924
6990		WITHERSLACK HALL	4-6-0	OC	Sdn		1948
34039		BOSCASTLE	4-6-2	3C	Bton		1946
(45231)	5231	3RD (VOLUNTEER) BATTALION,					
		THE WORCESTERSHIRE AND SHERWOOD					
		FORESTERS REGIMENT	4-6-0	OC	AW	1286	1936
(61264)	1264		4-6-0	OC	NB	26165	1947
(61306)	1306	MAYFLOWER	4-6-0	OC	NBQ	26207	1948
(62660)	506	BUTLER-HENDERSON	4-4-0	IC	Gorton		1920
(68088)			0-4-0T	IC	Dar	1205	1923
(69523)	4744		0-6-2T	IC	NB	22600	1921
71000		DUKE OF GLOUCESTER	4-6-2	3C	Crewe		1954
92212			2-10-0	OC	Sdn		1959

D4067		No.1802/B4	0-6-0DE		Dar		1961
(D9516) 56		8311/36	0-6-0DH		Sdn		1964
D9523 25		8311/25	0-6-0DH		Sdn		1964
No.3			0-4-0ST	OC	HL	3581	1924
		ROBERT NELSON No.4	0-6-0ST	IC	HE	1800	1936
		HILDA	0-4-0ST	OC	P	1963	1938
No.9		1802/B1	0-6-0ST	IC	HE	3825	1954
D4279		ARTHUR WRIGHT	0-4-0DM		JF	4210079	1952
No.1		QWAG	4wDM		RH	371971	1954
No.3	N.C.B. 11 1963	ALEN GRICE	4wDH		TH/S	134C	1964
51m			2w-2PMR		Wkm	(693	1932?)

HUNT & CO (HINCKLEY) LTD. LONDON ROAD WORKS. HINCKLEY.
Gauge : 4'8½". (SP 437937)

		0-4-0ST	OC	WB	2648	1941	Pvd
HUNTSMAN		0-6-0ST	OC	WB	2655	1942	Pvd

KETTON PORTLAND CEMENT CO LTD. KETTON CEMENT WORKS.
Gauge : 4'8½". (SK 982056, 987057) (Subsidiary of Thos W.Ward Ltd.)

No.1	0-4-0DH		JF	4220007	1960
No.4	0-6-0DH		JF	4240012	1961
	0-6-0DH		TH	293V	1980

LEICESTERSHIRE COUNTY COUNCIL, LEICESTERSHIRE MUSEUM OF TECHNOLOGY, ABBEY MEADOWS, CORPORATION ROAD. LEICESTER.
Gauge : 4'8½". (SK 589067)

		0-4-0ST	OC	BE	314	1906
No.2		0-4-0F	OC	AB	1815	1924
	MARS II	0-4-0ST	OC	RSHN	7493	1948

Gauge : 2'0".

	4wPM		MR	5038	1929
	4wPM		MR	5260	1931

MARKET BOSWORTH LIGHT RAILWAY, SHACKERSTONE STATION, MARKET BOSWORTH.
Gauge : 4'8½". (SK 379066)

(D2245)	No.2	0-6-0DM		(RSH	7864	1956
				(DC	2577	1956
M 50397		4w-4DMR		PR		1957 Dsm
		0-4-0WT	OC	EB		1906
N.C.B. 11		0-4-0ST	OC	HE	1493	1925
		0-6-0ST	OC	HL	3837	1934
21		0-6-0ST	OC	HL	3931	1938
V 47	HERBERT	0-4-0ST	OC	P	2012	1941
	LAMPORT No.3	0-6-0ST	OC	WB	2670	1942
No.3		0-6-0T	OC	RSHN	7537	1949
No.4		0-6-0T	OC	RSHN	7684	1951
No.7		0-4-0ST	OC	P	2130	1951
	NORTH GAWBER No.6 AREA	0-6-0T	OC	HC	1857	1952
2		0-6-0ST	OC	WB	3059	1953
	PEEPING TOM	4wDM		RH	235513	1945
No.347747	THE 1211 SQUADRON	0-6-0DM		RH	347747	1957
PVO 465		2-2wDM		David Brown		
9033	3	2w-2PMR		Wkm	6857	1954
9024	8	2w-2PMR		Wkm	7090	1955

HERBERT MORRIS LTD, NORTH ROAD, LOUGHBOROUGH.
Gauge : 4'8½".   (SK 537209)

1		4wDM	MR	2026	1920
2		4wDM	MR	4219	1935 Dsm

NATIONAL COAL BOARD.

                For full details see Section Four.

REDLAND ROADSTONE LTD.
Barrow-Upon-Soar Rail-Loading Terminal.
Gauge : 4'8½".   (SK 587168)

(D2867)	DIANNE	0-4-0DH	YE	2850	1961
	BUNTY	0-4-0DH	TH	146C	1964
		Rebuild of 0-4-0DM	JF	4210018	1950

Mountsorrel Granite Quarries.
Gauge : 4'8½".   (SK 578152)   RTC.

(02004)		0-4-0DH	YE	2815	1960 Dsm

RUTLAND RAILWAY MUSEUM, COTTESMORE IRON ORE MINES SIDINGS, ASHWELL ROAD,
COTTESMORE, near OAKHAM.
Gauge : 4'8½".   (SK 887137)

	UPPINGHAM	0-4-0ST	OC	P	1257	1912
	DORA	0-4-0ST	OC	AE	1973	1927
		0-4-0ST	OC	AB	1931	1927
	ELIZABETH	0-4-0ST	OC	P	1759	1928
YARD No.440	SINGAPORE	0-4-0ST	OC	HL	3865	1936
	COAL PRODUCTS No.6	0-6-0ST	IC	HE	2868	1943
		Rebuilt		HE	3883	1963
	SIR THOMAS ROYDEN	0-4-0ST	OC	AB	2088	1940
24		0-6-0ST	IC	HE	2411	1941
No.2	SALMON	0-6-0ST	OC	AB	2139	1942
		0-4-0ST	OC	P	2110	1950
No.3	1954/88	0-4-0DM		AB	352	1941
No.3		4wDM		RH	306092	1950
		4wDM		RH	305302	1951
		4wDM		FH	3887	1958
No.57		0-6-0DE		YE	2872	1962
DB 965072		2w-2PMR		Wkm	7587	1957 +

                + Currently stored elsewhere.

SEVERN TRENT WATER AUTHORITY, NOTTINGHAM DIVISION, near LEICESTER.
Gauge : 2'0".   (   )

19	U 193	4wDM	RH	7002/0567/6	1967

STANTON & STAVELEY LTD, HOLWELL WORKS, ASFORDBY HILL, MELTON MOWBRAY.
(Subsidiary of British Steel Corporation.)
Gauge : 4'8½".   (SK 725199)

19		4wDH	S	10019	1960
36	9411/70	4wDH	S	10036	1960

J.VERNON, CHURCH FARM, NEWBOLD VERDON.
Gauge : 1'10¼".   (SK 442038)

	PAMELA	0-4-0ST	OC	HE	920	1906	
	SYBIL MARY	0-4-0ST	OC	HE	921	1906	Dsm
24		4wDM		RH	382820	1955	

J.N.WALTON, MIDLANDS STEAM CENTRE, 106A, DERBY ROAD, LOUGHBOROUGH.
Gauge : 2'0".   (   )

SEA LION	2-4-0T	OC	WB	1484	1896

WELLAND VALLEY VINTAGE TRACTION CLUB, GLEBE ROAD, MARKET HARBOROUGH.
Gauge : 3'0".   (SP 742868)

KETTERING FURNACES No.8	0-6-0ST	OC	MW	1675	1906

LINCOLNSHIRE

ANGLIAN WATER AUTHORITY, LINCOLNSHIRE RIVER DIVISION, SOUTHREY, near LINCOLN.
Gauge : 2'0".   (TF 141664)

No.3	4wDM	RH	421432	1959
No.4	4wDM	RH	421433	1959

Locos are used on river bank work, etc, as required.

AVELING BARFORD HOLDINGS, SOUTH PARADE WORKS, GRANTHAM.
Gauge : 3'0".   (SK 917348)

2-2-0WT	G	AP	1607	1880	OOU

BRITISH SUGAR CORPORATION LTD.
Bardney Factory.
Gauge : 4'8½".   (TF 112686)

0-4-0DM	RH	327974	1954

Spalding Factory.
Gauge : 4'8½".   (TF 258247)

0-4-0DM	RH	304469	1951

BUTLINS LTD, SKEGNESS HOLIDAY CAMP.
Gauge : 2'0".   (TF 573669)

1863	C.P.HUNTINGTON	4-2-4DH	S/O	Chance	64/5030/24	1964

R.DALE, LOUTH.
Gauge : 2'0".   (   )

4wDM	MR	9264	1947

T.W.F.HALL, NORTH INGS FARM, DORRINGTON, near RUSKINGTON.
Gauge : 2'0".   (TF 098527)

4	6	4wDM	MR	7403	1939
3		4wDM	MR	7493	1940
		4wDM	MR	9655	1951
No.1		4wDM	RH	371937	1956

HAMBLETON BROS, SCRAP DEALERS, BRIGG ROAD, CAISTOR.
Gauge : 4'8½".   (TA 111025)

QC 2	0-4-0DM	JF	421004o	1951	OOU
QC 1	0-4-0DM	JF	4200046	1949	OOU
511 83 05	4wDM	RH	393306	1956	Dsm

H.M. DETENTION CENTRE, NORTH SEA CAMP, FREISTON, near BOSTON.
Gauge : 2'0".   (TF 387398, 398410, 399414)

4wDM	L	10994	1939	
4wDM	L	33650	1949	Dsm
4wDM	L	33651	1949	Dsm
4wDM	LB	51917	1960	Dsm
4wDM	LB	55413	1967	OOU
4wDM	LB	56371	1970	

JOHN LEE & SON (GRANTHAM) LTD, OLD WHARF ROAD, GRANTHAM.
Gauge : 4'8½".   (SK 908355)   RTC.

12608	4wDM	RH	321737	1952	OOU

LEY'S GEORGE FISCHER (LINCOLN) LTD, IRONFOUNDERS, NORTH HYKEHAM.
Gauge : 4'8½".   (SK 933672)

BAGNALL	4wDH	WB	3208	1961
	2-2wBE	WR	T8070	1979

LINCOLNSHIRE ASSOCIATION, MUSEUM OF LINCOLNSHIRE LIFE, BURTON ROAD, LINCOLN.
Gauge : 2'6".   (SK 972723)

4wPM	RP	52124	1918

Gauge : 2'3".

4wDM	RH	192888	1939

LYSAGHTS SPORTS & SOCIAL CLUB, HOLTON LE MOOR.
Gauge : 2'6".   (TF 094973)

CANNONBALL	4wDM S/O	RH	175403	1935

ROBIN PEARMAN.
Gauge : 2'6".   (   )

2	4wPM	MR	3849	1927

          Loco stored at a secret location.

RUSTON BUCYRUS LTD, BEEVOR STREET, LINCOLN.
Gauge : 4'8½".   (SK 966707)

T.374	4wDM R/R	Unimog

SMITH-CLAYTON FORGE LTD, TOWER WORKS, LINCOLN.
Gauge : 4'8½".   (SK 995713)

| | | | | 4wDM | | RH | 463154 | 1961 | |

J.H.P.WRIGHT, FENLAND AIRFIELD, HOLBEACH ST. JOHNS.
Gauge : 1'9".   (TF 333179)

| | | | 4-6-2DM | S/O | HC | D611 | 1938 |
| 6203 | QUEEN ELIZABETH | | 4-6-2DM | S/O | HC | D612 | 1938 |

MERSEYSIDE

J.G.ASHURST & SON, SCRAP MERCHANTS, VIST ROAD, NEWTON-LE-WILLOWS.
Yard (SJ 573966) with locos occasionally present for scrap or resale.

BRITISH INSULATED CALLENDERS CABLES LTD, PRESCOT.
Gauge : 4'8½".   (SJ 471924)

| | B.I.C.C. | | 0-4-0DH | | NB | 27653 | 1956 | |

B.I.C.C. METALS LTD, REFINERIES UNIT, PRESCOT.
Gauge : 2'6".   (SJ 470923)   RTC.

| WMD 2 | ENA SHARPLES | 4wDH | RH | 518493 | 1966 | OOU |

CAMMELL LAIRD SHIPBUILDERS LTD, BIRKENHEAD.
Gauge : 4'8½".   (SJ 328874, 330878)

| | 0-4-0DH | YE | 2805 | 1960 |
| | 4wDH | TH | 123V | 1963 |

CENTRAL ELECTRICITY GENERATING BOARD, BOLD POWER STATION, ST. HELENS.
Gauge : 4'8½".   (SJ 542934)   RTC.

| | 0-4-0DM | JF | 4210085 | 1953 | OOU |

C.EDMUNDSON, LIVERPOOL.
Gauge : 2'0".   (   )

| | 4wDM | MR | 4803 | 1934 |
| | 4wDM | MR | 7201 | 1937 |

FIBREGLASS LTD, INSULATION DIVISION, RAVENHEAD WORKS, ST. HELENS.
Gauge : 4'8½".   (SJ 504944)   (Subsidiary of Pilkington Bros Ltd.)

| J.B.M. | 0-4-0DE | YE | 2653 | 1957 |
| A.H.D. | 0-4-0DE | YE | 2730 | 1958 |

FORD MOTOR CO LTD, HALEWOOD FACTORY, LIVERPOOL.
Gauge : 4'8½".   (SJ 450845)

	0-4-0DH	YE	2807	1960
	0-4-0DH	YE	2675	1961
	0-4-0DH	YE	2679	1962

G.E.C. INDUSTRIAL LOCOMOTIVES LTD, VULCAN WORKS, NEWTON-LE-WILLOWS.
Gauge : 4'8½".   (SJ 587941)

ELIZABETH	0-4-ODM	(VF	D100	1949
		(DC	2271	1949
	0-6-ODM	RH	310088	1951

KNOWSLEY SAFARI PARK, KNOWSLEY HALL, near PRESCOT.
Gauge : 1'3".   (SJ   )

DUKE OF EDINBURGH	4-6-2DM	S/O	H.N.Barlow	1950

LAKESIDE MINIATURE RAILWAY, MARINE LAKE, SOUTHPORT.
Gauge : 1'3".   (SD 331174)

DUKE OF EDINBURGH	4-6-ODE	S/O	H.N.Barlow	1948
PRINCE CHARLES	4-6-ODE	S/O	H.N.Barlow	1954
GOLDEN JUBILEE				
1911-1961	4-6wDE		H.N.Barlow	1963
PRINCESS ANNE	6w-6wDH		SL	1971

LANCASHIRE FUEL CO LTD.
Rathbone Road Depot, Liverpool.
Gauge : 4'8½".   (SJ 385901)

LORD BERT	0-4-ODM	AB	383	1951
	0-6-ODE	HE	4550	1955

Southport Coal Concentration Depot.
Gauge : 4'8½".   (SD 344169)

3 NFD PR330	0-4-ODM	DC	2723	1961

LEVER BROS PORT SUNLIGHT LTD, PORT SUNLIGHT.
Gauge : 4'8½".   (SJ 341833)

LORD LEVERHULME	0-4-ODM	AB	388	1953
TRENCHARD	0-4-ODM	AB	401	1956

LOWTON METALS LTD, HAYDOCK, near ST. HELENS.
Gauge : 4'8½".   (SJ 577983)

(D2858)	0-4-ODH	YE	2817	1960

MERSEYSIDE METROPOLITAN COUNTY COUNCIL, MERSEYSIDE COUNTY MUSEUM,
WILLIAM BROWN STREET, LIVERPOOL.
Gauge : 4'8½".   (SJ 349908)

	LION	0-4-2	IC	Todd,Kitson & Laird	1838
No.1		0-6-OST	OC	AE 1465	1904
No.3		4w-4wRER			1892

NATIONAL COAL BOARD.

            For full details see Section Four.

NORWEST-HOLST, PLANT DEPOT, NETHERTON, LIVERPOOL.
Gauge : 1'0".   (SJ 365991)

	2w-2BE	Iso		
	2w-2BE	Iso		
	2w-2BE	Iso		
	2w-2BE	Iso		
	2w-2BE	Iso		

Locos present in yard between contracts.

PILKINGTON·BROS LTD, GLASS MANUFACTURERS.
Cowley Hill Works, St. Helens.
Gauge : 4'8½".   (SJ 514965)

CITY ROAD	0-4-0DE	YE	2626 1956
DONCASTER	0-4-0DE	YE	2654 1957
COWLEY HILL	0-4-0DE	YE	2687 1958
ST. ASAPH	0-4-0DE	YE	2781 1960
QUEENBOROUGH	0-4-0DE	YE	2782 1960

Sheet Works, St. Helens.
Gauge : 4'8½".   (SJ 512949)   RTC.

DH1	0-4-0DH	YE	2677 1960	OOU
DH2	0-4-0DH	YE	2820 1960	OOU

REA BULK HANDLING LTD, BIDSTON ORE DOCK & DUKE STREET WHARF, WALLASEY.
Gauge : 4'8½".   (SJ 300910, 306903, 311901)

PEGASUS	0-4-0DM	(VF	D99 1949	+
		(DC	2270 1949	
THESEUS	0-4-0DM	(VF	D138 1951	
		(DC	2272 1951	
WABANA	0-4-0DM	(RSH	7814 1953	
		(DC	2500 1953	
NARVIK	0-6-0DM	(VF	D229 1953	
		(DC	2501 1953	
PEPEL	0-6-0DM	(RSH	7857 1955	
		(DC	2569 1955	
W.H.SALTHOUSE	0-4-0DM	(RSH	7923 1959	
		(DC	2590 1959	
LABRADOR	0-4-0DE	YE	2732 1959	
KATHLEEN NICHOLLS	0-4-0DM	Bg/DC	2724 1963	
DOROTHY LIGHTFOOT	0-4-0DM	Bg/DC	2725 1963	

+ On permanent hire to W.J.Lee Ltd.

READS LTD, DRUM & CASK MAKERS, BULL LANE, AINTREE, LIVERPOOL.
Gauge : 4'8½".   (SJ 361971)   RTC.

	4wDM	RH	312432 1951	OOU

ROCKWARE GLASS CO LTD, ATLAS GLASS WORKS, ST. HELENS.
Gauge : 4'8½".   (SJ 519959)

400/703	POCKET NOOK I	0-6-0DM	HC	D1038 1958
400/704		0-6-0DM	HC	D1039 1958

SOUTHPORT PIER RAILWAY, SOUTHPORT.
Gauge : 60cm.   (SD 335176)

	SILVER BELLE	4-6-0DE		H.N.Barlow	1953	OOU
	ENGLISH ROSE	4w-4wDH		SL	23 1973	

STEAMPORT TRANSPORT MUSEUM, DERBY ROAD, SOUTHPORT.
Gauge : 4'8½".   (SD 341170)

5193		2-6-2T	OC	Sdn		1934
44806	MAGPIE	4-6-0	OC	Derby		1944
(47298)	7298	0-6-0T	IC	HE	1463	1924
76079		2-6-0	OC	Hor		1957
24081		4w-4wDE		Crewe		1960
M28361M		4w-4wRER		Derby		c1939
42	CECIL RAIKES	0-6-4T	IC	BP	2605	1885
No.1	WALESWOOD	0-4-0ST	OC	HC	750	1906
	LUCY	0-6-0ST	OC	AE	1568	1909
	EFFICIENT	0-4-0ST	OC	AB	1598	1918
No.9	KINSLEY	0-6-0ST	IC	HE	1954	1939
3		0-4-0ST	OC	P	1990	1940
75621		0-4-0ST	OC	P	1999	1941
	ST. MONANS	4wVBT	VCG	S	9373	1947
5		0-6-0ST	OC	P	2153	1954
No.1		4wBE		GB	2000	1945
	PERSIL	0-4-0DM		JF	4160001	1952
RS 142	TREVITHICK	0-4-0DE		RH	418598	1957

Rebuilt S.C.W.

U.G. GLASS CONTAINERS LTD, PEASLEY GLASSWORKS, PEASLEY CROSS LANE, ST. HELENS.
Gauge : 4'8½".   (SJ 517948)

378/625	PEASLEY		0-4-0DM	(VF	D296	1956
				(DC	2582	1956
378/626	SHERDLEY	378/625	4wDM	FH	3964	1961

WHITE MOSS PEAT CO LTD, SIMONSWOOD MOSS, PERIMETER ROAD, KIRKBY, LIVERPOOL.
Gauge : 2'0".   (SJ 442996)

	4wDM	MR	8696 1941
	4wDM	MR	20058 1949

WINGROVE & ROGERS LTD, ACORNFIELD ROAD, KIRKBY TRADING ESTATE.
(SJ   )

New WR locos under construction usually present.

NORFOLK

ANGLIAN BUILDING PRODUCTS LTD, ATLAS WORKS, LENWADE.
Gauge : 4'8½".   (TF 114179)

(2118)	0-6-0DM	Sdn	1959

ALAN BLOOM, BRESSINGHAM HALL, near DISS.
Gauge : 4'8½".   (TM 080806)

(30102)	102	GRANVILLE	0-4-0T	OC	9E	406	1893
(32662)	662	MARTELLO	0-6-0T	IC	Bton		1875
(41966)	80	THUNDERSLEY	4-4-2T	OC	RS	3367	1909
(42500)	2500		2-6-4T	3C	Derby		1934
(46100)	6100	ROYAL SCOT	4-6-0	3C	Derby		1930
(46233)	6233	DUCHESS OF SUTHERLAND	4-6-2	4C	Crewe		1938
(65567)	1217E		0-6-0	IC	Str		1905
70013		OLIVER CROMWELL	4-6-2	OC	Crewe		1951
377		KING HAAKON 7	2-6-0	OC	Nohab	1164	1919
5865		PEER GYNT	2-10-0	OC	Schichau		1944
141 R 73			2-8-2	OC	Lima	8939	1945 +
No.1			0-4-0T	OC	N	4444	1892
No.25			0-4-0ST	OC	N	5087	1896
		BLUEBOTTLE	0-4-0F	OC	AB	1472	1916
6841		WILLIAM FRANCIS	0-4-0+0-4-0	4C	BP	6841	1937
		MILLFIELD	0-4-0CT	OC	RSHN	7070	1942

        + Carries worksplate 9045 1946 in error.

Gauge : 1'11".   Nursery & Fen Railways.

No.316	GWYNEDD	0-4-0STT	OC	HE	316	1883	
No.994	GEORGE SHOLTO	0-4-0ST	OC	HE	994	1909	
No.1	EIGIAU	0-4-0WT	OC	OK	5668	1912	
		0-4-2ST	OC	KS	2395	1917	Dsm
	(BRONLLWYD)	0-6-0WT	OC	HC	1643	1930	+
12120		4wDM		MR	22210	1964	

        + Constructed from KS 2395 & HC 1643.

Gauge : 1'3".   Waveney Valley Railway.

No.1662	ROSENKAVALIER	4-6-2	OC	Krupp	1662	1937
No.1663	MÄNNERTREU	4-6-2	OC	Krupp	1663	1937
4472	FLYING SCOTSMAN	4-6-2	3C	W.P.Stewart		1976
		4wDM		Bressingham		1979

BOULTON & PAUL LTD, ENGINEERS, NORWICH WORKS, RIVERSIDE, NORWICH.
Gauge : 4'8½".   (TG 239079)

	0-4-0DM		Bg	3509 1958

BOULTON & PAUL (STEEL CONSTRUCTION) LTD, ENGINEERS, PREFLEX WORKS, LENWADE,
Gauge : 4'8¼".   (TG 113179)

	4wDM		RH	466625 1962

BRITISH INDUSTRIAL SAND LTD, QUARRIES DIVISION, MIDDLETON TOWERS SAND PITS,
LEZIATE, MIDDLETON, near KINGS LYNN,
Gauge : 4'8½".   (TF 672182, 674178)

(D2054)		0-6-0DM		Don		1959
3		4wDM		FH	3678	1953 OOU
46	14	4wDM		FH	3910	1959

BRITISH SUGAR CORPORATION LTD.
Cantley Factory.
Gauge : 4'8½".   (TG 386034)

1		0-6-0DM		RH	304468 1950
2		0-6-0DM		RH	395301 1956

Kings Lynn Factory, South Lynn.
Gauge : 4'8½".   (TF 609178)

	JAMIE	0-4-0DH		JF	4220025	1963
1	ROBBIE	0-4-0DH		JF	4220030	1964

Wissington Factory.
Gauge : 4'8½".   (TF 664973)

IVOR	0-4-0DH		JF	4220033	1965
	0-4-0DH		EEV	D1123	1966

G.T.CUSHING, STEAM MUSEUM, THURSFORD GREEN, THURSFORD, near FAKENHAM.
Gauge : 1'10¼".   (TF 980345)

CACKLER	0-4-0ST	OC	HE	671	1898

DOW CHEMICAL CO LTD, CROSSBANK ROAD, KINGS LYNN.
Gauge : 4'8½".   (TF 613215)

(07013)	0-6-0DE		RH	480698	1962

FRENCH KIER CONSTRUCTION LTD, PLANT DEPOT, SETCHEY, near KINGS LYNN.
Gauge : 2'0".   (TF 632136)

Ref						
LE/12/60		4wBE		WR	C6762	1963
LE/12/61		4wBE		WR	C6763	1963
LE/12/62		4wBE		WR	C6761	1963
LE/12/63		4wBE		WR	C6764	1963
LE/12/64		4wBE		WR	C6765	1963
LE/12/66		4wBE		WR	C6767	1963
LE/12/67		4wBE		WR	C6768	1963
LE/12/74		4wBE		WR	E6806	1965
LE/8-12/76		4wBE		WR	M7555	1972
LE/8-12/77		4wBE		WR	M7554	1972
LE/8-12/78		4wBE		WR	M7553	1972
LM-1	17575	4wBE		CE	B0987.1	1976
LM-2	17576	4wBE		CE	B0987	1976
LM-5	17593	4wBE		CE	B0987.5	1976
LM-7	17594	4wBE		CE	B0987	1976

Locos present in yard between contracts.

SID GEORGE, SAND, GRAVEL & DEMOLITION CONTRACTOR, BLACKBOROUGH END, MIDDLETON,
near KINGS LYNN.
Gauge : 1'11½".   (TF 667150)

EPV 785	2-2wDM		Potter	1977

PETER HILL, CAISTER CASTLE, near YARMOUTH.
Gauge : 4'8½".   (TG 504123)

42	RHONDDA	0-6-0ST	IC	MW	2010	1921

A.KING & SONS (METAL MERCHANTS), NORWICH SCRAPYARD, BESSEMER ROAD, NORWICH. (Closed)
Gauge : 4'8½".   (TG 226056)

(D2956)	A48L	0-4-0DM		AB	398	1956	OOU

<u>M.MAYES & C.FISHER, YAXHAM STATION YARD, YAXHAM, near DEREHAM.</u>
Gauge : 1'11½".   (TG 003102)

Y.P.L.R. No.1		2-2wVBT	VCG	Potter		1970
No.2	RUSTY	4wDM		L	32801	1948
No.3	PEST	4wDM		L	40011	1954
No.4	GOOFY	4wDM		OK	7688	c1936
No.5		4wDM		OK	7728	1935

<u>NATIONAL COAL BOARD, NORWICH COAL CONCENTRATION DEPOT, QUEENS ROAD, NORWICH.</u>
Gauge : 4'8½".   (TG 230077)

2325		0-6-0DM	(RSH	8184	1961
			(DC	2706	1961

<u>NORTH NORFOLK RAILWAY CO LTD.</u>
Gauge : 4'8½".   Locos are kept at :-

Sheringham.	(TG 156430)
Weybourne.	(TG 118419)

61572		4-6-0	IC	BP	6488	1928	
(65462)	7564	0-6-0	IC	Str		1912	
(E 79960)	79960	4wDMR		WMD		1958	
E 79963		4wDMR		WMD		1958	
3052	CAR No.91	4w-4wRER		MC		1932	
	PONY	0-4-0ST	OC	HL	2918	1912	
		0-6-0F	OC	WB	2370	1929	
45	COLWYN	0-6-0ST	IC	K	5470	1933	
	WISSINGTON	0-6-0ST	IC	HC	1700	1938	
5	JOHN D. HAMMER	0-6-0ST	OC	P	1970	1939	
No.1982	RING HAW	0-6-0ST	IC	HE	1982	1940	
	HARLAXTON	0-6-0T	OC	AB	2107	1941	
No.4		0-6-0ST	OC	WB	2680	1944	
		0-6-0T	OC	RSH	7845	1955	
	DOCTOR HARRY	0-4-0DM		JF	4100001	1945	
		0-4-0DM		JF	4210080	1953	
N.C.B. 10 1963		0-4-0DH		EES	8431	1963	
M. & G.N. No.1		2w-2PMR		Wkm	1521	1934	Dsm
No.2		2w-2PMR		Wkm	1522	1934	
960243	THE BUG	2w-2PMR		Wkm	1642	1934	Dsm

<u>STRUMPSHAW HALL STEAM MUSEUM, STRUMPSHAW HALL, near ACLE.</u>
Gauge : 1'11½".   (TF 345065)

No.6	GINETTE MARIE	0-4-0WT	OC	Jung	7509	1937

Gauge : 1'3".

2	CAGNEY	4-4-0	OC	P.McGarigle	1902

<u>Mr. WALKER, WOLFERTON STATION, near KINGS LYNN.</u>
Gauge : 4'8½".   (TF 661286)

960220		2w-2PMR	Wkm	1933

NORTHAMPTONSHIRE

BILLING AQUADROME LTD, BILLING, near NORTHAMPTON.
Gauge : 2'0".   (SP 808615)

S 87741	OLIVER		4wDM	S/O	MR	9869	1953
1380			4wDM		RH		

BRITISH LEYLAND U.K. LTD, WELLINGBOROUGH FOUNDRIES.
Gauge : 4'8½".   (SP 908676)

	4wDM		RH	386875 1955

BRITISH STEEL CORPORATION, TUBES DIVISION, CORBY GROUP.
Bessemer Plant Store, Corby Works, Corby.
Gauge : 4'8½".   (SP 904897)

47	(D9533)	8311/26	0-6-0DH		Sdn	1965	Dsm
48	(D9542)	8311/27	0-6-0DH		Sdn	1965	Dsm
49	(D9547)	8311/28	0-6-0DH		Sdn	1965	Dsm
51	(D9539)	8311/30	0-6-0DH		Sdn	1965	OOU
54	(D9553)	8311/34	0-6-0DH		Sdn	1965	OOU
55	(D9507)	8311/35	0-6-0DH		Sdn	1964	OOU
57	(D9532)	8311/37	0-6-0DH		Sdn	1965	Dsm
58	(D9554)	8311/38	0-6-0DH		Sdn	1965	OOU
60	(D9510)	8411/23	0-6-0DH		Sdn	1964	Dsm
63	(D9512)	8411/24	0-6-0DH		Sdn	1964	Dsm
66	(D9541)		0-6-0DH		Sdn	1965	Dsm
160	(D9538)		0-6-0DH		Sdn	1965	Dsm
20		8311/20	6wDH		RR	10273 1968	OOU
(30)		8311/1	0-8-0DH		YE	2894 1962	Dsm
37		8311/18	6wDH		RR	10275 1969	OOU
D2			0-4-0DH		NB	27079 1950	Dsm
D3			0-6-0DH		NB	27407 1954	Dsm
D4			0-6-0DH		NB	27408 1954	Dsm
5			0-6-0DH		NB	27409 1954	OOU
D8			0-6-0DH		NB	27871 1960	Dsm
D9			0-6-0DH		NB	28051 1962	OOU
D10			0-6-0DH		NB	28052 1962	OOU
D11			0-6-0DH		EEV	D913 1964	Dsm
D12			0-6-0DH		EEV	D914 1964	OOU
D13			0-6-0DH		EEV	D915 1964	OOU
D15			0-6-0DH		EEV	D1048 1965	OOU
D16			0-6-0DH		GECT	5366 1972	Dsm
D17			0-6-0DH		EEV	D1050 1965	OOU
D18			0-6-0DH		EEV	D1051 1965	OOU
D20			0-6-0DH		EEV	D1053 1965	OOU
D21			0-6-0DH		EEV	3970 1969	OOU
D22			0-6-0DH		EEV	3971 1969	OOU
D23			0-6-0DH		EEV	5354 1971	OOU
D24			0-6-0DH		EEV	5355 1971	Dsm
D25			0-6-0DH		EEV	5356 1971	Dsm
D26			0-6-0DH		EEV	5357 1971	OOU
D30			0-6-0DH		GECT	5367 1972	OOU
D31			0-6-0DH		GECT	5387 1973	OOU

B.S.C. Minerals, Pen Green Engineering Workshop.
Gauge : 4'8½".   (SP 897902)

(D9537)	52	8311/32	0-6-0DH		Sdn	1965	OOU
	32		6wDH		RR	10265 1967	OOU

BRITISH STEEL CORPORATION, TUBES DIVISION, CORBY WORKS GROUP, CORBY WORKS, CORBY.
Gauge : 4'8½".   (SP 909899)

B.S.C. 1		0-6-0DH	EEV	D1049	1965	@
B.S.C. 2		0-6-0DH	GECT	5395	1974	
D 27		0-6-0DH	EEV	5358	1971	
D 28		0-6-0DH	GECT	5365	1972	
D 29		0-6-0DH	GECT	5394	1974	+
D 32		0-6-0DH	GECT	5388	1973	
D 33		0-6-0DH	EEV	D916	1964	
D 35		0-6-0DH	GECT	5407	1975	
		0-6-0DH	EEV	D1052	1965	Dsm
		0-6-0DH	GECT	5408	1975	Dsm
		0-4-0DM	AB	476	1963	Dsm
		0-4-0WE	RSHN	7005	1940	Dsm
		0-4-0WE	HL	3820	1934	Dsm
No.3		4wWE	GB	420365/1	1974	OOU
No.4		4wWE	GB	420365/2	1974	OOU

@ Carries plates GECT 5408 1975 in error.
+ Carries plates EEV D1052 1965 in error.

GEORGE COHEN, SONS & CO LTD, COBORN WORKS, near KETTERING. (Closed)
Gauge : 4'8½".   (SP 850775)

243	B7	0-4-0DM	JF	22890	1939	OOU
		4wDM	MR	9922	1959	OOU

CORBY DISTRICT COUNCIL, WEST GLEBE PARK, COTTINGHAM ROAD, CORBY.
Gauge : 4'8½".   (SP 886890)

14	0-6-0ST OC	HL	3827	1934	

DAVENTRY BOROUGH COUNCIL, NEW STREET RECREATION GROUND, DAVENTRY.
Gauge : 4'8½".   (SP 574624)

0-6-0ST OC	WB	2654	1942	

FARTHINGSTONE SILOS LTD, NUNN MILLS ROAD, FAR COTTON, NORTHAMPTON.
Gauge : 4'8½".   (SP 763597)

SPEEDY	0-4-0DM	AB	361	1942	
SMOKY	0-4-0DM	AB	363	1942	
HOTWHEELS	0-6-0DM	AB	422	1958	

THOMAS E. GRAY & CO LTD, REFRACTORY MANUFACTURERS, ISEBROOK QUARRY, BURTON LATIMER.
Gauge : 4'8½".   (SP 887752)

	0-4-0DM	HE	2070	1940	
BUNTY	0-4-0DM	HE	4263	1952	

Gauge : 2'0".   (SP 892756)   RTC.

	4wDM	MR	5881	1935	OOU
THUNDERBIRD II	4wDM	MR	9411	1948	Dsm

SIR ROBERT McALPINE & SONS LTD, PLANT DEPOT, KETTERING.
Gauge : 2'0".   (SP 865769)

PLANT A 34818	0-4-0BE	WR	2065	1941	
PLANT A 26120	0-4-0BE	WR	6309	1961	
PLANT A 2613-	0-4-0BE	WR	6310	1961	

	PLANT A 066925	4wBE		WR	J7206 1969
2	PLANT A 056915	4wBE		WR	J7208 1969
3		4wBE		WR	J7271 1969
2	PLANT A 056917	4wBE		WR	J7272 1969
6	PLANT A 056920	4wBE		WR	J7273 1969
1	PLANT A 066923	4wBE		WR	J7274 1969
4	PLANT A 116917	4wBE		WR	J7275 1969
	PLANT A 037001	4wBE		WR	K7282 1970

Locos present in yard between contracts.

MINISTRY OF DEFENCE, ARMY DEPARTMENT, YARDLEY CHASE DEPOT.

For full details see Section Five.

NORTHAMPTONSHIRE IRONSTONE RAILWAY TRUST LTD, HUNSBURY HILL SITE.
Gauge : 4'8½".   (SP 735584)

No.14	BRILL	0-4-0ST	OC	MW	1795 1912
9365	BELVEDERE	4wVBT	VCG	S	9365 1946
		4wVBT	VCG	S	9369 1946
16		0-4-0DM		HE	2087 1940
	EXPRESS	4wDM		RH	235511 1945
		4wDM		RH	299100 1950

Gauge : Metre.

No.86	8315/86	0-6-0ST	OC	P	1871 1934
No.87	8315/87	0-6-0ST	OC	P	2029 1942
A 16 W		2w-2PMR		Wkm	6887 1954

Gauge : 2'0".

		4wPM		L	14006 1940
T 8		4wDM		MR	8731 1941
22		4wDM		MR	8756 1942
T 13		4wDM		MR	8969 1945
T 11		4wDM		MR	9711 1952

THE NORTHAMPTONSHIRE LOCOMOTIVE GROUP, QUARRY, near PITSFORD.
Gauge : Metre.   (SP 755672)

| 8315/85 | BANSHEE | 0-6-0ST | OC | P | 1870 1934 |

OVERSTONE SOLARIUM LIGHT RAILWAY, OVERSTONE SOLARIUM, SYWELL.
Gauge : 2'0".   (SP 819654)

| | | 4wDM | | MR | 8727 1941 |

SHANKS & McEWAN (ENGLAND) LTD, SLAG CONTRACTORS, CORBY STEELWORKS.
Gauge : 4'8½".   (SP 901910)

| 21 90 01 | | 0-4-0DH | | AB | 499 1965 |
| 21 90 02 | | 0-4-0DH | | RH | 504565 1965 |

WAGON REPAIRS LTD, NEILSON ROAD, FINEDON ROAD INDUSTRIAL ESTATE, WELLINGBOROUGH.
Gauge : 4'8½".   (SP 902700)

L 10		4wDM		RH	224345 1945
L 4	CHATSWORTH	4wDM		RH	235517 1945
L 8		4wDM		RH	393303 1956

WICKSTEED PARK LAKESIDE RAILWAY, KETTERING.
Gauge : 2'0".   (SP 883770)

	LADY OF THE LAKE	0-4-0DM	S/0		Bg	2042	1931
	KING ARTHUR	0-4-0DH	S/0		Bg	2043	1931
	CHEYENNE	4wDM	S/0		MR	22224	1966

Gauge : 4'8½".   (SP 879773)

	0-4-0ST	OC		AB	2323	1952	Pvd

K.WOOLMER, BAKERS LANE, WOODFORD, KETTERING.
Gauge : 2'0".   (   )

4wDM		L	36743	1951	Pvd

NORTHUMBERLAND

BRITISH STEEL CORPORATION, SCUNTHORPE DIVISION, BEAUMONT FLUOR MINE,
ALLENHEADS. (Closed)
Gauge : 2'0".   (NY 860453)

	0-4-0BE		WR	J7056	1969	OOU
	0-4-0BE		WR	M7544	1972	OOU
1	0-4-0BE		WR	7656	1973	OOU
2	4wBE		CE	B0134A	1973	OOU
3	4wBE		CE	B0134B	1973	OOU
4	4wBE		CE	B475	1975	OOU
	4wBE		CE	B0495	1975	OOU
4	4wBE		CE	B1599A	1978	OOU
	4wBE		CE	B1599B	1978	OOU
	BE		WR		1978	OOU
6	0-4-0BE		WR	T8013	1979	OOU
	4wBE		CE	B1854	1979	OOU

CENTRAL ELECTRICITY GENERATING BOARD, BLYTH POWER STATION, NORTH BLYTH.
Gauge : 4'8½".   (NZ 300836, 302833)

No.27	0-4-0DH		NB	27762	1958
No.28	0-4-0DH		NB	27874	1958
No.31	0-6-0DH		JF	4240014	1962
	0-6-0DH		AB	487	1964

G.F.HORSEMAN, SLAGGYFORD LIGHT RAILWAY, THE ISLAND, SLAGGYFORD, near ALSTON.
Gauge : 2'0".   (NY 680523)

	4wDM		HE	2577	1942
3236	0-4-0DM	S/0	Bg	3236	1947

HUGHES BOLCKOW & CO LTD, SHIPBREAKERS, NORTH BLYTH.
Gauge : 4'8½".   (NZ 311825)

	A.M.W. No.212	0-4-0DM		JF	22959	1941

MINERALS INDUSTRIES LTD, SCRAITHOLE MINE, CARR SHIELD, near NENTHEAD.
Gauge : 1'6".   (NY 803468)

		0-4-0BE		WR	7177 1967

JOHN MOFFITT, "HUNDAY", NATIONAL TRACTOR & FARM MUSEUM, PEEPY FARM, NEWTON, STOCKSFIELD.
Gauge : 2'6".   (NZ 042652)

DALMUNZIE		4wPM	MR	2014 1920
YARD No.85		0-4-0DM	HE	2250 1940

NATIONAL COAL BOARD.
> For full details see Section Four.

SOUTH TYNEDALE RAILWAY PRESERVATION SOCIETY, SLAGGYFORD STATION.
Gauge : 4'8½".   (NY 676524)

No.13	THE BARRA	0-4-0ST OC	HL	3732 1928

NORTH YORKSHIRE

ALNE BRICK CO LTD.
Alne, near Easingwold.
Gauge : 2'0".   (SE 522663)

	4wDM	MR	8694 1943

Hemingbrough Brickworks, near Selby.
Gauge : 2'0".   (SE 674313)

		4wDM	MR	7494 1940	
No.4	DRUID	4wDM	MR	8644 1941	
		4wDM	MR	8746 1943	Dsm
		4wDM	HE	2959 1944	OOU
No.2	44	4wDM	MR	9778 1953	OOU

BRITISH OIL & CAKE MILLS LTD, OLYMPIA WORKS, SELBY.
Gauge : 4'8½".   (SE 624326, 625327)

0-4-0DM		JF	4200003 1946
4wDM R/R		S&H	7501 1966

BRITISH SUGAR CORPORATION LTD.
Poppleton Factory, York.
Gauge : 4'8½".   (SE 576531)

0-4-0DM	RH	395304 1956

Selby Factory.
Gauge : 4'8½".   (SE 627323, 629322)

0-4-0DM	RH	327964 1953

BUTLINS LTD, FILEY HOLIDAY CAMP.
Gauge : 2'0".   (TA 118775)

| 141 | C.P.HUNTINGDON | 4-2-4DH | S/O | Chance | 76/50141-24 | 1976 |

DERWENT VALLEY RAILWAY CO, LAYERTHORPE STATION, YORK. (Closed)
Gauge : 4'8½".   (SE 612521)

No.1	(D2298)	LORD WENLOCK	0-6-0DM		(RSH	8157	1960
No.2		CLAUDE THOMPSON	0-4-0DM		(DC	2679	1960
					JF	4210142	1958

ELMET INDUSTRIAL (TRACTORS) LTD, THE NOOK, SOUTH MILFORD.
Gauge : 2'0".   (SE 493316)

		4wDM		MR	5213	1930
		4wDM		MR	8979	1946
		4wDM		RH	252864	1947

Locos hired out to various concerns.

LIGHTWATER VALLEY LEISURE LTD, LIGHTWATER VALLEY FARM, RIPON.
Gauge : 1'3".   (   )

	LITTLE GIANT	4-4-2	OC	BL	10	1905 +
111	YVETTE	4-4-0	OC	W.Younger		1942
	ROYAL SCOT	4-6-0	OC	Carland		
1326	BLACOLVESLEY	4-4-4PM	S/O	BL		1909
278	7	2-8-0DH	S/O	SL	17.6.79	1979

+ Currently under renovation elsewhere.

MINISTRY OF DEFENCE, ARMY DEPARTMENT, HESSAY DEPOT.
For full details see Section Five.

NATIONAL RAILWAY MUSEUM, LEEMAN ROAD, YORK.
Gauge : 4'8½".   (SE 594519)

		THE AGENORIA	0-4-0	VC	Foster Rastrick		1829
1868	49	COLUMBINE	2-2-2	OC	Crewe	1	1845
No.3			0-4-0	IC	Bury		1846
66		AEROLITE	2-2-4T	IC	KTH	281	1851
910			2-4-0	IC	Ghd		1875
82		BOXHILL	0-6-0T	IC	Bton		1880
214		GLADSTONE	0-4-2	IC	Bton		1882
790		HARDWICKE	2-4-0	IC	Crewe	3286	1892
1621			4-4-0	IC	Ghd		1893
563			4-4-0	OC	9E	380	1893
673			4-2-2	IC	Derby		1897
990		HENRY OAKLEY	4-4-2	OC	Don	769	1898
251			4-4-2	OC	Don	991	1902
2818			2-8-0	OC	Sdn	2122	1905
(30245)	245		0-4-4T	IC	9E	501	1905
(30925)	925	CHELTENHAM	4-4-0	3C	Elh		1934
(31737)	737		4-4-0	IC	Afd		1901
35029		ELLERMAN LINES	4-6-2	3C	Elh		1949
(41000)	1000		4-4-0	3C	Derby		1902
(42700)	2700		2-6-0	OC	Hor		1926
(44767)	4767	GEORGE STEPHENSON	4-6-0	OC	Crewe		1947
46229		DUCHESS OF HAMILTON	4-6-2	4C	Crewe		1938
(50621)	1008		2-4-2T	IC	Hor	1	1889

(60022)	4468	MALLARD	4-6-2	3C	Don	1870	1938
(60800)	4771	GREEN ARROW	2-6-2	3C	Don	1837	1936
(62785)	490		2-4-0	IC	Str	836	1894
(65894)	2392		0-6-0	IC	Dar		1923
(68633)	87		0-6-0T	IC	Str	1249	1904
(68846)	1247		0-6-0ST	IC	SS	4492	1899
69023		JOEM	0-6-0T	IC	Dar	2151	1951
D2860			0-4-0DH		YE	2843	1960
03090			0-6-0DM		Don		1960
D5500			2w-2-2w+2w-2-2wDE		BT	71	1957
20050			4w-4wDE		(EE	2347	1957
					(VF	D375	1957
(DS 75)	75S		4wRE		Siemens	6	1898
E5001			4w-4wRE		Don		1958
26020			4w-4wWE		MV	1027	1951
(26500)	No.1		4w-4wWE/RE		BE		1905
84001			4w-4wWE		NB	27793	1960
APT-E	PC1/PC2		4w-4w-4wArtic GTE		Derby		1972
8143	1293		4w-4RE		MV/MC		1925
11179	3131		4w-4RE		EE/Elh		1938
28249			4w-4wRE		Oerlikon M.C.		1915
960209			2w-2PMR		Wkm	899	1933
		HODBARROW	0-4-0ST	OC	HE	299	1882
		WOOLMER	0-6-0ST	OC	AE	1572	1910
607			4-8-4	OC	VF	4674	1935
No.15		EUSTACE FORTH	0-4-0ST	OC	RSHN	7063	1942
		IMPERIAL No.1	0-4-0F	OC	AB	2373	1956
5			4wVBT	VCG	S	9629	1957
		ROCKET	0-2-2	OC	Loco.Ent	No.2	1979
		JOHN ALCOCK	0-6-0DM		HE	1697	1932
			0-4-0DE		AW	D21	1933
240			0-4-0DM		(EE	847	1934
					(DC	2047	1934
			0-6-0DM		RSHN	7746	1954

Gauge : 2'6".

(822)	No.1	THE EARL	0-6-0T	OC	BP	3496	1902

Gauge : 2'0".

1		CHALONER	0-4-0VBT	VC	DeW		1877
			4wDM		RH	187105	1937
809			2w-2-2-2wRE		EE	809	1930

Gauge : 1'11½".

No.1	K		0-4-0+0-4-0T	4C	BP	5292	1909
No.2			4wPM		MR	1377	1918

Gauge : 1'6".

	WREN	0-4-0STT	OC	BP	2825	1887

Some locomotives are usually under renovation, or stored, in the Museum Annexe at Leeman Road Goods Station and at the N.C.L. Depot, York. Certain locos will be used on 'Special Runs' and also exhibited at other sites.

NORTH BAY RAILWAY, NORTHSTEAD MANOR GARDENS, NORTH BAY, SCARBOROUGH.
Gauge : 1'8".   (TA 035898)

1931	NEPTUNE	4-6-2DM	S/O	HC	D565	1931
1932	TRITON	4-6-2DM	S/O	HC	D573	1932

NORTH YORKSHIRE MOORS RAILWAY PRESERVATION SOCIETY.
Gauge : 4'8½".   Locos are kept at :-

Goathland.	(NZ 836013)
Grosmont.	(NZ 828049, 828053)
Levisham.	(NZ 818909)
New Bridge.	(NZ 803854)
Pickering.	(NZ 797842)

6619			0-6-2T	IC		Sdn		1928	
(30841)	No.841	GREENE KING	4-6-0	OC		Elh		1936	
34027		TAW VALLEY	4-6-2	3C		Bton		1946	
(45428)	5428	ERIC TREACY	4-6-0	OC		AW	1483	1937	
(62005)	2005		2-6-0	OC		NBQ	26609	1949	
(63395)	2238		0-8-0	OC		Dar		1918	
63460			0-8-0	3C		Dar		1919	
75014			4-6-0	OC		Sdn		1951	
80135			2-6-4T	OC		Bton		1956	
92134			2-10-0	OC		Crewe		1957	
D821		GREYHOUND	4w-4wDH			Sdn		1960	
D2207			0-6-0DM			(VF	D208	1953	
						(DC	2482	1953	
D5032			4w-4wDE			Crewe		1959	
D7029			4w-4wDH			BPH	7923	1962	
(D9520)	45	8311/24 24	0-6-0DH			Sdn		1964	
(D9529)	61	8411/20	0-6-0DH			Sdn		1965	
(E 50341)	D10		4w-4wDMR			GRC		1957	
Sc 51118	D11		4w-4wDMR			GRC		1957	
No.29			0-6-2T	IC		K	4263	1904	
No.5			0-6-2T	IC		RS	3377	1909	
No.20		JENNIFER	0-6-0T	OC		HC	1743	1942	
No.3180		ANTWERP	0-6-0ST	IC		HE	3180	1944	
No.31		METEOR	0-6-0T	IC		RSH	7609	1950	
No.1		MIRVALE	0-4-0ST	OC		HC	1882	1955	
No.47		MOORBARROW	0-6-0ST	OC		RSH	7849	1955	
No.21			0-4-0DM			JF	4210094	1954	
		STANTON No.44	0-4-0DE			YE	2622	1956	
No.1		FRED	2w-2PMR			Wkm	578	1932	
No.2		PAULINE	2w-2PMR			Wkm			
3	DB 965053	FRANK	2w-2PMR			Wkm	7576	1956	
4	DB 965108	NELSON	2w-2PMR			Wkm	7623	1957	
5		GRAHAM	2w-2PMR			Wkm	593	1932	
6	7	KEN	2w-2PMR			Wkm	1724	1934	
7			2w-2PMR			Wkm	7565	1956	
			2w-2PMR			Wkm	417	1931	DsmT
			2w-2PMR			Wkm	1305	1933	DsmT
			2w-2PMR			Wkm	1523	1934	DsmT

HENRY OAKLAND & SONS LTD. (HEPWORTH CERAMICS), ESCRICK TILE WORKS, near YORK.
Gauge : 2'0".   (SE 625403)   RTC.

No.2	DICK		4wDM		MR	7179	1937	Dsm

ROWNTREE-MACKINTOSH LTD, CHOCOLATE MANUFACTURERS, YORK.
Gauge : 4'8½".   (SE 605539)

No.1		0-4-0DE		RH	423661 1958
No.2		4wDM		RH	432479 1959
No.4		0-6-0DH		TH	285V 1979

SETTLE LIMESTONE LTD, HORTON QUARRY, HORTON-IN-RIBBLESDALE.
Gauge : 4'8½".   (SD 802722)   (Member of Tarmac Group.)

	RAYLEIGH	0-4-0DE		RH	312985 1952	
RS 223	DAVY	0-4-0DE		YE	2610 1956	
	RAMSAY	0-4-0DE		RH	402800 1957	Dsm

TILCON LTD, SWINDEN LIMEWORKS, GRASSINGTON, near SKIPTON.
Gauge : 4'8½".   (SD 983614)

(08054)		0-6-0DE		Dar	1953
12083		0-6-0DE		Derby	1950
(15231)		0-6-0DE		Afd	1951 oOU
2		4wDM		FH	3893 1958 OOU

YORKSHIRE DALES RAILWAY CO LTD, STEAM & VINTAGE TRANSPORT CENTRE, EMBSAY.
Gauge : 4'8½".   (SE 007533)

48151		2-8-0	OC	Crewe	1942 +	
		0-4-0ST	OC	P	1159 1908	
20	DOROTHY	0-6-0ST	OC	HC	1450 1922	
	AIREDALE	0-6-0ST	IC	HE	1440 1923	
	ANN	4wVBT	VCG	S	7232 1927	
		0-6-0ST	OC	HE	1810 1937	Dsm
	SLOUGH ESTATES LTD No.5	0-6-0ST	OC	HC	1709 1939	
S 112		0-6-0ST	IC	HE	2414 1941	
S 119	BEATRICE	0-6-0ST	IC	HE	2705 1945	
	FOLESHILL	0-4-0ST	OC	P	2085 1948	
No.140		0-6-0T	OC	HC	1821 1948	
No.3	CITY LINK	0-4-0ST	OC	YE	2474 1949	
No.2		0-4-0ST	OC	RSHN	7661 1950	
No.22		0-4-0ST	OC	AB	2320 1952	
	PRIMROSE No.2	0-6-0ST	IC	HE	3715 1952	
	DARFIELD No.1	0-6-0ST	IC	HE	3783 1953	
No.69		0-6-0ST	IC	HE	3785 1953	
	MONCKTON No.1	0-6-0ST	IC	HE	3788 1953	
MDE 15		4wDM		Bg/DC	2136 1938	
	H.W.ROBINSON	0-4-0DM		JF	4100003 1946	
		4wDM		RH	294263 1950	
887		4wDM		RH	394009 1955	
DB 965095		2w-2PMR		Wkm	7610 1957	

          + Currently under renovation at Wakefield.

Gauge : 2'0".

		4wPM	L	9993 1938 +
Y.W.R. L2		4wPM	L	10225 1938

          + Currently under renovation at Barnoldswick, near Colne.

YORKSHIRE GRAIN DRIERS LTD, HARVEST MILLS, DUNNINGTON STATION, YORK.
Gauge : 4'8½".   (SE 674516)

	CHURCHILL	0-4-0DM		JF	4100005 1947 OOU

YORKSHIRE WATER AUTHORITY, RIVERS DIVISION, RICCALL PLANT DEPOT.
Gauge : 2'0".   (SE 608373)

24		4wDM		MR	7498 1941
35		4wDM		MR	8698 1941

Locos used on river bank work, etc, as required.

NOTTINGHAMSHIRE

BLUE CIRCLE INDUSTRIES LTD, KILVINGTON GYPSUM WORKS, near NEWARK.
Gauge : 4'8½".   (SK 798435)

(D2865)	0-4-0DH		YE	2848 1961

Gauge : 3'0".   (SK 797435)

	0-6-0DM		RH	281290 1949
No.1	0-6-0DM		RH	281291 1949 OOU
	4wDH		MR	102T007 1974
	4wDH		SMH	102T016 1976

BOOTS PURE DRUG CO LTD, MANUFACTURING CHEMISTS, BEESTON, NOTTINGHAM.
Gauge : 4'8½".   (SK 545363)   RTC.

No.2	P.N.B. 292	0-4-0F	OC	AB	2008 1935
	B 7211	0-4-0DE		RH	384139 1955

BRITISH GYPSUM LTD.
Gotham Works, Gotham.
Gauge : 2'2".   (SK 539292)   (Underground)   RTC.

	4wDM		RH	168790 1933 Dsm
	4wDM		RH	175417 1936 Dsm

Rushcliffe Works, East Leake.
Gauge : 4'8½".   (SK 553280)

	4wDM		RH	236364 1946
	4wDM		RH	398616 1956

BRITISH SUGAR CORPORATION LTD.
Colwick Factory.
Gauge : 4'8½".   (SK 617400)

No.1	0-4-0DM		RH	319292 1953
No.2	0-4-0DM		RH	375718 1955

Newark Factory, Kelham, near Newark.
Gauge : 4'8½".   (SK 795554)   RTC.

SIR ALFRED WOOD	0-6-0DM		RH	319294 1953 OOU

CENTRAL ELECTRICITY GENERATING BOARD.
High Marnham Power Station, near Newark.
Gauge : 4'8½".   (SK 802712)

No.2   0-4-0DH   AB   441 1959

Staythorpe Power Station, near Newark.
Gauge : 4'8½".   (SK 763538, 764534)

No.2   NANCY   4wDM   RH   263001 1949   OOU
No.4   0-4-0DE   RH   312986 1952
No.5A   0-4-0DE   RH   420137 1958
No.1B   0-4-0DE   RH   421435 1958
No.2B   0-4-0DE   RH   449754 1961

J.R.CLARK, PLANT DEALER, WIGSLEY WOOD SCRAPYARD.
Gauge : 4'8½".   (SK 845708)

    A 1092   4wDM   RH   224354 1945   OOU
    SF 364   4wDM   RH   338415 1953   OOU

J.CRAVEN, EAST VIEW, MAIN STREET, WALESBY, NEWARK.
Gauge : 3'0".   (SK 684707)

                                2w-2PMR   Wkm   9673 1964

Gauge : 2'0".

                                4wDM   RH   441944 1960
                                4wDM   B.House c1974

GLASS BULBS LTD, HARWORTH.
Gauge : 4'8½".   (SK 624909)

                                4wDM   RH   299099 1950
                                4wDH   RH   513141 1966   OOU

HOVERINGHAM GRAVELS (MIDLANDS) LTD, HOVERINGHAM WORKSHOPS, NOTTINGHAM.
Gauge : 2'0".   (SK 701481)

    H 85   4wDM   HE   7178 1971   OOU

M.A.G.JACOB, 53, RIBBLESDALE ROAD, SAWLEY, LONG EATON.
Gauge : 2'0".   (   )

    L 5   4wDM   RH   370555 1953

MINISTRY OF DEFENCE, ARMY DEPARTMENT.
Chilwell Depot, Nottingham.
            For full details see Section Five.

Ruddington Depot.
            For full details see Section Five.

MINISTRY OF DEFENCE, ROYAL ORDNANCE FACTORY, RANSKILL. (Closed)
Gauge : 4'8½".   (SK 673867)   RTC.

No.11   R.O.F. No.6   0-4-0DM   JF   22988 1942   OOU

NATIONAL COAL BOARD.

For full details see Section Four.

NEWARK DISTRICT COUNCIL, MILLGATE MUSEUM OF SOCIAL AND FOLK HISTORY, NEWARK.
Gauge : 4'8½".   (   )

	0-6-0ST IC	HC	1682	1937

SEVERN TRENT WATER AUTHORITY, STOKE BARDOLPH SEWAGE WORKS, NOTTINGHAM.
Gauge : 2'0".   (SK 633421)

U 197	4wDM	MR	22129	1962

SEVERN TRENT WATER AUTHORITY, TRENT RIVER MANAGEMENT DIVISION,
ORSTON ROAD EAST DEPOT, WEST BRIDGFORD.
Gauge : 2'0".   (SK 585385)

Locos, used on river bank work, etc, as required are here for repair occasionally.

STEETLEY REFRACTORIES LTD, CERAMICS DIVISION, RHODESIA, near WORKSOP.
Gauge : 4'8½".   (SK 554792)   RTC.

105000	981	OUGHTIBRIDGE No.1	0-4-0DE	YE	2652	1957	OOU

P.TUXFORD, PLANT YARD, PODDERS LANE, MAPPERLEY PLAINS, NOTTINGHAM.
Gauge : 2'6".   (SK 604444)

0-4-0DM	HE	3411	1947	OOU

A.J.WILSON, 6, TRENTDALE ROAD, CARLTON, NOTTINGHAM.
Gauge : 2'0".   (SK 607408)

THE WASP	4wPM	Wilson	c1969

OXFORDSHIRE

BLUE CIRCLE INDUSTRIES LTD, OXFORD WORKS, SHIPTON-ON-CHERWELL.
Gauge : 4'8½".   (SP 482175)

0-4-0DH	JF	4220037	1966
4wDH	TH	213V	1969

PETER COURT PLANT HIRE, BUTCHERS MEADOW, BALSCOTT, near BANBURY.
Gauge : 2'0".   (SP 387408)

4wDM	OK	6931	OOU
4wDM	OK	7734	OOU

GREATWESTERN PRESERVATIONS LTD, DIDCOT RAILWAY CENTRE.
Gauge : 4'8½".   (SU 524906)

1340	TROJAN	0-4-0ST	OC	AE	1386	1897
1363		0-6-0ST	OC	Sdn	2377	1910
1466		0-4-2T	IC	Sdn		1936
3650		0-6-0PT	IC	Sdn		1939
3738		0-6-0PT	IC	Sdn		1937
3822		2-8-0	OC	Sdn		1940
4144		2-6-2T	OC	Sdn		1946
4942	MAINDY HALL	4-6-0	OC	Sdn		1929
5029	NUNNEY CASTLE	4-6-0	4C	Sdn		1934
5051	EARL BATHURST/					
	DRYSLLWYN CASTLE	4-6-0	4C	Sdn		1936
5322		2-6-0	OC	Sdn		1917
5572		2-6-2T	OC	Sdn		1929
5900	HINDERTON HALL	4-6-0	OC	Sdn		1931
6106		2-6-2T	OC	Sdn		1931
6697		0-6-2T	IC	AW	985	1928
6998	BURTON AGNES HALL	4-6-0	OC	Sdn		1949
7202		2-8-2T	OC	Sdn		1934
7808	COOKHAM MANOR	4-6-0	OC	Sdn		1938
34051	WINSTON CHURCHILL	4-6-2	3C	Bton		1946
92220	EVENING STAR	2-10-0	OC	Sdn		1960
D7018		4w-4wDH		BPH	7912	1961
(W22W) No.22		4w-4wDMR		AEC		1941
(5)		0-4-0DH		Derby C.& W		1960
No.5		0-4-0WT	OC	GE		1857
2	PONTYBEREM	0-6-0ST	OC	AE	1421	1900
No.1	BONNIE PRINCE CHARLIE	0-4-0ST	OC	RSH	7544	1949
47		0-6-0ST	OC	RSH	7800	1954
		0-6-0DH		HE	5238	1962
(A 21 W)		2w-2PMR		Wkm	6892	1954 DsmT
(B 37 W	PWM 3963)	2w-2PMR		Wkm	6948	1955 DsmT

R.HILTON, "POPLARS", NORTH MORETON, DIDCOT.
Gauge : 3'0".   (SU 552904)

No.1	(ED 10)	0-4-0ST	OC	WB	1889	1911

Gauge : 1'11½".

	KIDBROOKE	0-4-0ST	OC	WB	2043	1917

ALAN KEEF LTD, COTE FARM, COTE, near BAMPTON.
Gauge : 4'8½".   (SP 351030)

	4wDM		FH	2893 1944

Gauge : Metre.

CAMBRAI	0-6-0T	OC	Corpet	493 1888
THE ROCK	0-4-0DM		HE	2419 1941

Gauge : 3'2¼".

4wDM		RH	266561 1948 Dsm

Gauge : 3'0".

No.1 13/500	LADY MORRISON	0-4-2ST	OC	KS	3024 1916 Dsm
	DINMOR	4wDM		JF	3900011 1947
3		4wDM		MR	10118 1949

Gauge : 4'0".

		4wDM	MR	9909 1958

Gauge : 2'0".

			4wDM	MR	5243 1930 + Dsm
3			2-4wBE	WR	832 1935
			4wDM	JF	21294 1936	Dsm
	CILGWYN		4wDM	RH	175414 1936
			4wDM	MR	7057 1938	Dsm
			4wDM	MR	7304 1938	Dsm
			4wDM	MR	8681 1941	Dsm
	DIGGER		4wDM	MR	8882 1944
18 U 191	T.R.A. No.16 OWL	4wDM	RH	283513 1949
			0-4-0DM	HE	5222 1958
16			4wDM	MR	11141 1960
	SUE		4wDM	RH	476106 1964
			4wDM	MR	22239 1965
1215			4wDM	MR	22253 1965
1			4wDM	MR	11311 1966
EL 16			4wBE	CE	5667 1969
			4wBE	CE	5667 1969
			4wDM	OK	7595

          + In use as a workbench.

Gauge : 1'9".

	OLD SPARKY	4wDM	RH	487963 1963

Gauge : 1'3".

		2w-2DMR	Minirail	c1976

          Also other locos usually present for rebuild, resale & hire.

LEISURE TRACK, COTSWOLD WILD LIFE PARK, BURFORD.
Gauge : 2'0".   (SP 237084)

No.3 97513		4wDM	MR	9976 1954
S 87632 ADAM		4wDM S/O	MR	9978 1954

MINISTRY OF DEFENCE, ARMY DEPARTMENT.
Bicester Central Workshops.
          For full details see Section Five.

Bicester Military Railway.
          For full details see Section Five.

NUFFIELD SALVAGE CO LTD, OLD HENLEY ROAD, EWELME, WALLINGFORD.
Gauge : 4'8½".   (SU 650896)

215	0-4-0DM	HE	3393 1946	Dsm
216	0-4-0DM	HE	3394 1946	Dsm
219	0-4-0DM	HE	3397 1946	Dsm
221	0-4-0DM	HE	3131 1944	Dsm

Gauge : 2'6".

	4wDM	RH	221624 1943	OOU
	4wDM	RH	242919 1947	OOU

PLEASURE-RAIL LTD, BLENHEIM PALACE RAILWAY, WOODSTOCK.
Gauge : 1'3".   (SP 444163)

5751	SIR WINSTON CHURCHILL	4-6-2	OC		G&S	9	1946
		2-6-4PE	S/0		S.Batty		c1955
		6w-4PM			G&S		1960
20	TRACY JO	2-6-2DM	S/0		G&S	20	1964
	DOCTOR DIESEL	4w-4wDE			Minirail		1966

J.M.WALKER, LINDSEY FARM, HIGH COGGES, near WITNEY.
Gauge : 4'8½".   (SP 380089)

		0-4-0ST	OC	P	1438 1916

Locos for resale occasionally present.

SHROPSHIRE

BRITISH SUGAR CORPORATION LTD, ALLSCOTT FACTORY.
Gauge : 4'8½".   (SJ 605124)

(D2302)		0-6-0DM		(RSH	8161 1960
				(DC	2683 1960
AR 18		0-4-0DM		RH	304474 1951

CAMBRIAN RAILWAYS SOCIETY LTD, OSWESTRY.
Gauge : 4'8½".   (SJ 294297)

7822	FOXCOTE MANOR	4-6-0	OC	Sdn	1950
No.8		0-4-0ST	OC	BP	1827 1879
		0-6-0ST	OC	AB	885 1900
	ADAM	0-4-0ST	OC	P	1430 1916
140/589/0001		0-4-0ST	OC	P	2131 1951
	NORMA	0-6-0ST	IC	HE	3770 1952
3		0-6-0DM		HE	3526 1947
322		4wDM		FH	3541 1952
No.1		0-4-0DM		HC	D843 1954

R.G.HONEYCHURCH, KINNERLEY STATION.
Gauge : 2'6".   (SJ 337198)

	YARD No.P.9260	0-4-0DM		HE	2248 1940 OOU

Gauge : 1'11½".

		4wDM		RH	327904 1951 OOU

IRONBRIDGE GORGE MUSEUM TRUST LTD.
Blists Hill Industrial Museum, near Ironbridge.
Gauge : 4'8½".   (SJ 696037)

49395		0-8-0	IC	Crewe	5662 1921

Coalbrookdale Museum.
Gauge : 4'8½".   (SJ 668049)

```
5                              0-4-OST   OC      Coalbrookdale c1865
                               0-4-OVBT  VCG          S      6185 1925 +
                               0-4-OVBT  VCG          S      6155 1925
                               Rebuilt from 0-4-OST OC  MW

        + Rebuild of 6  0-4-OST OC   C'dale  c1865.
```

MINISTRY OF DEFENCE, ARMY DEPARTMENT, DONNINGTON DEPOT.
            For full details see Section Five.

NORTH WEST WATER AUTHORITY, LLANFORDA HALL WATERWORKS, OSWESTRY.
Gauge : 2'0".   (SJ 277295)

```
81.01                          4wDM                 RH    452294 1960
81.02                          4wDM                 RH    260712 1948
81.03      L.C.W.W. 18         4wDM                 HE      6299 1964
81.04                          4wDM                 HE      6298 1964
81 A 06                        4wDM                 RH    496039 1963
81 A 07                        4wDM                 RH    496038 1963
```

THE ROYAL AIR FORCE MUSEUM, c/o R.A.F. COSFORD.
Gauge : 60cm.   (SJ 787053)

```
                               4wDM                 RH    370571 1954
```

SEVERN VALLEY RAILWAY CO LTD.
Gauge : 4'8½".   Locos are kept at :-

```
            Arley, Hereford & Worcester.           (SO 800764)
            Bewdley, Hereford & Worcester.         (SO 793753)
            Bridgnorth.                            (SO 715926)
            Hampton Lodge.                         (SO 744863)
            Highley.                               (SO 749831)
```

```
3020       CORNWALL           2-2-2    OC      Crewe            1858
813                           0-6-OST  IC      HC       555     1901
1501                          0-6-OPT  OC      Sdn              1949
2857                          2-8-0    OC      Sdn              1918
3205                          0-6-0    IC      Sdn              1946
3612                          0-6-OPT  IC      Sdn              1939    Dsm
4141                          2-6-2T   OC      Sdn              1946
4150                          2-6-2T   OC      Sdn              1947
4566                          2-6-2T   OC      Sdn              1924
4930       HAGLEY HALL        4-6-0    OC      Sdn              1929
5164                          2-6-2T   OC      Sdn              1936
5764                          0-6-OPT  IC      Sdn              1929
6960       RAVENINGHAM HALL   4-6-0    OC      Sdn              1944
(7325) 9303                   2-6-0    OC      Sdn              1932
7714                          0-6-OPT  IC      KS      4449     1930
7802       BRADLEY MANOR      4-6-0    OC      Sdn              1938
7812       ERLESTOKE MANOR    4-6-0    OC      Sdn              1939
7819       HINTON MANOR       4-6-0    OC      Sdn              1939
(42968) 2968                  2-6-0    OC      Crewe            1934
43106                         2-6-0    OC      Dar     2148     1951
(45000) 5000                  4-6-0    OC      Crewe   216      1934
45110      R.A.F. BIGGIN HILL 4-6-0    OC      VF      4653     1935
(45690) 5690 LEANDER          4-6-0    3C      Crewe   288      1936
                              Rebuilt  Derby            1973
(46443) 6443                  2-6-0    OC      Crewe            1950
46521                         2-6-0    OC      Sdn              1953
```

47383			0-6-0T	IC	VF	3954	1926	
(48773)	8233		2-8-0	OC	NB	24607	1940	
(61994)	3442	THE GREAT MARQUESS	2-6-0	3C	Dar	1761	1938	
75069			4-6-0	OC	Sdn		1955	
78019			2-6-0	OC	Dar		1954	
80079			2-6-4T	OC	Bton		1954	
D1013		WESTERN RANGER	6w-6wDH		Sdn		1962	
D1062		WESTERN COURIER	6w-6wDH		Crewe		1963	
		THE LADY ARMAGHDALE	0-6-0T	IC	HE	686	1898	
2047		WARWICKSHIRE	0-6-0ST	IC	MW	2047	1926	
No.4			0-4-0ST	OC	P	1738	1928	
600		GORDON	2-10-0	OC	NB	25437	1943	
No.17		HIGHFLYER	0-4-0DM		JF	22912	1940	
			0-4-0DM		RH	319290	1953	
		WILLIAM	0-4-0DM		RH	408297	1957	
1			0-4-0DM		RH	414301	1957	
11509		ALAN	0-4-0DM		RH	414304	1957	
D2961			0-4-0DE		RH	418596	1957	
		ARCHIBALD	0-4-0DM		RH	418789	1957	
(PT 2P)			2w-2PMR		Wkm	1580	1934	
PWM 2827			2w-2PMR		Wkm	5005	1949	
PWM 2830			2w-2PMR		Wkm	5008	1949	
PWM 3189			2w-2PMR		Wkm	5019	1948	DsmT
PWM 3774			2w-2PMR		Wkm	6653	1953	Dsm
DB 965054			2w-2PMR		Wkm	7577	1957	DsmT
(PT 1P	TP 49P)		2w-2PMR		Wkm	7690	1957	
(9021)	6		2w-2PMR		Wkm	8085	1958	
9032			2w-2PMR		Wkm	8200	1958	

TELFORD DEVELOPMENT CORPORATION, TOWN PARK, TELFORD.
Gauge : 2'0".   (SJ 700077)

	THOMAS	4wVBT VCG	Kierstead Ltd/ Alan Keef	1979	
1		4wPM	D.Skinner	c1975	

TELFORD HORSEHAY STEAM TRUST, THE OLD LOCO SHED, HORSEHAY, TELFORD.
Gauge : 4'8½".   (SJ 675073)

	PETER	0-6-0ST	OC	AB	782	1896
		4wVBT	VCG	S	9535	1952
5619		0-6-2T	IC	Sdn		1925
1	TOM	0-4-0DH		NB	27414	1954
		2w-2PMR		Wkm		

D.L.WALKER, THE FIRS, HOPE PARK, MINSTERLEY.
Gauge : 1'11½".   (SJ 329008)

	DOROTHEA	0-4-0ST	OC	HE	763	1901

SOMERSET

A.R.C.(SOUTHERN) LTD, WHATLEY QUARRY, near FROME.
Gauge : 4'8½".   (ST 733479)

1	2162	4wDH	TH/S	133C	1963 +
2	2166	4wDH	TH/S	136C	1964
3	2323	4wDH	TH	152V	1965
4		4wDH	TH	200V	1968

+ Carries 133V on plate in error.

BREAN CENTRAL MINIATURE RAILWAY, MID SOMERSET LEISURE CENTRE, COAST ROAD, BERROW.
Gauge : 1'6".   (   )

	4w-4wPM	A.R.Deacon	1975

BRITISH CELLOPHANE LTD, BRIDGWATER.
Gauge : 4'8½".   (ST 309382)

D2133	0-6-0DM	Sdn	1960

BUTLINS LTD, MINEHEAD HOLIDAY CAMP.
Gauge : 2'0".   (SS 984464)

31	C.P.HUNTINGDON	4-2-4DH	S/O	Chance	64-5031-24 1964

EAST SOMERSET RAILWAY CO LTD, WEST CRANMORE RAILWAY STATION, SHEPTON MALLET.
Gauge : 4'8½".   (ST 664429)

6634			0-6-2T	IC	Sdn		1928
(31323)	323	BLUEBELL	0-6-0T	IC	Afd		1910
32110			0-6-0T	IC	Bton		1877
47493			0-6-0T	IC	VF	4195	1927
75029		THE GREEN KNIGHT	4-6-0	OC	Sdn		1954
92203		BLACK PRINCE	2-10-0	OC	Sdn		1959
4101			0-4-0CT	OC	D	4101	1901
1398		LORD FISHER	0-4-0ST	OC	AB	1398	1915
		GLENFIELD	0-4-0ST	OC	AB	1719	1920
68005			0-6-0ST	IC	RSH	7169	1944

FISONS LTD, HORTICULTURE DIVISION, ECLIPSE PEAT WORKS, ASHCOTT ROAD, MEARE,
GLASTONBURY & SHAPWICK HEATH SITE.
Gauge : 2'0".   (ST 451406, 454406, 435404)

1	4wDM	LB	53726	1963
2	4wDM	LB	50888	1959
3	4wDM	LB	51989	1960
4	4wDM	LB	55070	1966
	4wDM	L	42319	1956
6	4wDM	L	37658	1952
7	4wDM	L	42494	1956
8	4wDM	L	37170	1951
9	4wDM	L	34758	1949
10	4wDM	L	10498	1938
	4wDM	Eclipse		
12	4wDM	L	26366	1944

FOSTER YEOMAN QUARRIES LTD, MEREHEAD STONE TERMINAL, TORR WORKS, SHEPTON MALLET.
Gauge : 4'8½".   (ST 693426)

1010		WESTERN CAMPAIGNER	6w-6wDH		Sdn		1962	Pvd
(D3002)	11	DULCOTE	0-6-0DE		Derby		1952	
(3003)	22	MEREHEAD	0-6-0DE		Derby		1952	Dsm
(08032)	33	MENDIP	0-6-0DE		Derby		1954	
				Rebuilt	Derby		1975	
44		WESTERN YEOMAN II	4w-4wDE		GM	798083-1	1980	

MENDIP DISTRICT COUNCIL, WELSHMILL ADVENTURE PLAYGROUND, FROME.
Gauge : 4'8½".   (ST 778486)

			4wVBT VCG	S	9387	1948

MILK MARKETING BOARD, CHARD JUNCTION.
Gauge : 4'8½".   (ST 341047)

			0-4-0DM	RH	304470	1951

MINISTRY OF DEFENCE, ROYAL ORDNANCE FACTORY, PURITON.
Gauge : 4'8½".   (ST   )

6320	R.O.F. BRIDGWATER No.1	0-4-0DH	AB	578	1972
6321	R.O.F. BRIDGWATER No.2	0-4-0DH	AB	579	1972

Gauge : 2'6".   (ST   )

3582		4wBE	GB	1877	1943
3583		4wBE	GB	1698	1941
3584		4wBE	GB	1699	1941
		4wDM	SMH	40SPF522	1981

SOMERSET & DORSET RAILWAY MUSEUM TRUST, WASHFORD STATION.
Gauge : 4'8½".   (ST 044412)

53808	88		2-8-0	OC	RS	3894	1925	
		ISABEL	0-6-0ST	OC	HL	3437	1919	
		KILMERSDON	0-4-0ST	OC	P	1788	1929	
No.1			0-4-0F	OC	WB	2473	1932	
24			4wDM		RH	210479	1941	+
900855			2w-2PMR		Wkm	6967	1954	
			2w-2PMR		Wkm			@

          + Carries plate RH 306089.
          @ Currently stored elsewhere.

P.C.VALLINS, SLADES HILL NURSERY, TEMPLECOMBE.
Gauge : 2'0".   (   )

			4wPM	L	9256	1937
20			4wPM	L	18557	1942

WESTONZOYLAND ENGINE GROUP, WESTONZOYLAND PUMPING STATION, near BRIDGWATER.
Gauge : 2'0".   (   )

87030	30		4wDM	MR	40S310	1968

WEST SOMERSET RAILWAY CO.
Gauge : 4'8½".   Locos are kept at :-

Bishops Lydeard.   (ST 164290)
Minehead.   (SS 975463)
Williton Goods Yard.   (ST 085416)

4561		2-6-2T	OC	Sdn		1924
5542		2-6-2T	OC	Sdn		1928
6412	THE FLOCKTON FLYER	0-6-0PT	IC	Sdn		1934
32678 5		0-6-0T	IC	Bton		1880
D2994		0-6-0DE		RH	480695	1962
D7017		4w-4wDH		BPH	7911	1961
D9526		0-6-0DH		Sdn		1964
(D9551) 50	8311/29 29	0-6-0DH		Sdn		1965
(W)50413		4w-4wDMR		PR		1957
(W)50414		4w-4wDMR		PR		1957
1163	WHITEHEAD	0-4-0ST	OC	P	1163	1908
2994	VULCAN	0-6-0ST	OC	WB	2994	1950
2996	VICTOR	0-6-0ST	OC	WB	2996	1951
		4wDM		RH	183062	1937

SOUTH YORKSHIRE

A.D.H. LTD, DEMOLITION PLANT SALES, WALLACE ROAD, SHEFFIELD.
Gauge : 4'8½".   (SK 348890)

4wDM	RH	252828	1948

ALLIED STEEL & WIRE LTD, REINFORCEMENT STEEL SERVICES DIVISION, McCALLS WORKS,
MEADOWHALL ROAD, SHEFFIELD.
Gauge : 4'8½".   (SK 391916)

22/5
0-6-0DH	YE	2896	1964

AMALGAMATED CONSTRUCTION CO LTD, MINING & CIVIL ENGINEERS, HOYLEMILL ROAD,
STAIRFOOT, BARNSLEY.
Gauge : 2'0".   (   )

4wDM	RH	249565	1948
4wDM	RH	481552	1962

Locos currently on contract work at Thorne Colliery, South Yorkshire.

BATCHELORS FOODS LTD, WADSLEY BRIDGE, SHEFFIELD.
Gauge : 4'8½".   (SK 326920)   RTC.

S116-1
0-4-0DH	TH	102C	1960	
Rebuild of 0-4-0DM	JF	4200019	1947	OOU

BEATSON, CLARK & CO LTD, STAIRFOOT GLASS WORKS, BARNSLEY.
Gauge : 4'8½".   (SE 369061)

4wDM	FH	3883	1958

C.F.BOOTH LTD, SCRAP MERCHANTS.
Clarence Metal Works, Armer Street, Rotherham.
Gauge : 4'8½".   (SK 421924)

D2854		0-4-0DH	YE	2813 1960	
(D2958)		0-4-0DM	RH	390777 1956	
		0-4-0DM	JF	4210007 1949	OOU

Also other locos for scrap occasionally present.

York Road Station Sidings, Doncaster.
Gauge : 4'8½".   (SE 567041)   RTC.

N.C. No.3	PLANT No.15	0-4-0DM	HC	D965 1956	OOU

Also other locos for scrap occasionally present.

P.BRIDDEN, SHEFFIELD.
Gauge : 2'0".   (   )

4wDM	OK	(3444 1930?)

BRITISH FUEL CO, NUNNERY COAL DEPOT, SHEFFIELD.
Gauge : 4'8½".   (SK 368877)

MARIA	0-4-0DE	RH	384144 1955

BRITISH STEEL CORPORATION, B.S.C.(CHEMICALS) LTD, CARBONISATION & ELECTRODE
COATING WORKS GROUP.
Brookhouse Works, Beighton, Sheffield.   (Closed)
Gauge : 4'8½".   (SK 448843, 449842)   RTC.

1	0-4-0WE	WSO	4424/1 1946	OOU
	0-4-0WE	WSO	7848 1960	OOU

Orgreave Works, Orgreave, Sheffield.
Gauge : 4'8½".   (SK 426874)

	0-6-0DE	YE	2528 1953	OOU
2444/16	0-6-0DE	YE	2753 1959	
2444/17	0-6-0DE	YE	2754 1960	
2444/18	0-6-0DE	YE	2866 1962	
2444/20	0-6-0DE	YE	2670 1958	
4044/21	4wDH	RR	10245 1966	Dsm
4044/22	4wDH	RR	10246 1966	OOU
4044/23	0-6-0DE	YE	2708 1959	
	0-4-0WE	WSO	4571 1947	OOU
	0-4-0WE	WSO	8266 1962	

BRITISH STEEL CORPORATION, SHEFFIELD DIVISION, PROFIT CENTRES, B.S.C. FORGES,
FOUNDRIES & ENGINEERING, RIVER DON & ASSOCIATED WORKS, RIVER DON WORKS,
BRIGHTSIDE LANE, BRIGHTSIDE, SHEFFIELD.
Gauge : 4'8½".   (SK 382902, 387898)

No.1	30	4wBE	Bg/MV	2186 1940	+
B.S.C. No.33	9/6729	0-4-0DM	HC	D817 1954	OOU
B.S.C. No.37	9/6776 No.44	0-4-0DM	HC	D1157 1959	OOU
B.S.C. No.39		0-4-0DM	HC	D996 1957	
B.S.C. No.40	PLANT No.0911255	4wDH	S	10002 1959	
B.S.C. No.41	PLANT No.0911420	4wDH	S	10009 1959	
B.S.C. No.42		4wDH	S	10026 1960	@
B.S.C. No.43		4wDH	S	10018 1960	

			4wDH		S	10025 1960	
			0-4-0DM		HC	D994 1957	Dsm

+ Used at West Treatment Plant.
@ Carries plate S 10043.

BRITISH STEEL CORPORATION, SHEFFIELD DIVISION, STEELWORKS GROUP, ROTHERHAM WORKS.
Aldwarke Works, Rotherham.
Gauge : 4'8½".   (SK 449951, 449953, 451954)

No.	Name/No.	Name	Type		Builder	Works No. Year	Notes
10	624/66		0-4-0DE		YE	2682 1958	
11		HENRY	0-4-0DE		YE	2801 1961	OOU
		WILLIAM	0-4-0DE		YE	2880 1962	OOU
		EDWARD	0-4-0DE		YE	2953 1965	OOU
			4w-4wDH		TH	197-198V 1968	
No.37			0-6-0DH		HE	7002 1971	
40			0-6-0DH		HE	7357 1973	
B.S.C. 41			0-6-0DH		HE	8805 1978	
84			0-4-0DE		(BP	7877 1960	+
					(BT	330 1960	
85			0-4-0DE		(BP	7878 1960	
					(BT	331 1960	OOU
86			0-4-0DE		(BP	7939 1960	OOU
					(BT	332 1960	
87			0-4-0DE		(BP	7940 1960	OOU
					(BT	333 1960	
88			0-4-0DE		(BP	7941 1960	+
					(BT	334 1960	
89			0-4-0DE		(BP	7942 1961	
					(BT	335 1961	
91			0-4-0DE		(BP	7944 1961	
					(BT	337 1961	
92			0-4-0DE		(BP	7945 1961	
					(BT	338 1961	
93			0-6-0DE		YE	2889 1962	
94			0-6-0DE		YE	2890 1962	
95			0-6-0DE		YE	2891 1963	
96			0-6-0DE		YE	2905 1963	
97			0-6-0DE		YE	2906 1963	
98			0-6-0DE		YE	2907 1963	
PWG 284R			4wDM	R/R	Unimog	1975	

+ Locos can work in tandem.

Templeborough Works, Rotherham.
Gauge : 4'8½".   (SK 410915, 411916, 414918, 416915, 417920, 419912)

No.	No.	Name	Type		Builder	Works No. Year	Notes
			0-4-0ST	OC	RSH	7020 1941	+ Dsm
1		SHEFFIELD	0-4-0DE		YE	2481 1950	
2		ROTHERHAM	0-4-0DE		YE	2480 1950	
7	624/63		0-4-0DE		YE	2664 1957	
8	624/64		0-4-0DE		YE	2680 1958	
9			0-4-0DE		YE	2681 1958	
20	624/67		0-4-0DE		YE	2688 1959	
21	624/68		0-4-0DE		YE	2710 1959	
22	624/69		0-4-0DE		YE	2785 1960	
23			0-4-0DE		YE	2786 1960	
24	624/71		0-4-0DE		YE	2860 1962	
25	624/72		0-4-0DE		YE	2861 1962	
31	624/83		0-6-0DE		YE	2904 1964	
32	624/84		0-6-0DE		YE	2935 1964	
33	624/85		0-6-0DE		YE	2946 1965	
34	624/86		0-6-0DE		YE	2947 1965	

36		0-6-0DH	HE	7001 1971	
38		0-6-0DH	HE	7003 1971	
39	624/95	0-6-0DH	HE	7004 1971	
No.42		0-6-0DH	HE	8902 1978	

+ Frame used for weighbridge testing.
Locos also shunt McCalls Ickles Works (SK 418916) and the Ickles Railway
Materials and Industrial Products Department of B.S.C's Forges, Foundries and
Engineering Group.

BRITISH STEEL CORPORATION, SHEFFIELD DIVISION, STEELWORKS GROUP,
STOCKSBRIDGE AND TINSLEY PARK WORKS.
Stocksbridge Railway, Stocksbridge, Sheffield. (Stocksbridge Railway Co.)
Gauge : 4'8½". (SK 260990, 267987)

		0-6-0DE	YE	2608 1956	Dsm
	"TONY"	2w-2DMR	Wkm	9688 1965	

Stocksbridge Works, Stocksbridge, Sheffield.
Gauge : 4'8½". (SK 260990, 267992)

No.23	656/38/01	0-6-0DE	YE	2606 1955	
No.24		0-6-0DE	YE	2607 1956	OOU
No.25		0-6-0DE	YE	2667 1957	
No.26		0-6-0DE	YE	2671 1957	
No.27		0-6-0DE	YE	2720 1958	
No.28		0-6-0DE	YE	2721 1958	
No.29		0-6-0DE	YE	2722 1958	
No.30		0-6-0DE	YE	2750 1959	
No.31		0-6-0DE	YE	2751 1959	
No.32		0-6-0DE	YE	2752 1959	
No.33		0-6-0DE	YE	2740 1959	
4		0-6-0DE	YE	2594 1956	
12		0-6-0DE	YE	2635 1957	
32		0-6-0DE	YE	2736 1959	
	522 17 01	0-6-0DE	YE	2739 1959	
23		0-6-0DE	YE	2798 1961	

Tinsley Park Works, Greenland Road, Sheffield.
Gauge : 4'8½". (SK 401889, 401897)

43		0-4-0DM	HC	D1156 1959	
50	522 12 01	0-6-0DH	TH/S	119C 1962	Dsm
51	522 13 01	0-6-0DH	TH/S	122C 1963	OOU
52		0-6-0DH	TH/S	124C 1963	
53		0-6-0DH	TH	261V 1976	
54		0-6-0DH	S	10107 1963	
55		0-6-0DH	S	10108 1963	
56		0-6-0DH	S	10111 1963	
57		0-6-0DH	S	10053 1961	
No.116		0-6-0DH	S	10112 1963	
62		0-6-0DH	S	10114 1963	
63		0-6-0DH	S	10115 1963	
64		0-6-0DH	RR	10173 1963	
91		0-6-0DH	S	10091 1962	

BRITISH TISSUES LTD, OUGHTIBRIDGE DIVISION, SPRING GROVE PAPER MILLS,
OUGHTIBRIDGE, near SHEFFIELD.
Gauge : 4'8½". (SK 303941)

M.P.P. 6		0-4-0DH	JF	22499 1938

CEMENTATION MINING LTD, PLANT DEPOT, BENTLEY WORKS, DONCASTER.
Gauge : 2'0".   (SE 563056)

2	70108	0-4-OBE	WR	J7184	1969
A		0-4-OBE	WR	M7479	1972
1		4wBE	CE	B0132	1973
4		4wBE	CE	B0132B	1973
		4wBE	CE	B0145A	1973
3		4wBE	CE	B0145B	1973
5		4wBE	CE	B0145C	1973
	70217	4wBE	CE	B0145D	1973
		4wBE	CE	B0167	1974

Gauge : 1'6".

No.1	0-4-OBE	WR	J7292	1969

Locos present in yard between contracts.

CENTRAL ELECTRICITY GENERATING BOARD, WOODHEAD TUNNEL & DUNFORD BRIDGE.
Gauge : 2'0".   (SK 114998, 156022)

4wDM	RH	444208	1961
4wBE	CE	5843	1971

COALITE & CHEMICAL PRODUCTS LTD, GRIMETHORPE WORKS, BARNSLEY.
Gauge : 4'8½".   (SE 403084)

1	4wDH	RR	10258	1966
2	4wDH	RR	10259	1966

GEORGE COHEN, SONS & CO LTD, TINSLEY WORKS, SHEFFIELD ROAD, SHEFFIELD.
Gauge : 4'8½".   (SK 401913)

4wDM	RH	513142	1967

CRODA HYDROCARBONS LTD, KILNHURST WORKS.
Gauge : 4'8½".   (SK 464982)

0-4-ODH	RR	10204	1965

DAVY LOEWY ENGINEERING CO LTD, PRINCE OF WALES ROAD, DARNALL, SHEFFIELD.
Gauge : 4'8½".   (SK 395875)

7600	4wDH	TH	189C	1967

DONCASTER COALITE CO LTD, ASKERN.
Gauge : 4'8½".   (SE 559134)

1	4wDH	S	10164	1963
No.2	4wDH	RR	10179	1964
	0-6-ODH	TH	238V	1971

DONCASTER METROPOLITAN COUNCIL, MUSEUM OF SOUTH YORKSHIRE LIFE,
CUSWORTH HALL, DONCASTER.
Gauge : 2'0".   (SE 547039)

0-4-ODM	HE	2008	1939

ALLAN FINLAY, HARTWOOD EXPORT MACHINERY LTD, HANGMANSTONE DEPOT, SHEFFIELD ROAD, BIRDWELL, near BARNSLEY.
Yard (SE 348009) with locos for scrap occasionally present.

FIRTH BROWN LTD, STEEL MANUFACTURERS, ATLAS WORKS, SHEFFIELD.
Gauge : 4'8½". (SK 368888)

A1		4wDH		TH	145V 1964
A2		4wDH		TH	155V 1965
A3		4wDH		TH	162V 1966
A4		4wDH		TH	229V 1970
A6		4wDH		TH	251V 1974

FISONS LTD, HORTICULTURE DIVISION, HATFIELDS MOOR, near DONCASTER.
Gauge : 3'0". (SE 713084)

02-15	4wDM		MR	10159 1949
	4wDM		MR	40S302 1967
	4wDM		MR	40S378 1971
	4wDM		HE	7366 1974
No.2	4wDM		HE	7367 1974
	4wDM		Diema	3543 1974

FLATHER BRIGHT STEELS, TINSLEY WORKS, SHEFFIELD ROAD, TINSLEY.
Gauge : 1'6". (SK 401913)

1112	M.U.F.C.	4wBE	GB	6061 1961

HADFIELDS LTD, EAST HECLA WORKS, VULCAN ROAD, TINSLEY, SHEFFIELD.
Gauge : 4'8½". (SK 395912)

No.18131	4wDH		TH	117V 1962
18167	4wDH		TH	121V 1962
No.18459	4wDH		TH	143V 1964
19856	0-4-0DH		TH	116C 1962
	Rebuild of 0-4-0DM		JF	4200007 1946

HADFIELDS LTD, LEEDS ROAD WORKS, ATTERCLIFFE, SHEFFIELD. (Closed)
Gauge : 4'8½". (SK 383888, 386893)

621 / 3	LOCO	0-4-0DH		NB	27733 1957 .OOU
22/1		0-4-0DE		RH	424840 1960 OOU
No.2		0-4-0DH		EEV	D1125 1966 OOU

HALLAMSHIRE RAILWAY SOCIETY, PENISTONE STATION GOODS YARD.
Gauge : 4'8½". (SE 243035)

No.7		0-4-0ST	OC	HC	1689 1937
	CARBONISATION No.1	0-6-0F	OC	RSH	7847 1955
		0-4-0DM		HE	2088 1940
621/4	D1 IVOR LOCO	4wDM		RH	212653 1942

S.HARRISON & SONS (TRANSPORT) LTD, 310, SHEFFIELD ROAD, SHEFFIELD.
Gauge : 4'8½". (SK 399910)

		0-4-0ST	OC	AB	2217 1947 Pvd
No.3	WEE-YORKIE	0-4-0ST	OC	AB	2360 1954 Pvd
	CATHRYN	0-6-0T	OC	HC	1884 1955 Pvd

WALTER HESELWOOD LTD, METAL MERCHANTS, STEVENSON ROAD, SHEFFIELD.
Yard (SK 375893) with locos for scrap occasionally present.

PETER HICKSON, DONCASTER.
Gauge : 2'0".   (   )

4wDM	MR	+
4wDM	MR	+
4wDM	MR	+

+ Three of 5359 1931; 8875, 8877, 8885 of 1944.

THOMAS HILL (ROTHERHAM) LTD, VANGUARD WORKS, HOOTON ROAD, KILNHURST.
Gauge : 4'8½".   (SK 465975)

2	BILSDALE	0-6-0DE	YE	2665	1957
No.2		4wDH	S	10021	1959
22		0-6-0DE	YE	2794	1961
2		0-6-0DE	YE	2878	1963
		4wDH	RR	10260	1966
20		4wDH	TH	187V	1967

Locos under construction and repair usually present; locos are
also hired out to various concerns.

J.H.TRACTORS LTD, TICKHILL PLANT HIRE, PLANT DEPOT, APY LANE, TICKHILL, near DONCASTER.
Gauge : 2'0".   (SK 583931)

MBS 002	4wBE	WR	4817	1951
MBS 004	4wBE	WR	4819	1951
MBS 008	0-4-0BE	WR		
MBS 010	0-4-0BE	WR		
MBS 213	0-4-0BE	WR		
MBS 323	4wBE	WR	5115	1953
MBS 324	4wBE	WR	5316	1955
MBS 346	4wBE	WR	H7066	1968
MBS 347	4wBE	WR	H7067	1968
MBS 348	0-4-0BE	WR	H7049	1968
MBS 492	0-4-0BE	WR		
MBS 494	0-4-0BE	WR	6131	1959
MBS 520	0-4-0BE	WR	7068	1968
MBS 521	4wBE	WR	1199	1938
GT 63	0-4-0BE	WR	E6946	1965
8	0-4-0BE	WR		

Gauge : 1'6".

MBS 432	0-4-0BE	WR	H7185	1968	
MBS 433	0-4-0BE	WR	4320	1950	
MBS 493	0-4-0BE	WR			
	0-4-0BE	WR	5157	1953	Dsm
	0-4-0BE	WR	5244	1954	Dsm
	0-4-0BE	WR	6703	1962	Dsm

Locos present in yard between contracts.

THOMAS MARSHALL & CO (LOXLEY) LTD, STORRS BRIDGE FIRECLAY MINE, LOXLEY.
Gauge : 1'4".   (SK 288902)   (Underground)

4wBE	GB	2782	1957

MOORSIDE MINING CO LTD, MOORSIDE COLLIERY, MOSBORO MOOR, MOSBOROUGH, SHEFFIELD.
Gauge : 2'0".   (SK 420817)

		0-4-0DM	HC	(DM750 1949?) OOU

NATIONAL COAL BOARD.
                    For full details see Section Four.

OH STEELFOUNDERS AND ENGINEERS LTD, ALSING ROAD, TINSLEY, SHEFFIELD.
Gauge : 4'8½".   (SK 394910)

16892		4wBE	GB	2389 1952

ROE BROS & CO LTD, IRON & STEEL MERCHANTS, BLACKBURN ROAD, SHEFFIELD.
Gauge : 4'8½".   (SK 388924)

5909		4wDM	RH	412430 1957 OOU
	ELIZABETH	4wDM	RH	466630 1962

THE SHEFFIELD TRADES HISTORICAL SOCIETY, WORTLEY TOP FORGE, near SHEFFIELD.
Gauge : 2'0".   (SK 299999)

		4wPM	FH	

W. & F. SMITH LTD, IRON & STEEL MERCHANTS, ECCLESFIELD EAST SIDINGS.
Yard (SK 367941) with locos for resale occasionally present.

SPROTBOROUGH CASTINGS LTD, SPROTBROUGH WORKS, DONCASTER.
Gauge : 4'8½".   (SE 553032)   RTC.   (Subsidiary of T.W.Ward.)

2766		4wDM	JF	4120001 1951 OOU

T.J.THOMSON & SON LTD, MANGHAM WORKS, TAYLORS LANE, PARKGATE, ROTHERHAM.
Gauge : 2'0".   (SK 434951)

		4wDM	RH	187102 1937 OOU

TINSLEY WIRE (SHEFFIELD) LTD, ATTERCLIFFE COMMON, SHEFFIELD.
Gauge : 4'8½".   (SK 393902)

		4wDM	FH	3677 1954

TRACKWORK ASSOCIATES LTD, c/o ANGLOPILE LTD, FINNINGLEY.
Gauge : 4'8½".   (SK 673997)

	CHARLES	4wDM	RH	417889 1958

THOS W. WARD LTD, DEMOLITION CONTRACTORS,
Templeborough Contractors Plant Depot, Sheffield Road, Sheffield.
Yard (SK 404913) with locos for resale or hire occasionally present.

Tinsley Scrap Depot, Attercliffe Common, Sheffield.
Gauge : 4'8½".   (SK 392901)

		Type				Builder	No.	Year	
		0-4-0DM				JF	22975	1942	
		0-4-0DM				JF	4210075	1952	OOU
		4wDM				RH	412428	1957	OOU
		4wDM				RH	458952	1960	
D2		0-4-0DH				AB	448	1961	

          Also other locos for scrap occasionally present.

YORKSHIRE WATER AUTHORITY, SOUTHERN DIVISION, SHEFFIELD WATER POLLUTION CONTROL AREA,
BLACKBURN MEADOWS SEWAGE WORKS, SHEFFIELD.
Gauge : 4'8½".   (SK 396916)

| 2 | 4005 | 0-4-0DE | RH | 434774 | 1961 |
| 044/201 | | 4wDH | TH | 265V | 1976 |

STAFFORDSHIRE

ALTON TOWERS RAILWAY, ALTON TOWERS, near LEEK.
Gauge : 2'0".   (SK 075437)

| | ALTONIA | 0-4-0DM | S/O | | Bg | 1769 | 1929 |

Gauge : 1'3".

| | JOHN | 4-4-2 | OC | A.Barnes | 103 | 1924 |

BAGULEY-DREWRY LTD, BURTON-ON-TRENT.
(SK 243225)
          New BD locos under construction, and other locos under repair,
          occasionally present.

M.BAMFORD, CHILDRENS PLAYGROUND, ROCESTER.
Gauge : 4'8½".   (SK 108394)

| No.67 | | 0-4-0ST | OC | AB | 2352 | 1954 |

BASS MUSEUM, HORNINGLOW STREET, BURTON-ON-TRENT.
Gauge : 4'8½".   (SK 248234)

| No.9 | | 0-4-0ST | OC | NR | 5907 | 1901 |
| 20 | | 4wDM | | KC | | 1926 |

BRITISH INDUSTRIAL SAND LTD, MONEYSTONE QUARRY, OAKAMOOR.
Gauge : 4'8½".   (SK 047451)

| | BRIGHTSIDE | 0-4-0DH | | YE | 2672 | 1959 |
| (07003) | | 0-6-0DE | | RH | 480688 | 1962 |

BRITISH REINFORCED CONCRETE ENGINEERING LTD, STAFFORD.
Gauge : 4'8½".   (SJ 933218)

4wDH	TH/S	103C	1960

BRITISH STEEL CORPORATION, SCUNTHORPE DIVISION, SHELTON WORKS, ETRURIA, STOKE.
Gauge : 4'8½".   (SJ 867477, 874477)

TIGER	0-4-0DH	NB	27651	1956	Dsm
	0-4-0DE	YE	2601	1956	OOU
9/7352	0-4-0DH	NB	27937	1959	Dsm
JANUS	0-6-0DE	YE	2772	1960	
SQUIRREL	0-4-0DE	YE	2780	1960	
WEASEL	0-4-0DE	YE	2783	1960	
ATLAS	0-6-0DE	YE	2787	1961	
LUDSTONE	0-6-0DE	YE	2868	1962	
BADGER	0-4-0DE	YE	2869	1962	
OTTER	0-4-0DE	YE	2873	1964	OOU
EARL GRANVILLE	0-4-0DE	YE	2885	1963	
LORD FARINGDON	0-4-0DH	AB	606	1976	
LORD LEVESON	0-6-0DH	AB	607	1976	

BROOKFIELD FOUNDRY & ENGINEERING CO, CALIFORNIA WORKS, STOKE-ON-TRENT.
Gauge : 4'8½".   (SJ 881442)   RTC.

0-4-0ST	OC	KS	4388	1926	OOU
0-6-0PT	OC	WB	2613	1940	OOU

CENTRAL ELECTRICITY GENERATING BOARD, MEAFORD 'B' POWER STATION.
Gauge : 4'8½".   (SJ 888368)

No.1	0-4-0DH	AB	440	1958
No.2	0-4-0DH	AB	442	1959
No.3	0-4-0DH	AB	443	1959
No.4	0-6-0DH	AB	486	1964

CHASEWATER LIGHT RAILWAY PRESERVATION SOCIETY, NORTON CANES.
Gauge : 4'8½".   (SK 034070)

No.11	ALFRED PAGET	0-4-0ST	OC	N	2937	1882
		0-6-0ST	OC	HC	431	1895
		0-4-0ST	OC	P	917	1902
	ASBESTOS	0-4-0ST	OC	HL	2780	1909
		0-4-0ST	OC	AB	1223	1911
No.2	THE COLONEL	0-4-0ST	OC	P	1351	1914
	INVICTA	0-4-0ST	OC	AB	2220	1946
10	WHIT No.4	0-6-0T	OC	HC	1822	1949
1		4wPM		MR	1947	1919
21		4wDM		KC	1612	1929
No.1		4wDM		FH	2914	1944
		0-4-0DE		RH	458641	1963

SAM G. CLOWES, WALLGRANGE FARM, NEWCASTLE ROAD, LEEK.
Yard (SJ 978549) with locos for resale occasionally present.

CRODA SYNTHETIC CHEMICALS LTD, FOUR ASHES WORKS.
Gauge : 4'8½".   (SJ 917085)   R.S.W.

MP1	0-4-0F	OC	AB	1944	1927

Property Services Agency, Fauld (Scropton) Depot. (Closed)
Gauge : 2'0".   (SK 193286)   RTC.

180		4wBE	GB	1608 1939	OOU
181		4wBE	GB	1609 1939	OOU
233		4wBE	GB	1838 1942	OOU
234		4wBE	GB	1839 1942	OOU

Quality Testing Unit, Cold Meece, near Swynnerton.

For full details see Section Five.

FAIRCLOUGH CIVIL ENGINEERING LTD, TUNNELLING DIVISION, PLANT DEPOT, SWYNNERTON.
Gauge : 2'0".   (SJ 850325)

S151	193.022	4wBE	CE	5955A	1972
S152		4wBE	CE	5955B	1972
S153		4wBE	CE	5955C	1972
S154		4wBE	CE	5955D	1972
S177		4wBE	CE	B0111A	1973
S178		4wBE	CE	B0111B	1973
S179		4wBE	CE	B0111C	1973
S191	193.023	4wBE	CE	B0131A	1973
S192		4wBE	CE	B0131B	1973
		4wBE	CE	B0147	1973
S200		4wBE	CE	B0152A	1973
S206		4wBE	CE	B0152	1973
S205		4wBE	CE	B0152/2A	1973
S206		4wBE	CE	B0152/2B	1973
S207	193.007	4wBE	CE	B0152-1	1973
S208		4wBE	CE	B0152/2	1973
S213	193.024	4wBE	CE	B0183A	1974
S214		4wBE	CE	B0183B	1974
S232		4wBE	CE	B0459A	1975
S233		4wBE	CE	B0459B	1975
S234		4wBE	CE	B0459C	1975
S237		4wBE	CE	B0471A	1975
S238		4wBE	CE	B0471B	1975
S239		4wBE	CE	B0471C	1975
S240		4wBE	CE	B0471D	1975
S241	193.026	4wBE	CE	B0471E	1975
S242	193.027	4wBE	CE	B0471F	1975
S260	192.021	4wBE	CE	B0465	1974
S261	193.028	4wBE	CE	B0941A	1976
S262		4wBE	CE	B0941B	1976
S263	193.029	4wBE	CE	B0941C	1976
S264	193.030	4wBE	CE	B0948.1	1976
S265	193.031	4wBE	CE	B0948.2	1976
S270		4wBE	CE	B0952.1	1976
S271	193.032	4wBE	CE	B0952.2	1976
S274		4wBE	CE	B0957A	1976
S275	193.033	4wBE	CE	B0957B	1976
S276		4wBE	CE	B0957C	1976
S278		4wBE	CE	B0958.1	1976
S279		4wBE	CE	B0958A	1976
S287		4wBE	CE	B1551A	1977
S288		4wBE	CE	B1551	1977
S289		4wBE	CE	B1551B	1977
S290		4wBE	CE	B1551C	1977
S291	192.011	4wBE	CE	B1552	1977

Gauge : 1'6".

S103		4wBE	CE	5792A	1970
S104	193.013	4wBE	CE	5792B	1970
S105		4wBE	CE	5792C	1970
S106	193.014	4wBE	CE	5792D	1970
S128		4wBE	CE	5882A	1971
S129		4wBE	CE	5882B	1971
S130		4wBE	CE	5882C	1971
S131		4wBE	CE	5882D	1971
S137		4wBE	CE	5911A	1972
S138		4wBE	CE	5911B	1972
S139		4wBE	CE	5911C	1972
S141		4wBE	CE	5926A	1972
S142		4wBE	CE	5926C	1972
S143		4wBE	CE	5926D	1972
S146	192.015	4wBE	CE	5926/A	1972
S147		4wBE	CE	5926/2	1972

Locos present in yard between contracts.

FOXFIELD LIGHT RAILWAY SOCIETY, DILHORNE, near STOKE-ON-TRENT.
Gauge : 4'8½".   (SJ 976446)

No.35	RHIWNANT	0-6-0ST	IC	MW	1317	1895	
	HENRY CORT	0-4-0ST	OC	P	933	1903	
No.12	TOPHAM	0-4-0ST	OC	WB	2193	1922	
3	CRANFORD	0-6-0ST	OC	AE	1919	1924	
No.6	LEWISHAM	0-6-0ST	OC	WB	2221	1927	
No.1		0-4-0F	OC	AB	1984	1930	
No.3	J.T.DALY	0-4-0ST	OC	WB	2450	1931	
	ROBERT	0-6-0ST	OC	AE	2068	1933	
1		0-4-0ST	OC	P	1803	1933	
	LITTLE BARFORD	0-4-0ST	OC	AB	2069	1939	
	HAWARDEN	0-4-0ST	OC	WB	2623	1940	
No.15	ROKER	0-4-0CT	OC	RSH	7006	1940	
11		0-4-0ST	OC	P	2081	1947	
7	WIMBLEBURY	0-6-0ST	IC	HE	3839	1956	
	HELEN	4wDM		MR	2262	1923	
		4wDM		RH	242915	1946	+
	CORONATION	0-4-0DH		NB	27097	1953	
RS 154		0-4-0DM		RH	395305	1956	
(900332)		2w-2PMR		Wkm	497	1932	Dsm
(900379)	PT2	2w-2PMR		Wkm	580	1932	DsmT
(PWM 2204)		2w-2PMR		Wkm	4121	1946	Dsm
(PWM 2807	B 170 W)	2w-2PMR		Wkm	4985	1949	Dsm
TR 3		2w-2PMR		Wkm	6884	1954	Dsm
TR 5		2w-2PMR		Wkm	6900	1954	Dsm
	F.L.R.	2w-2PMR		Wkm	7139	1955	
(A 34 W)		2w-2PMR		Wkm	8501	1960	

+ Rebuilt from 2'0" gauge.

G.E.C. TURBINE GENERATORS LTD, MAIN WORKS, STAFFORD.
Gauge : 4'8½".   (SJ 932221)

11044		4wBE	EE	788	1930
	TONKA	0-4-0DE	RH	424841	1960

LICHFIELD DISTRICT COUNCIL, BEACON PARK, LICHFIELD.
Gauge : 4'8½".   (SK 110097)

	4wDM	FH	3809	1963

J.MURPHY & SONS LTD, PLANT DEPOT, HAWKS GREEN LANE, CANNOCK.
Gauge : 2'0".   (SJ 994108)

		4wBE	WR	N7605 1973
		4wBE	WR	N7606 1973
		4wBE	WR	N7607 1973
		0-4-0BE	WR	7617 1973
		4wBE	WR	N7620 1973
		4wBE	WR	N7621 1973
		0-4-0BE	WR	M7550 1972

Gauge : 1'6".

		0-4-0BE	WR	M7548 1972

Locos present in yard between contracts.

NATIONAL COAL BOARD.   See also entry under Greater London.

For full details see Section Four.

NORTH STAFFORDSHIRE RAILWAY CO (1978) LTD, CHURNET VALLEY STEAM RAILWAY MUSEUM, CHEDDLETON STATION, near LEEK.
Gauge : 4'8½".   (SJ 983519)

44422					
80136		0-6-0	IC	Derby	1927
No.9	JOSIAH WEDGWOOD	2-6-4T	OC	Bton	1956
		0-6-0ST	IC	HE 3777 1952
		0-4-0DM		RSHN 6980 1940

NORTH STAFFS & CHESHIRE TRACTION ENGINE CLUB, KLONDYKE MILL, DRAYCOTT-IN-THE-CLAY.
Gauge : 1'11½".   (SK 156289)

No.4	PALMERSTON	0-4-0STT OC	GE	1864

ROM RIVER REINFORCEMENT CO LTD, TRENT VALLEY TRADING ESTATE, LICHFIELD.
Gauge : 4'8½".   (SK 133102)

	0-4-0DH	NB	27940 1959

SEVERN TRENT WATER AUTHORITY, UPPER TRENT WATER RECLAMATION DIVISION.
Gauge : 2'0".   Locos are kept at :-

BL	Blithe Valley Works, Checkley, near Uttoxeter.	(SK 033373)
BU	Burslem Works, Burslem.	(SJ 863487)
BW	Burslem Workshops, Westport Road, Burslem.	(SJ 865502)
N	Newstead Works, Trentham.	(SJ 894404)
T	Tunstall Works, Stoke-On-Trent.	(SJ 854506)

87025	25	4wDM	L	11410 1939	BW	+ OOU
87028	28	4wDM	L	26288 1944	BU	
	92	4wDM	L	39005 1952	BW	OOU
	920	4wDM	L	39419 1953	T	
		4wDM	MR	22238 1965		
		4wDM	MR	40S308 1968		
		4wDM	SMH	104060G 1976	N	
		4wDM	SMH	104063G 1976	N	

+ Incorporates parts of L 3908.

STAFFORD BOROUGH COUNCIL, VICTORIA GARDENS, STAFFORD.
Gauge : 2'0".   (SJ 919229)

ISABEL	0-4-0ST OC	WB	1491 1898

STAFFORDSHIRE COUNTY MUSEUM, SHUGBOROUGH HALL ESTATE, GREAT HAYWOOD, STAFFORD.
Gauge : 4'8½"   (SJ 992215)

(G.W.R. 252)		0-6-0	IC	EW		1855	Dsm
1439		0-4-0ST	IC	Crewe	842	1865	
No.6		0-4-0ST	OC	R.Heath		c1886	
	MOSS BAY	0-4-0ST	OC	KS	4167	1920	
No.2		0-6-2T	IC	Stoke		1923	
No.2	(BEL 2)	4wBE		Stoke		1917	
		0-4-0PM		Bg	800	1920	

TILSLEY DIESEL INDUSTRY LTD, STONEWALL INDUSTRIAL ESTATE, SILVERDALE.
Yard (SJ 829466) with locos for resale occasionally present.

TILSLEY & LOVATT LTD, NEWSTEAD TRADING ESTATE, TRENTHAM.
Yard (SJ 887409) with locos for resale & repair occasionally present.

TRENTHAM GARDENS LTD, MINIATURE RAILWAY, TRENTHAM, STOKE.
Gauge : 2'0".   (SJ 864406)

BRORA	0-4-0PM	S/o		Bg	1797 1930
			Rebuilt	Bg	2083 1934
GOLSPIE	0-4-0DM	S/o		Bg	2085 1934
DUNROBIN	0-6-0DM	S/o		Bg	3014 1938

WAGON REPAIRS LTD, FENTON, STOKE-ON-TRENT.
Gauge : 4'8½".   (SJ 881439)

L20	YARD No.4859	4wDM		FH	3746 1955
L19	50	4wDM		FH	3960 1961

SUFFOLK

BRITISH SUGAR CORPORATION LTD,
Bury St. Edmunds Factory,
Gauge : 4'8½".   (TL 862654)

HERCULES	0-4-0DM		RH	281271 1950
	0-4-0DH		RH	497753 1963

Ipswich Factory, Sproughton, Ipswich.
Gauge : 4'8½".   (TM 137448)

0-4-0DE		RH	408304 1957

DEPARTMENT OF THE ENVIRONMENT.
Havergate Island, Orford Ness.
Gauge : 2'0".   (TM 415480)

4wDM		MR	22209 1964
4wDM		MR	22211 1964

<u>Site near Aldeburgh, Orford Ness.</u>
Gauge : 2'0".   (TM 462545)

		4wDM	MR	22212	1964	OOU

<u>EAST ANGLIAN TRAMWAY MUSEUM SOCIETY, CARLTON COLVILLE, LOWESTOFT.</u>
Gauge : 4'8½".   (TM 505903)

ARMY 9035		2w-2PMR	Wkm	8195	1958

Gauge : 2'0".

No.1		4wDM	MR	5902	1934
No.2		4wDM	MR	5912	1934

<u>FELIXSTOWE DOCK & RAILWAY CO LTD, FELIXSTOWE.</u>
Gauge : 4'8½".   (TM 283330)

D3489	COLONEL TOMLINE	0-6-0DE	Dar		1958

<u>R.FINBOW, "CAITHNESS", BACTON.</u>
Gauge : 4'8½".   (TM 064682)

FRY		4wVBT VCG	S	7492	1928

<u>RANSOMES, SIMS & JEFFERIES LTD, NACTON FOUNDRY, IPSWICH.</u>
Gauge : 4'8½".   (TM 204418)

RANSOMES		4wDM	RH	466629	1962

<u>SUFFOLK ROADSTONE LTD, ROADSTONE TERMINAL, PRINCES STREET, IPSWICH.</u>
Gauge : 4'8½".   (TM 157439)

		4wDM	RH	237927	1946	OOU

SURREY

<u>BRITISH INDUSTRIAL SAND LTD, HOLMETHORPE, REDHILL.</u>
Gauge : 4'8½":   (TQ 288517)

1		0-4-0DM	RSH	7901	1958
2		0-4-0DM	Bg/DC	2159	1941
3		0-4-0DM	AB	332	1938

<u>BROCKHAM MUSEUM ASSOCIATION, BROCKHAM QUARRY, CHALKPIT LANE, BETCHWORTH,</u>
<u>near DORKING.</u>
Gauge : 3'2¼".   (TQ 198510)

No.4	TOWNSEND HOOK	0-4-0T	OC	FJ	172L	1880
10	MONTY	4wDM		OK	7269	1936

Gauge : 3'0".

2	SCALDWELL	0-6-0ST	OC	P	1316	1913

Gauge : 2'11".

| | | | 4wDM | | MR | 10161 | 1950 |

Gauge : 2'0".

4		POLAR BEAR	2-4-0T	OC	WB	1781	1905
5		PETER	0-4-0ST	OC	WB	2067	1918
23			4wPM		MR	872	1918
				Rebuilt	MR	3720	1925
15			4wDM		MR	1320	1918
No.2	14		4wDM		RH	166024	1933
16		PELDON	4wDM		JF	21295	1936
17		REDLAND	4wDM		OK	6193	1937
No.7	11	THE MAJOR	4wDM		OK	7741	1937
12			4wDM		R&R	80	1937
37		WD 904	2w-2PMR		Wkm	3403	1943
13			4wDM		HE	3097	1944
2	50		4wBE		WR	5031	1953

Gauge : 1'10".

| 23 | | | 0-4-0T | IC | Spence | | 1920 |

J.L.BUTLER, 5, HEATH RISE, GROVE HEATH, RIPLEY.
Gauge : 60cm.   (TQ 046557)

1		COVERTCOAT	0-4-0ST	OC	HE	679	1898
		ARCHER	4wDM		MR	4709	1936
		BARGEE	4wDM		MR	8540	1940
3			0-4-0DM		Dtz	19531	c1941
		MAC	4wDM		MR	8994	1946

J.CROSSKEY, SURREY LIGHT RAILWAY.
Gauge : 2'0".   (   )

2		4wDM		RH	174535	1936
4		4wDM		RH	177642	1936
22		4wDM		RH	226302	1944
		4wDM		HE	3621	1947
		4wDM		RH	277265	1949
		4wDM		MR	20073	1950
		4wDM		RH	264252	1952
6		4wBE		WR	D6912	1964
		2w-2PM		Rhiwbach			Dsm

Locos are kept at a private location.

DELTA CONSTRUCTION LTD, PLANT DEPOT, GODALMING.
Gauge : 2'0".   (   )

| L1 | | 4wBE | | WR | 4818 | 1951 |

Gauge : 1'6".

EL2		0-4-0BE		WR	N7641	1974
EL3		0-4-0BE		WR	N7660	1974
EL4		0-4-0BE		WR	N7615	1973
EL5		0-4-0BE		WR	N7616	1973

Locos present in yard between contracts.

J.EWING, 25, PARK AVENUE, CAMBERLEY.
Gauge : 2'0".   (   )

| | | | 4wDM | FH | 4008 1963 |

J.W.HARDWICK, SONS & CO LTD, SCRAP MERCHANTS, WEST EWELL.
Yard (TQ 198645) with locos for scrap or resale occasionally present.

E.J.H.P.INGRAM, 17, CANNON GROVE, FETCHAM.
Gauge : 60cm.   (   )

| | | | 4wDM | MR | 7153 1937 |

JOHNSTON CONSTRUCTION, PLANT YARD, CHERTSEY.
Yard (   ) with locos present between contracts.
            See entry under Warwickshire for loco fleet details.

J.B.LATHAM, "CHANNINGS", KETTLEWELL HILL, WOKING.
Gauge : 3'2¼".   (TQ 003598)

| No.5 | WILLIAM FINLAY | 0-4-0T | OC | FJ | 173L 1880 |

R.MARNER, HORSEHILL FARM, NORWOOD HILL, HORLEY.
Gauge : 2'0".   (   )

| 27 | | | 4wDM | MR | 5863 1934 |

NATIONAL RIFLE ASSOCIATION, CENTURY RANGE, BISLEY, BROOKWOOD.
Gauge : 2'0".   (SU 941585)

| | | | 4wDM | LB | 52579 1961 |

P.D.NICHOLSON, 17, CROSSLANDS ROAD, WEST EWELL.
Gauge : 2'0".   (TQ 207636)

| No.1830 | PLUTO | | 4wPM | FH | 1830 1933 |

P.RAMPTON & FRIENDS, BURGATE FARM, VANN LANE, HAMBLEDON.
Gauge : 60cm.   (TQ 001381)

1	SABERO	0-6-0T	OC	Couillet	1140 1895
2	SAMELICES	0-6-0T	OC	Couillet	1209 1898
3	OLLEROS	0-6-0T	OC	Couillet	1318 1940
6	LA HERRERA	0-6-0T	OC	Sabero	c1937
7	SOTILLOS	0-6-2T	OC	Borsig	6022 1906
101		0-4-2T	OC	Hen	16073 1918
102		0-4-0T	OC	Hen	16043 1918
103		0-4-0T	OC	Hen	16045 1918
		2-6-2+2-6-2	4C	Hano	10634 1928
	RENISHAW 4	0-4-4-0T	VCG	AE	2057 1931
	RENISHAW 5	0-4-4-0T	OC	WB	2545 1936
		0-4-2T		WB	2895 1948
No.18		4wDE		DK	c1918

Gauge : 55cm.

| | SANTA ANA | 0-4-2ST | OC | HC | 640 1903 |

            Some of the locos are stored at another, unknown, location.

REDLAND BRICKS LTD.
North Holmwood Brickworks, Dorking. (Closed)
Gauge : 2'0".   (TQ 173474)

4wBE	WR	4998	1953

Nutbourne Brickworks, Hambledon, near Godalming.
Gauge : 2'0".   (SU 973375)

4wDM	MR	7199	1937
4wDM	MR	8678	1941
4wDM	MR	21513	1958

WILLIAM F. REES LTD, HIGH STREET, OLD WOKING.
Gauge : 1'0".   (TQ 015569)

2w-2BE	Iso
2w-2BE	Iso

Suppliers of contractors locos.

A.STREETER & CO LTD, PLANT DEPOT, CATTESHALL WHARF, WARRAMILL ROAD, GODALMING.
Repair depot at WARSOP ELECTRICAL CO LTD, GUILDFORD.
Gauge : 2'0".   (SU 983446, 988505)

	4wDM	Jung	5869	Dsm
	4wBE	WR	D6803 1964	
5	0-4-0BE	WR	F7026 1966	
2	0-4-0BE	WR		
3	0-4-0BE	WR		
	0-4-0BE	WR		
	0-4-0BE	WR		
	4wBE	WR		
	4wBE	CE	5940A 1972	
	4wBE	CE	5940B 1972	
	4wBE	CE	5940C 1972	
2	4wBE	CE	5940D 1972	
	4wBE	CE	5961A 1972	
	4wBE	CE	5961B 1972	
	4wBE	CE	5961C 1972	
1	4wBE	CE	5961D 1972	
4	4wBE	CE	B0107A 1973	
	4wBE	CE	B0107B 1973	
	4wBE	CE	B0107C 1973	
009	4wBE	CE	B0113A 1973	
	4wBE	CE	B0113B 1973	
	4wBE	CE	B0119A 1973	
	4wBE	CE	B0119B 1973	
	4wBE	CE	B0119C 1973	
No.20	4wBE	CE	B0119D 1973	
	4wBE	CE	B0142A 1973	
	4wBE	CE	B0142B 1973	
	4wBE	CE	B0142C 1973	
	4wBE	CE	B0166 1974	

Locos present in yard and repair depot between contracts.

SURREY & HAMPSHIRE CANAL SOCIETY, ASH EMBANKMENT SITE, ALDERSHOT,(HANTS). (Closed)
Gauge : 2'0".   (SU 921564)

1	4wDM	HE	1944 1939
	4wDM	MR	22070 1960

I.SUTCLIFFE, BOURNE VALLEY GRIT NARROW GAUGE RAILWAY.
Gauge : 2'0".   (   )

			4wDM	MR	7128 1938

THAMES WATER AUTHORITY, METROPOLITAN WATER DIVISION, CHARLTON ROAD DEPOT,
SHEPPERTON.
Gauge : 2'0".   (TQ 083694)

IVAN		4wPM	FH	3317 1957

WEY VALLEY LIGHT RAILWAY, FARNHAM DISTRICT SCOUTS SERVICE TEAM, FARNHAM.
Gauge : 2'0".   (SU 848473)

3	4wPM	FH	1767 1931
	4wDM	OK	3685 1931
	4wPM	OK	4588 1932
	4wDM	OK	6504 1936
	4wDM	MR	5713 1936
6	4wDM	FH	2528 1941
2	4wPM	Thakeham	1950
1	4wPM	Bredonvale	c1950
960228	2w-2PMR	Wkm	1309 1933
	2w-2PM	Wkm	2981 1941 +
4	4wDM	Wkm	3031 1941
7	2w-2PM	Wkm	3032 1941
5	2w-2PM	Wkm	3287 1943

+ Converted to a passenger coach.

TYNE & WEAR

BOWES RAILWAY CO LTD, SPRINGWELL.
Gauge : 4'8½".   (NZ 285589)

No.22	No.85	0-4-0ST	OC	AB	2274 1949
	W.S.T.	0-4-0ST	OC	AB	2361 1954
101		4wDM		FH	3922 1959

BRITISH STEEL CORPORATION, B.S.C. SECTIONS, JARROW WORKS, WESTERN ROAD, JARROW.
Gauge : 4'8½".   (NZ 324655)

60	0-4-0DH	S	10067 1961	
61	0-4-0DH	S	10069 1961	
62	0-4-0DH	S	10084 1961	OOU
63	0-4-0DH	S	10066 1961	OOU

CENTRAL ELECTRICITY GENERATING BOARD.
Dunston Power Station, near Gateshead. (Closed)
Gauge : 4'8½".   (NZ 221628)

No.55	0-4-0DE	RH	381751 1955	OOU
No.30	0-4-0DE	RH	412707 1957	OOU

Stella North Power Station, Newburn-On-Tyne.
Gauge : 4'8½".   (NZ 177646)

No.24		4wDH	TH/S	188C	1967
No.25		4wDH	TH/S	177C	1967

Stella South Power Station, near Blaydon.
Gauge : 4'8½".   (NZ 174643)

	T.I.C. No.31	0-4-0DE	RH	412714	1957
		0-4-0DH	JF	4240013	1962	Dsm
	THORPE MARSH No.2	0-6-0DH	JF	4240015	1962
1		0-6-0DH	JF	4240020	1964

CLARK HAWTHORN LTD, WILLINGTON QUAY, WALLSEND.
Gauge : 4'8½".   (NZ 315661)

1459	4wDM	FH	3880	1958

JONES & BAILEY LTD, COAL STAITHES, HARTON.
Gauge : 2'0".   (NZ 359670)

4wDM	MR	22237	1965

LOCOMOTION ENTERPRISES (1975) LTD, BOWES RAILWAY, SPRINGWELL, GATESHEAD.
(NZ   )

Manufacturers of Replica Locomotives.

MINISTRY OF DEFENCE, ROYAL ORDNANCE FACTORY, BIRTLEY.

For full details see Section Five.

NATIONAL COAL BOARD.

For full details see Section Four.

N.E.I. PARSONS LTD, HEATON WORKS, NEWCASTLE-UPON-TYNE.
Gauge : 4'8½".   (NZ 277655)

No.1	0-4-0DM	HE	4212	1950
No.2	0-4-0DM .	HE	4522	1953	·OOU
3	0-4-0DH	NB	27656	1957

PORT OF SUNDERLAND AUTHORITY, SOUTH DOCKS, SUNDERLAND.
Gauge : 4'8½".   (NZ 410580)

P.S.A. No.21	0-4-0DE	RH	395294	1956
P.S.A. No.22	0-4-0DE	RH	416210	1959

PORT OF TYNE AUTHORITY.
Albert edward Dock, Percy Main, North Shields.
Gauge : 4'8½".   (NZ 344671)

No.32	0-4-0DE	RH	416208	1957
No.35	0-4-0DE	RH	418600	1957

Tyne Dock, South Shields.
Gauge : 4'8½".   (NZ 350658, 355652)   RTC.

58	0-4-0DE	RH	381755	1955

RAINE & CO LTD, DELTA IRON & STEEL WORKS, BLAYDON.
Gauge : 4'8½".   (NZ 211633)

1		0-4-0DE	(BT	339 1961	
			(BP	7946 1961	
		0-4-0DE	(BT	443 1962	
			(BP	7873 1962	Dsm

RIBBLESDALE CEMENT LTD, NEWCASTLE DEPOT, RAILWAY STREET, NEWCASTLE-UPON-TYNE.
Gauge : 4'8½".   (NZ 244635)

	RIBBLESDALE No.7	0-4-0DM	AB	394 1955

ROWNTREE-MACKINTOSH LTD, CHOCOLATE MANUFACTURERS, FAWDON FACTORY, NEWCASTLE.
Gauge : 4'8½".   (NZ 220687)

		4wDM	RH	421419 1958
No.3		4wDM	RH	441934 1960

LESLIE SANDERSON LTD, WEST LINES TRADING ESTATE, BIRTLEY.
Yard (NZ 264548) with locos for resale occasionally present.

SHEPHERDS SCRAP METALS (NEWCASTLE) LTD, SCRAP DEALERS, ST. PETERS STATION, NEWCASTLE.
Gauge : 4'8½".   (NZ 275637)

25		4wDM	RH	275884 1949	
5898	BMA 25605	4wDM	RH	441933 1960	OOU
		4wDH	TH/S	112C 1961	

SWAN HUNTER SHIPBUILDERS LTD. (Member Company of British Shipbuilders.)
Carville & Neptune Yards, Wallsend.
Gauge : 4'8½".   (NZ 298653)

	LION	0-4-0DH	RH	468042 1963

Hebburn Shipyard, Hebburn.
Gauge : 4'8½".   (NZ 306654, 309655)

	TRIUMPH	0-4-0DM	RH	304472 1951 Dsm
	APOLLO	0-4-0DM	RH	319288 1953

Naval Yard, Walker.
Gauge : 4'8½".   (NZ 296638)

MTC 3472	TIGER	0-4-0DH	RH	437368 1959

TANFIELD RAILWAY PRESERVATION SOCIETY, MARLEY HILL.
Gauge : 4'8½".   (NZ 207573)

	HOLWELL No.3	0-4-0ST	OC	BH	266 1873
No.3		0-4-0ST	OC	RWH	2009 1884
No.3		0-4-0WT	OC	EB	37 1898
		0-6-0ST	OC	AB	1015 1904
112	CYCLOPS	0-4-0ST	OC	HL	2711 1907
No.6		0-4-2ST	OC	AB	1193 1910
		0-4-0ST	OC	HL	2859 1911
No.17		0-6-0T	OC	AB	1338 1913
32		0-4-0ST	OC	AB	1659 1920
	COAL PRODUCTS No.3	0-6-0ST	OC	HL	3575 1923

	STAGSHAW	0-6-0ST	OC		HL	3513	1927	
	CORONATION	0-4-0ST	OC		HC	1672	1937	
		0-4-0CT	OC		RSHN	7007	1940	
49		0-6-0ST	IC		RSH	7098	1944	
	SIR CECIL A. COCHRANE	0-4-0ST	OC		RSH	7409	1948	
38		0-6-0T	OC		HC	1823	1949	
No.4		4wVBT	VCG		S	9559	1953	
No.44	9103/44	0-6-0ST	OC		RSH	7760	1953	
No.38		0-6-0ST	OC		RSH	7763	1954	
21		0-4-0ST	OC		RSHN	7796	1954	
No.16		0-6-0ST	OC		RSH	7944	1957	
		0-4-0DE			AW	D22	1933	

T.J.THOMSON & SON LTD, TYNE YARD, DUNSTON, near GATESHEAD.
Gauge : 4'8½".   (NZ 225627)

	CHURCHILL	0-4-0DM		RH	281270	1951	
No.3	P.S.A. No.20	0-4-0DM		RH	327969	1954	OOU

TYNE & WEAR METROPOLITAN COUNTY COUNCIL, MIDDLE ENGINE LANE STORE, near CHIRTON.
Gauge : 4'8½".   (NZ 323693)

		0-4-0	VC	RS	A4	1826
41		0-6-0PT	IC	K	2509	1883
10		0-6-0DM		Consett	No.10	1958
DE 900730	(3267)	4w-4wRE				1904

TYNE WEAR METRO, NEWCASTLE-UPON-TYNE.
Gauge : 4'8½".   Locos are kept at :-

	Newbridge Street Depot.		(NZ 254647)	
	South Gosforth Car Sheds.		(NZ 250686)	

		0-4-0DM	HE	4264	1952
WL 1		0-6-0DE	BT	801	1977
WL 2		0-6-0DE	BT	802	1977
WL 3		0-6-0DE	BT	803	1977
WL 4		0-6-0DE	BT	804	1977
WL 5		0-6-0DE	BT	805	1977

WASHINGTON DEVELOPMENT CORPORATION, "F" PIT MUSEUM.
Gauge : 2'0".   (NZ 303574)

0-4-0DM	RH	392157	1956

WARWICKSHIRE

BLUE CIRCLE INDUSTRIES LTD, HARBURY CEMENT WORKS.
Gauge : 4'8½".   (SP 394587)

0-4-0DH	JF	4220008	1959

CENTRAL ELECTRICITY GENERATING BOARD, HAMS HALL POWER STATION, near COLESHILL.
Gauge : 4'8½".   (SP 206915)

1		0-6-0DH	EEV	D1137	1966
2		0-6-0DH	EEV	D1227	1967
3		0-6-0DH	EEV	D1228	1967

GEORGE COHEN, SONS & CO LTD, KINGSBURY.
Gauge : 4'8½".   (SP 219969)

A.M.W. No.241		0-4-0DM	JF	22995	1943

Also other locos for scrap occasionally present.

K.FENWICK.
Gauge : 1'11½".   (   )

D		4wDM	RH	182137	1936
		4wDM	RH	226278	1942
		4wDM	RH	229631	1944

Present location unknown.

FORD MOTOR CO LTD, PRINCES DRIVE, LEAMINGTON.
Gauge : 2'0".   (SP 314652)

4	P 37829	4wDM	FH	3916	1959

G.E.C. MACHINES LTD, RUGBY.
Gauge : 4'8½".   (SP 506765)

No.9510	MAZDA	0-4-0DE	RH	268881	1949	OOU
No.12342		0-4-0DE	RH	423657	1958	

ISO SPEEDIC CO LTD, FABRICATIONS & ELECTRIC VEHICLES, CHARLES STREET, WARWICK.
(SP 294656)

New Iso locos under construction occasionally present.

JOHNSTON CONSTRUCTION LTD, PLANT DEPOT, WATLING STREET, SHAWELL.
Gauge : 2'0"?   (   )

		0-4-0BE	WR	7650	1973
		0-4-0BE	WR	7651	1973

Gauge : 1'0".

	2w-2BE	Iso	

Locos present in yard between contracts.
See also entry under Surrey.

LILLEY/WADDINGTON LTD, PLANT DEPOT, HAUNCHWOOD COLLIERY SITE, GALLEY COMMON, near NUNEATON.
Yard (SP 313917) with locos present between contracts.

See entry under Essex for loco fleet details.

MILLER BUCKLEY PLANT LTD, PLANT DEPOT, WATLING STREET, RUGBY.
Gauge : 1'6".   (SP 533788)

L 10		4wBE	CE	5827	1970
L 11		4wBE	CE	5920	1972
L 12		4wBE	CE	5965A	1973
L 13		4wBE	CE	5965B	1973
L 14		4wBE	CE	5965C	1973
L 15		4wBE	CE	B0109A	1973
L 16		4wBE	CE	B0109B	1973

   Locos present in yard between contracts.

MINISTRY OF DEFENCE, ARMY DEPARTMENT.
Kineton Depot.

   For full details see Section Five.

Long Marston Depot.

   For full details see Section Five,

MIXCONCRETE AGGREGATES LTD, SOUTH ROAD GRAVEL PITS, CHARLECOTE,
near STRATFORD-UPON-AVON.
Gauge : 2'0".   (SP 265575)   RTC.

T 2		4wDM	MR	8739	1942	OOU

NATIONAL COAL BOARD.

   For full details see Section Four.

RAILWAY PROJECT GROUP, PEYTO'S CLOSE, CHESTERTON, near WARWICK.
Gauge : 1'11½".   (SP 357582)

	0-6-0T	OC	AB	1578	1918

RUGBY PORTLAND CEMENT CO LTD.
New Bilton Works, near Rugby.
Gauge : 4'8½".   (SP 487756)

	4wDH	TH	173V	1966

Southam Works.
Gauge : 4'8½".   (SP 422640)

SOUTHAM 2	0-4-0DM	HC	D625	1942
	4wDH	TH	164V	1966

SEVERN LAMB LTD, WESTERN ROAD, STRATFORD-UPON-AVON.
(SP 197554)

   New miniature locos under construction & locos in for repair
   usually present.

SEVERN TRENT WATER AUTHORITY, TAME DIVISION, COLESHILL INCINERATION PLANT &
FILTER BEDS.

   For details see entry under West Midlands.

A.D.SMITH, OLDBERROW HOUSE, HENLEY-IN-ARDEN.
Gauge : 2'0".   (SP 121659)

2	LADY LUXBOROUGH	0-4-0ST OC		WB	2088 1919

WARWICK DISTRICT COUNCIL, NEWBOLD COMYN PARK, LEAMINGTON SPA.
Gauge : 4'8½".   (SP 333657)

75170		0-6-0ST IC	WB	2758 1944

WEST MIDLANDS

ALBRIGHT & WILSON LTD, CHEMICAL MANUFACTURERS, OLDBURY.
Gauge : 4'8½".   (SO 994883)

4wDM	FH	3686 1954

ALLEN ROWLAND & CO LTD, STATION WORKS, WARWICK ROAD, TYSELEY.
Gauge : 4'8½".   (SP 109840)

No.6	872/51	0-4-0DM	JF	4210116 1956	
No.7	872/52	0-4-0DM	JF	4210125 1957	OOU

JOHN BAGNALL & SONS LTD, LEABROOK IRONWORKS, WEDNESBURY.
Gauge : 4'8½".   (SO 979943)

4wDM	RH	269603 1949	OOU
4wDM	RH	398611 1957	
4wDM	FH	4016 1964	

BAILLIE CONTRACTING CO LTD, PLANT DEPOTS, 2061, COVENTRY ROAD, SHELDON, BIRMINGHAM &
TORRINGTON AVENUE, COVENTRY.
Gauge : 1'0".'   (SP 143840, 285778)

L 43	2w-2BE	Iso	
	2w-2BE	Iso	T51 1974
	2w-2BE	Iso	
	2w-2BE	Iso	

Locos present in yards between contracts.

BIRMINGHAM CORPORATION.
Childrens Playground, Hindlow Close, Duddeston, Birmingham.
Gauge : 4'8½".   (SP 087877)

0-6-0ST IC	MW	2015 1921

Museum Of Science & Industry, Newhall Street, Birmingham.
Gauge : 4'8½".   (SP 064874)

| 46235 | CITY OF BIRMINGHAM | 4-6-2 | 4C | Crewe | | 1939 | |

Gauge : 2'8".

| | SECUNDUS | 0-6-0WT | OC | Bellis & Seekings | | 1874 | |

Gauge : 2'0".

| 1 | LEONARD | 0-4-0ST | OC | WB | 2087 | 1919 | |
| No.56 | (LORNA DOONE) | 0-4-0ST | OC | KS | 4250 | 1922 | |

BRADLEY & FOSTER LTD, DARLASTON IRON WORKS, BENTLEY ROAD SOUTH, DARLASTON.
Gauge : 3'0".   (SO 984977)   RTC.

| | | 4wDM | | BD | 3750 | 1979 | OOU |

BRITISH LEYLAND U.K. LTD.
Cofton Hackett Factory.
Gauge : 4'8½".   (SP 011764)

| | COFTON | 4wDH | | TH | 276V | 1977 | |

Courthouse Green Works, Coventry.
Gauge : 4'8½".   (SP 353812)

| | | 4wDM | | RH | 224346 | 1945 | |

Longbridge Works, Birmingham.
Gauge : 4'8½".   (SP 009775, 011774)

(3011)	LICKEY	0-6-0DE		Derby		1952	
No.1		0-4-0DH		JF	4220034	1965	
	LONGBRIDGE	0-4-0DH		HE	6982	1970	
	FRANKLEY	4wDH		TH	283V	1978	

BRITISH STEEL CORPORATION, SHEFFIELD DIVISION, STEELWORKS GROUP,
BILSTON & WOLVERHAMPTON WORKS.
Bilston Works, Bilston.
Gauge : 4'8½".   (SO 942958)

6	MARGARET	0-4-0DE		YE	2663	1957	
9	ANDREW	0-4-0DE		YE	2796	1960	
10	GEORGE	0-4-0DE		YE	2797	1960	
12	MICHAEL	0-4-0DE		YE	2879	1962	

Wolverhampton Works, Horseley Fields, Wolverhampton.
Gauge : 4'8½".   (SO 927988)

| 9111/85 | STANTON No.45 | 0-4-0DE | | YE | 2623 | 1956 | |

BRITISH STEEL CORPORATION, TUBES DIVISION, GENERAL TUBES WORKS GROUP,
BROMFORD WORKS, WHEELWRIGHT ROAD, ERDINGTON, BIRMINGHAM.
Gauge · 4'8½".   (SP 113899)

9111/82	STANTON No.42	0-4-0DE		YE	2598	1956	
58		0-4-0DH		S	10098	1962	
59		0-4-0DH		S	10099	1962	
	BARABEL	0-4-0DH		RR	10202	1964	
	WILLIAM A.	0-4-0DH		EEV	D1279	1969	

C.BRYANT & SON LTD, PLANT DEPOT, DORIS ROAD, BIRMINGHAM.
Gauge : 1'6".   (SP 095866)

		4wBE	CE	B0122 1973 +

+ Plate reads CE0443.
Loco present in yard between contracts.

BRYMBO STEELS WORKS LTD, CABLE STREET MILLS, CABLE STREET, WOLVERHAMPTON.
Gauge : 2'0".   (SO 925977)   (A member of the G.K.N. Group of Companies.)

2	CM 9075	4wBE	WR	C6716 1963
	CM 9076	4wBE	WR	C6717 1963

CASHMORES GLYNWED DISTRIBUTION, GREAT BRIDGE.
Gauge : 4'8½".   (SO 981933)

	0-4-0DM	JF	22498 1939	Dsm
A.M.W. No.213	0-4-0DM	JF	22960 1941	OOU

CENTRAL ELECTRICITY GENERATING BOARD.
Nechells Power Station.
Gauge : 4'8½".   (SP 098896)

N.E. No.1	0-6-0DH	BD	3681 1972

Walsall Power Station, Birchills.
Gauge : 4'8½".   (SP 000998)   RTC.

3087	0-6-0DE	Derby	1954
	4wDM	RH	262997 1949

COURTAULDS LTD, COVENTRY.
Gauge : 4'8½".   (SP 336808)

HENRY	0-4-0ST OC	HL	2491 1901	Pvd

DUCTILE STEELS LTD, PLANETARY ROAD, WEDNESFIELD.
Gauge : 4'8½".   (SO 945997)   RTC.

	4wDM	RH	349039 1953	OOU

DUNLOP CO LTD, ERDINGTON, BIRMINGHAM.
Gauge : 4'8½".   (SP 127901)

	4wDM	MR	9932 1972

J.J.GALLAGHER & CO LTD, PLANT DEPOT, ARMOUR CLOSE, LITTLE GREEN LANE, BIRMINGHAM.
Gauge : 2'0".   (SP 097864)

	4wDM	MR	5868 1934

Gauge : 1'6".

	4wBE	CE	5956 1972
	4wBE	CE	B0922 1975
	4wBE	CE	B0922 1975

Locos present in yard between contracts.

R.A.GIBLIN LTD, BRIQUETTING STEEL SWARF & CAST IRON BORINGS, STAFFORD STREET, WEDNESBURY.
Gauge : 4'8½".   (SO 985947)

			4wDM	Bg/DC	2107 1937	OOU
L 14			4wDM	RH	235515 1945	OOU
L 7			4wDM	RH	349038 1954	OOU

HUNT BROS (OLDBURY) LTD, WEST BROMWICH LANE, OLDBURY.
Gauge : 1'3".   (SO 989899)

No.1	SUTTON BELLE	4-4-2	OC		Cannon	1933
				Rebuilt		1953
No.2	SUTTON FLYER	4-4-2	OC		Cannon	
No.3	PRINCE OF WALES	4-4-2	OC		BL	11 1914
		4w-4wPMR			Guest	

Locos in store.

J. & H.B. JACKSON LTD, SCRAP MERCHANTS, COVENTRY.
Gauge : 4'8½".   (SP 343804)   RTC.

		0-4-0DM	HC	D604 1936	Pvd

L.C.P. FUEL CO, PENSNETT TRADING ESTATE, SHUT END.
Gauge : 4'8½".   (SO 903898)

(02003)	PETER	0-4-0DH	YE	2812 1960
(D2868)	SAM	0-4-0DH	YE	2851 1961

L.C.P. HOLDINGS LTD, PENSNETT TRADING ESTATE, SHUT END.
Gauge : 4'8½".   (SO 901897)

2025	WINSTON CHURCHILL	0-6-0ST IC	MW	2025 1923	Pvd
		4wDM	RH	215755 1942	Pvd

F.H.LLOYD & CO LTD, JAMES BRIDGE STEELWORKS, WEDNESBURY.
Gauge : 4'8½".   (SO 992969)

VL 6		4wDM	RH	382824 1955	OOU

J.MARSHALL, 38, SPRING LANE, HOCKLEY HEATH.
Gauge : 2'0".   (SP 152725)

No.1	ODDSON	4wVBT G	Marshall	1970

MARTIN & CO (CONTRACTORS) LTD, PLANT DEPOT, 1195, BRISTOL ROAD SOUTH, NORTHFIELD, BIRMINGHAM.
Gauge : 1'6".   (SP 015786)

		0-4-0BE	WR	M7546 1972

Loco present in yard between contracts.

METROPOLITAN-CAMMELL CARRIAGE & WAGON CO LTD, WASHWOOD HEATH.
Gauge : 4'8½".   (SP 103889)

9611/61	STANTON No.40	0-4-0DE	YE	2596	1955
HF 0591		0-4-0DM	JF	4210112	1956
HF 0793		0-4-0DM	JF	4210133	1957
9611/63	STANTON No.48	0-4-0DE	YE	2705	1958
	PETER	0-4-0DH	HE	7424	1978

Also new MC railcars under construction usually present.

NATIONAL COAL BOARD.

For full details see Section Four.

N.E.I. CLARKE CHAPMAN - JOHN THOMPSON LTD, DUDLEY PORT.
Gauge : 4'8½".   (SO 973925)

4wDM	RH	408496	1957

ROUND OAK STEEL WORKS LTD, BRIERLEY HILL.
Gauge : 4'8½".   (SO 919881)   (Subsidiary of British Steel Corporation.)

No.1	0-4-0DE	YE	2593	1955	
No.2	0-4-0DE	YE	2614	1957	
No.3	0-4-0DE	YE	2662	1957	
No.4	0-4-0DE	YE	2774	1959	
No.5	0-4-0DE	YE	2784	1960	
No.6	0-4-0DE	YE	2795	1960	Dsm
No.7	0-4-0DE	YE	2821	1961	
No.8	0-4-0DE	YE	2881	1962	
No.9	0-4-0DE	YE	2882	1962	
No.10	0-4-0DE	YE	2883	1963	
No.11	0-4-0DE	YE	2532	1953	

SEVERN TRENT WATER AUTHORITY, TAME DIVISION.
Gauge : 2'0".   Locos are kept at :-

Coleshill Filter Beds, Warwickshire.	(SP 196914)	
Coleshill Incineration Plant, Warwickshire.	(SP 192916)	
Lagoon Works, Water Orton.	(SP 156917, 159913)	
Minworth Main Depot & Workshops.	(SP 164926)	

87023		4wDM	MR	22236	1965	OOU
87031		4wDM	MR	40S383	1971	
87032	32	4wDM	MR	40S412	1973	
87033	33	4wDM	MR	40SD501	1975	
87034	34	4wDM	MR	40SD502	1975	
87035	35	4wDM	MR	40SD503	1975	
8739B		4wDM	SMH	40SD515	1979	
8740B		4wDM	SMH	40SD516	1979	

SHERIDAN CONTRACTORS LTD, PLANT DEPOT, BELMONT ROW, NECHELLS GREEN, BIRMINGHAM.
Gauge : 2'0".   (SP 081874)

4wBE	CE	B1890	1979
4wBE	CE	B1890	1979

Gauge : 1'6".

0-4-0BE	WR		1973 +
0-4-0BE	WR		1973 +
4wBE	CE	2200A	1979

+ These are two of 7642, 7643, 7652, 7659.
Locos present in yard between contracts.

SOUTH STAFFS WAGON CO LTD, BLOOMFIELD ROAD, PRINCES END, TIPTON.
Gauge : 4'8½". (SO 948934)

	0-4-0DE	RH	420140	1958
No.3	0-4-0DE	RH	461960	1962

STANDARD GAUGE STEAM TRUST, TYSELEY, BIRMINGHAM.
Gauge : 4'8½". (SP 105841)

4983	ALBERT HALL	4-6-0	OC	Sdn		1931
5043	EARL OF MOUNT EDGCUMBE	4-6-0	4C	Sdn		1936
5080	DEFIANT	4-6-0	4C	Sdn		1939
5637		0-6-2T	IC	Sdn		1925
7027	THORNBURY CASTLE	4-6-0	4C	Sdn		1949
7029	CLUN CASTLE	4-6-0	4C	Sdn		1950
7752		0-6-0PT	IC	NB	24040	1930
7760		0-6-0PT	IC	NB	24048	1930
9600		0-6-0PT	IC	Sdn		1945
(45593) 5593 KOLHAPUR		4-6-0	3C	NBQ	24151	1935
		0-4-0ST	OC	P	1722	1926
1		0-4-0ST	OC	P	2004	1941

GEORGE WARD (MOXLEY) LTD, BAGGOTT'S BRIDGE CLAYWORKS, DARLASTON.
Gauge : 2'0". (SO 965964) RTC.

4wPM	L	3834	1931	Dsm
4wDM	LB	52885	1962	OOU

WELLMAN, SMITH, OWEN ENGINEERING CORP LTD, DARLASTON.
(SO   )

New WSO locos under construction occasionally present.

W. & T. (LEISURE) LTD, DUDLEY ZOO MINIATURE RAILWAY, DUDLEY.
Gauge : 1'3". (SO 947911)

MICHAEL	4-4-2	OC	A.Barnes	105	1930
CLARA	0-4-2PM	S/O	G&S		1961

WEST SUSSEX

<u>J.M.BALDOCK, HOLLYCOMBE STEAM FAIR & RAILWAY, LIPHOOK.</u>
Gauge : 4'8½".   (SU 852295)

	Name	Wheels	Cyl	Builder	No.	Date
	ALBERT	0-4-0ST	OC	HL	2450	1899
	SIR VINCENT	4wWT	G	AP	8800	1917

Gauge : 2'0".

No.1	CALEDONIA	0-4-0WT	OC	AB	1995	1931
		4wDM		RH		1941 +

+ Either 203016 or 203019.

Gauge : 1'10¼".   (SU 851296)

	JERRY M.	0-4-0ST	OC	HE	638	1895

<u>BLUEBELL RAILWAY CO LTD.</u>
Gauge : 4'8½".   Locos are kept at :-

Horsted Keynes.   (TQ 372293)
Sheffield Park.   (TQ 403238)

		Name	Wheels	Cyl	Builder	No.	Date	Notes
(9017)	3217	EARL OF BERKELEY	4-4-0	IC	Sdn		1938	
30064			0-6-0T	OC	VIW	4432	1943	
(30096)	96	NORMANDY	0-4-0T	OC	9E	396	1893	
(30541)	541		0-6-0	IC	Elh		1939	
(30583)	488		4-4-2T	OC	N	3209	1885	
(30847)	847		4-6-0	OC	Elh		1936	
(30928)	No.928	STOWE	4-4-0	3C	Elh		1934	
(31027)	No.27	PRIMROSE	0-6-0T	IC	Afd		1910	
(31178)	1178		0-6-0T	IC	Afd		1910	
(31263)	263		0-4-4T	IC	Afd		1905	
(31592)	No.592		0-6-0	IC	Longhedge		1901	
(31618)	1618		2-6-0	OC	Bton		1928	
31638			2-6-0	OC	Afd		1931	
(32473)	473	BIRCH GROVE	0-6-2T	IC	Bton		1898	
(32636)	72	FENCHURCH	0-6-0T	IC	Bton		1872	
(32655)	55	STEPNEY	0-6-0T	IC	Bton		1875	
33001			0-6-0	IC	Bton		1942	
(34023)	21C123	BLACKMOOR VALE	4-6-2	3C	Bton		1946	
34059		SIR ARCHIBALD SINCLAIR	4-6-2	3C	Bton		1947	
(58850)	2650		0-6-0T	OC	Bow	181	1880	
73082		CAMELOT	4-6-0	OC	Derby		1955	
75027			4-6-0	OC	Sdn		1954	
80100			2-6-4T	OC	Bton		1955	
92240			2-10-0	OC	Crewe		1958	
No.3		BAXTER	0-4-0T	OC	FJ	158	1877	
	24	THE BLUE CIRCLE	2-2-0WT	G	AP	9449	1926	
		STAMFORD	0-6-0ST	OC	AE	1972	1927	
		BRITANNIA	4wPM		H	957	1926	
6944			2w-2PMR		Wkm		1932	Dsm
		PWM 3959	2w-2PMR		Wkm	6944	1955	
		PWM 3962	2w-2PMR		Wkm	6947	1955	
			2w-2PMR		Wkm	7445	1956	Dsm
			2w-2PMR		Wkm			DsmT
			2w-2PMR		Wkm			DsmT
			2w-2PMR		Syl	14384		

BLUE CIRCLE INDUSTRIES LTD, SHOREHAM WORKS, UPPER BEEDING.
Gauge : 4'8½".   (TQ 197086)

| | | | 0-4-0DM | | RH | 260754 | 1950 | |
| | | | 0-4-0DM | | RH | 260755 | 1950 | |

J.LEMON-BURTON, "PAYNESFIELD", ALBOURNE GREEN.
Gauge : 1'3".   (TQ 243179)

212		0-4-0	OC	R.H.Morse	82	1939	
		0-6-0	OC	J.Lemon-Burton			
		4wDM		L	51721	1960	

MIDHURST WHITES LTD, MIDHURST.
Gauge : 2'6".   (SU 877212)

	4wPM		MR	6035	1937	OOU
	4wDM		MR	8981	1946	OOU
	4wDM		MR	22235	1965	OOU

P.D.FUELS LTD, CRAWLEY COAL CONCENTRATION DEPOT.
Gauge : 4'8½".   (TQ 288388)

| (D2246) | 0-6-0DM | (RSH | 7865 | 1956 |
| | | (DC | 2578 | 1956 |

P.SMITH, 46, HIDE GARDENS, RUSTINGTON.
Gauge : 2'0".   (   )

| | 4wPM | | MR | 5297 | 1931 |

SOUTHERN INDUSTRIAL HISTORY CENTRE TRUST, CHALK PITS MUSEUM, HOUGHTON BRIDGE,
AMBERLEY, ARUNDEL.
Gauge : 2'0".   (   )

		0-4-0T	OC	Decauville	1126	1950
	C.C.S.W.	4wDM		FH	1980	1936
18		4wDM		RH	187081	1937
	THAKEHAM TILES No.3	4wDM		HE	2208	1941
LOD/758054		4wDM		HE	2536	1941
	THAKEHAM TILES No.4	4wDM		HE	3653	1948

Gauge : 2'6".

| 79/190 | | 4wDM | | OK | 5926 | 1935 |

Gauge : 1'8".

| | 4wDM | | L | 33937 | 1949 |

THAKEHAM TILES LTD, ROCK LANE, THAKEHAM, near STORRINGTON.
Gauge : 2'0".   (TQ 104151)   RTC.

| | 4wDM | | HE | 5258 | 1957 | Dsm |

WEST YORKSHIRE

W.H.ARNOTT YOUNG & CO LTD, SCRAP MERCHANTS, DUDLEY HILL STATION, BRADFORD.
Yard (SE 184308) with locos for scrap occasionally present.

BRADFORD METROPOLITAN CORPORATION, ECCLESHILL INDUSTRIAL MUSEUM.
Gauge : 4'8½".   (SE 184353)

	NELLIE		0-4-0ST	OC	HC	1435	1922	

J.BUCKLER, 123, HOWDENCLOUGH ROAD, BRUNTCLIFFE, near LEEDS.
Gauge : 1'10¼".   (SE 244272)

	ALAN GEORGE		0-4-0ST	OC	HE	606	1894	
SOU/2/11/49	SHOLTO		4wDM		HE	2433	1941	

CENTRAL ELECTRICITY GENERATING BOARD.
Elland Power Station.
Gauge : 4'8½".   (SE 119219)

ELLAND No.1	AUSTIN WALKER	0-4-0DM		HC	D1153	1959	
ELLAND No.2		0-4-0DH		JF	4220010	1960	

Kirkstall Power Station, Leeds.
Gauge : 4'8½".   (SE 268345)   RTC.

No.2		0-4-0DM		HC	D637	1948	OOU

Skelton Grange Power Station, near Leeds.
Gauge : 4'8½".   (SE 334312, 337309)   RTC.

No.11	J.JOHNSON	0-6-0DH	YE	2835	1961	OOU
12	W.MARSHALL	0-6-0DH	YE	2836	1961	OOU
13	R.PEEL	0-6-0DH	YE	2837	1961	OOU

Thornhill Power Station, Dewsbury.
Gauge : 4'8½".   (SE 229201)

		0-4-0DH	JF	4210002	1949	OOU
No.1		0-4-0DH	JF	4220038	1966	OOU

Wakefield Power Station.
Gauge : 4'8½".   (SE 348198)   RTC.

	0-4-0DH	YE	2673	1958	OOU
	4wDH	TH/S	138C	1964	OOU

CROSSLEY BROTHERS (SHIPLEY) LTD, SHIPLEY STATION.
Gauge : 4'8½".   (SE 148372)

	HARRY	0-4-0ST	OC	AB	1823	1924	
		4wDM		RH	200514	1940	Dsm
0040		4wDM		RH	284838	1950	
9	BETH	4wDM		RH	425483	1958	
602		4wDM		RH	417892	1959	OOU

Also other locos for scrap occasionally present.

GREENBAT LTD, ARMLEY ROAD, LEEDS.
(SE 284335)

New GB locos under construction occasionally present.

HUNSLET ENGINE CO LTD, JACK LANE, LEEDS.
Gauge : 4'8½".   (SE 305321)

		0-4-0DM	RH	327966	1954
		0-4-0DM	HE	4630	1956
		4wDM	HE	5176	1957
D1		0-6-0DM	HC	D1186	1959
	BRIAN	0-4-0DH	HE	7406	1977

Also locos under construction & repair present.

KEIGHLEY & WORTH VALLEY LIGHT RAILWAY LTD.
Gauge : 4'8½".   Locos are kept at :-

	Haworth.	(SE 034371)
	Keighley.	(SE 066412)
	Oakworth.	(SE 052389)
	Oxenhope.	(SE 032355)

4612		0-6-0PT	IC	Sdn		1942
(5775) L89		0-6-0PT	IC	Sdn		1929
(30072) 72		0-6-0T	OC	VIW	4446	1943
41241		2-6-2T	OC	Crewe		1949
42765		2-6-0	OC	Crewe	5757	1927
43924		0-6-0	IC	Derby		1920
45212		4-6-0	OC	AW	1253	1935
47279		0-6-0T	IC	VF	3736	1924
(48431) 8431		2-8-0	OC	Sdn		1944
51218		0-4-0ST	OC	Hor	811	1901
52044		0-6-0	IC	BP	2840	1887
68077		0-6-0ST	IC	AB	2215	1947
75078		4-6-0	OC	Sdn		1956
78022		2-6-0	OC	Dar		1954
80002		2-6-4T	OC	Derby		1952
D2511 BRM 5477		0-6-0DM		HC	D1202	1961
E79962 62		4wDMR		WMD		1958
(E79964)64		4wDMR		WMD		1958
	BELLEROPHON	0-6-0WT	OC	HF	C	1874
752		0-6-0ST	IC	BP	1989	1881
1210	SIR BERKELEY	0-6-0ST	IC	MW	1210	1891
	LORD MAYOR	0-4-0ST	OC	HC	402	1893
52		0-6-2T	IC	NR	5408	1899
31	HAMBURG	0-6-0T	IC	HC	679	1903
19		0-4-0ST	OC	Hor	1097	1910
67		0-6-0T	IC	HC	1369	1919
5820		2-8-0	OC	Lima	8758	1945
1931		2-8-0	OC	VF	5200	1945
	FRED	0-6-0ST	IC	RSH	7289	1945
118	BRUSSELLS	0-6-0ST	IC	HC	1782	1945
		0-4-0ST	OC	AB	2226	1946
No.57	SAMSON	0-6-0ST	IC	RSHN	7668	1950
62		0-6-0ST	IC	RSHN	7673	1950
63		0-6-0ST	IC	RSHN	7761	1954
32	HUSKISSON	0-6-0DM		HE	2699	1944
9/6728	E.S.C. No.32	0-4-0DM		HC	D816	1954
D0226	VULCAN	0-6-0DE		(EE	2345	1956
				(VF	D226	1956
9/6773	B.S.C. No.34	0-4-0DM		HC	D990	1957
9/6777	B.S.C. No.38	0-4-0DM		HC	D995	1957
		0-4-0DM		P	5003	1961

		2w-2PMR		Wkm		1932	Dsm
		2w-2PMR		Wkm	(7573	1956?)	Dsm
		2w-2PMR		Wkm			Dsm

LEEDS CITY MUSEUM, MUSEUM OF INDUSTRY AND SCIENCE, ARMLEY MILLS, CANAL ROAD, LEEDS.
Gauge : 3'0".   (SE 275342)

| | LORD GRANBY | 0-4-0ST | OC | HC | 633 | 1902 | |

Gauge : 2'0".

No.1	BARBER	0-6-2ST	OC	TG	441	1908	
		4wPM		MR	1369	1918	
		4wDM		RH	172901	1935	
		4wDM		L	29890	1946	Dsm
		4wDM		L	35421	1949	Dsm
7		4wDM		OK			
		4wDM		MR			

Gauge : 1'6".

| | JACK | 0-4-0WT | OC | HE | 684 | 1898 + |

+ Currently under renovation elsewhere.

P.N.LOWE, ABBEY LIGHT RAILWAY, BRIDGE ROAD, KIRKSTALL, LEEDS.
Gauge : 2'0".   (SE 262357)

No.41		4wDM		MR	5859	1934
YWA L3	A.M.W. No.197	4wDM		RH	198287	1940
	LOWECO	4wDM		L	20449	1942
		4wDM		HE	2463	1944

MIDDLETON RAILWAY TRUST, HUNSLET, LEEDS.
Gauge : 4'8½".   (SE 302309)

(68153) 59		4wVBT	VCG	S	8837	1933	
1310		0-4-0T	IC	Ghd	38	1891	
Nr.385		0-4-0WT	OC	Hartmann	2110	1895	
	WINDLE	0-4-0WT	OC	EB	53	1909	
	HENRY DE LACY II	0-4-0ST	OC	HC	1309	1917	
No.6	PERCY	0-4-0ST	OC	HL	3860	1935	
		0-4-0ST	OC	P	2003	1941	
	MATTHEW MURRAY	0-4-0ST	OC	WB	2702	1943	
		0-4-0ST	OC	P	2103	1950	
H.M.Co 1	MARY	0-4-0DM		HC	D577	1932	
	COURAGE	4wDM		HE	1786	1935	
		0-4-0DM		JF	3900002	1945	
	CARROLL	0-4-0DM		HC	D631	1946	
(DE 320467	DB 965049)	2w-2PMR		Wkm	7564	1956	
		2w-2PMR		Wkm			DsmT

MINISTRY OF DEFENCE, ROYAL ORDNANCE FACTORY, CROSSGATES, LEEDS.
Gauge : 4'8½".   (SE   )

| | | 4wDM | | FH | 3918 | 1959 |

Also uses M.O.D.,A.D. locos. For full details see Section Five.

MODERNA MODERNA LTD, MODERNA MILLS, MYTHOLMROYD.
Gauge : 2'0".   (SE 016259)

| | | 4wBE | | Kershaw | 1964 |

MONCKTON COKE & CHEMICAL CO LTD, ROYSTON, near BARNSLEY.
Gauge : 4'8½".   (SE 376120)

			4wWE		GB	(1379 1935?)
			4wWE		GB 420452-2 1979	

NATIONAL COAL BOARD.
          For full details see Section Four.

PLASMOR LTD, CONCRETE BLOCK MANUFACTURERS, WOMERSLEY ROAD, KNOTTINGLEY.
Gauge : 4'8½".   (SE 503228)

		4wDH		Plasmor	c1972

PROCOR (U.K.) LTD, HORBURY JUNCTION WORKS, WAKEFIELD.
Gauge : 4'8½".   (   )

5972	OLTON HALL	4-6-0	OC	Sdn	1937

R.N.REDMAN, ARTHINGTON STATION.
Gauge : 2'11".   (SE 257445)

		4wDM		HC	D571 1932

ROCKWARE GLASS LTD, HEADLANDS GLASS WORKS, KNOTTINGLEY.
Gauge : 4'8½".   (SE 496232)

		4wDM		RH	305307 1952	Dsm
		4wDM		RH	393301 1955	

WAGON REPAIRS LTD, HAWKINCROFT ROAD, off QUARRY HILL, HORBURY, WAKEFIELD.
Gauge : 4'8½".   (SE 285183)

4061		4wDM		RH	279594 1949	OOU

YORKSHIRE WATER AUTHORITY, WESTERN DIVISION.
Bradford Depot.
Gauge : 2'0".   (SE 168333)

		4wPM		FH	3627 1953
L2		4wDM		RH	

Esholt Purification Works.
Gauge : 4'8½".   (SE 185394, 185396)   RTC.

	ELIZABETH	0-4-0ST	OC	HC	1888 1958	
398		0-4-0DM		HC	D989 1957	OOU
	PRINCE OF WALES	0-4-0DH		HE	7159 1969	

North Bierley Purification Works, Oakenshaw.
Gauge : 2'0".   (SE 179277)

		4wDM		MR	8959 1945	OOU
L8		4wDM		HE	7195 1974	

WILTSHIRE

BLUE CIRCLE INDUSTRIES LTD, WESTBURY WORKS.
Gauge : 4'8½".   (ST 885527)

4		4wDH	FH	3996 1962	OOU
1		0-4-0DH	EEV	8449 1965	
		0-6-0DH	TH	278V 1978	

BRAYDON HALL, MINETY.
Gauge : 4'8½".   (SU 027901)

| No.4 | | 0-6-0ST IC | MW | 641 1877 | |

BRITISH LELY LTD, AGRICULTURAL MACHINERY MANUFACTURERS, WOOTTON BASSETT.
Gauge : 4'8½".   (SU 059824)

| | | 4wDM | RH | 183060 1937 | OOU |

BRITISH LEYLAND U.K. LTD, LEYLAND CARS, MANUFACTURING DIVISION, SWINDON BODY PLANT,
BRIDGE END ROAD, STRATTON ST. MARGARET, SWINDON.
Gauge : 4'8½".   (SU 167864)

SBL 1		0-4-0DM	JF	4210105 1955	
		0-4-0DM	JF	4210137 1958	
SBL 3		0-4-0DH	JF	4220009 1960	
SBL 4		0-4-0DH	JF	4220017 1961	
SBL 5		0-4-0DH	JF	4220018 1961	
1		0-4-0DH	JF	4220032 1965	

BRITISH RAIL ENGINEERING LTD, SWINDON WORKS.
Gauge : 4'8½".   (SU 140846)

| D1015 | WESTERN CHAMPION | 6w-6wDH | Sdn | 1962 | Pvd |
| (D5705) | TDB 968006 | 6w-4wDE | MV | 1958 | Pvd |

Also non-B.R. locos here occasionally for renovation, etc.

COOPERS (METALS) LTD, BRIDGE HOUSE, GIPSY LANE WORKS, GIPSY LANE, SWINDON.
Gauge : 4'8½".   (SU 165860)

| 3213 | 2 | 0-4-0DM | JF | 4210082 1953 | OOU |

S.G.HIBBERD, SALISBURY.
Gauge : 2'0".   (   )

| | | 4wDM | FH | 2525 1941 | |
| | | 4wDM | FH | 3787 1956 | |

KINGSTON MINERALS LTD, MONKS PARK MINE, CORSHAM.
Gauge : 2'6".   (ST 878682)

		4wDM	RH	359169 1953	
		4wDM	RH	398101 1956	
8		4wBE	GB	2920 1958	

Battery loco is underground.

LONGLEAT LIGHT RAILWAY, LONGLEAT, WARMINSTER.
Gauge : 1'3".   (ST 808432)

1	LENKA	4w-4DHR		SL	7322	1973 +
2	CEAWLIN	2-8-0DH	S/o	SL	R7	1975
3	DOUGAL	0-6-2T	OC	SL		1970

+ 4 wheel power unit incorporates the main frame of 4wDM L 7280 1936.

MINISTRY OF DEFENCE, AIR FORCE DEPARTMENT.
Chilmark Depot.
Gauge : 4'8½".   (ST 982302)

A.M.W. No.154	0-4-0DM		JF	22604 1939
	0-4-0DH		AB	482 1963

Also uses M.O.D.,A.D. locos. For full details see Section Five.

Gauge : 2'0",   (ST 976312)

A.M.W. No.189	4wDM	RH	200512 1940
	4wDH	BD	3698 1974
	4wDH	BD	3699 1974
	4wBE	BD	3702 1974
	4wBE	BD	3703 1974
	4wBE	BD	3704 1974

Dinton Depot.
Gauge : 2'0".   (SU 008308)

A.M.W. No.165	4wDM	RH	194784 1939
A.M.W. No.194	4wDM	RH	200516 1940
	4wDH	BD	3700 1974
	4wDH	BD	3701 1974

MINISTRY OF DEFENCE, ARMY DEPARTMENT.
Tidworth Depot.

For full details see Section Five.

Warminster Depot.

For full details see Section Five.

MINISTRY OF DEFENCE, NAVY DEPARTMENT, ROYAL NAVAL ARMAMENT DEPOT, DEAN HILL.
Gauge : 4'8½".   (SU 276266)

0-4-0DM	HE	3395 1946

Also uses M.O.D.,A.D. locos. For full details see Section Five.

Gauge : 2'6".

P 6495	4wDM	HE	6651 1965
P 6496	4wDM	HE	6652 1965
P 13350	4wDM	HE	6659 1965
P 13351	4wDM	HE	6660 1965
YARD No.P 26553	4wDM	HE	7495 1977

WILLIAM PIKE LTD, SCRAP MERCHANTS, CHEMICALS ROAD, WEST WILTS INDUSTRIAL ESTATE, WESTBURY.
Yard (ST 853528) with locos for resale occasionally present.

ROGERS & COOKE (SALISBURY) LTD. QUIDHAMPTON. (Subsidiary of E.C.C. Ltd.)
Gauge : 4'8½".   (SU 114314)

		4wDM	R/R	S&H	7502	1966

SWINDON & CRICKLADE RAILWAY SOCIETY. near SWINDON.
Gauge : 4'8½".   (SU 110897)

7903	FOREMARKE HALL	4-6-0	OC	Sdn		1949
		0-4-0DM		JF	21442	1936

THAMESDOWN BOROUGH COUNCIL, ARTS & RECREATION GROUP, MUSEUMS DIVISION,
GREAT WESTERN RAILWAY MUSEUM, FARINGDON ROAD, SWINDON.
Gauge : 7'0¼".   (SU 145846)

	NORTH STAR	2-2-2	IC	Sdn		1925 +

+ Replica, incorporating parts of the original loco, RS 150 1837.

Gauge : 4'8½".

2516		0-6-0	IC	Sdn	1557	1897
(3440) 3717	CITY OF TRURO	4-4-0	IC	Sdn	2000	1903
4003	LODE STAR	4-6-0	4C	Sdn	2231	1907
9400		0-6-0PT	IC	Sdn		1947

P.S.WEAVER, NEW FARM, LAYCOCK, near CORSHAM.
Gauge : 1'9".   (ST 899691)

	0-4-0VBT	VCG	P.Weaver		1978

SECTION 2 SCOTLAND

BORDERS   187

CENTRAL   187

DUMFRIES & GALLOWAY   189

FIFE   189

GRAMPIAN   192

HIGHLANDS   192

LOTHIAN   194

ORKNEY   +

STRATHCLYDE   195

TAYSIDE   206

WESTERN ISLES   206

+ No known locomotives exist.

BORDERS

DR.R.P.JACK, THE STATION, EDDLESTON.
Gauge : 2'0".   (NT 242471)

		0-4-0T	OC	AB	1871 1925
		4wDM		HE	2927 1944

NORIT-CLASMANN LTD, AUCHENCORTH MOSS, near LEADBURN.
Gauge : 2'0".   (NT 209529)

	4wDM	MR	40S343 1969

CENTRAL

B.P. OIL GRANGEMOUTH REFINERY LTD, GRANGEMOUTH REFINERY.
Gauge : 4'8½".   (NS 942817, 947809, 952822)

No.2	144-2	0-6-0DH	EEV	D917 1965
No.3	144-3	0-6-0DH	EEV	D1232 1968
9		0-6-0DH	HE	7304 1972
10		0-6-0DH	AB	600 1976
11		0-6-0DH	AB	649 1980
12		0-6-0DH	TH	290V 1980

BRITISH ALUMINIUM CO LTD, DAVIDS LOAN, FALKIRK.
Gauge : 4'8½".   (NS 892818)

No.1	0429	0-4-0DM	JF	22902 1943	OOU
No.2		0-4-0DM	JF	4100015 1948	OOU
		4wDH	EEV	D908 1964	

CALEDONIAN PEAT PRODUCTS LTD, GARDRUM MOSS, SHIELDHILL, near FALKIRK.
Gauge : 3'0".   (NS 885757)

LO 1	4wDM	RH	394022 1956
3	4wDM	RH	398088 1956
	4wDM	RH	

WILLIAM KERR LTD, SCRAP DEALER, RIVERBANK WORKS, STIRLING.
Gauge : 2'0".   (NS 805953)

5	4wDM	RH	339105 1953	OOU

    Also other locos for scrap occasionally present.

MINISTRY OF DEFENCE, ARMY DEPARTMENT, STIRLING DEPOT.
    For full details see Section Five.

NATIONAL COAL BOARD.
    For full details see Section Four.

RICHARDSONS MOSS LITTER CO LTD, LETHAM MOSS WORKS, near AIRTH STATION.
Gauge : 2'0".   (NS 871867)

	4wDM			MR	5402 1932	
No.2	4wDM			RH	223698 1944	
		Rebuilt	Richardsons	1976	OOU
	4wDM			MR	21505 1955	
	4wDM			LB	53225 1962	OOU

SCOTTISH RAILWAY PRESERVATION SOCIETY.
Bo'Ness Site.
Gauge : 4'8½".   (NT 003817)

1313		4-6-0	OC	Motala	586 1917
	SIR JOHN KING	0-4-0ST	OC	HL	3640 1926
	CLYDESMILL No.3	0-4-0ST	OC	AB	1937 1928
No.24		0-6-0T	OC	AB	2335 1953
BO'NESS AND KINNEIL					
RAILWAY D7 PIONEER	4wDM		RH	275883 1949
P 6687		0-4-0DE		RH	312984 1951

Grahamston Goods Depot, Falkirk.
Gauge : 4'8½".   (NS 891805)

(55189) 419		0-4-4T	IC	St Rollox	1908
(62712) No.246	MORAYSHIRE	4-4-0	3C	Dar	1391 1928
(65243) No.673	MAUDE	0-6-0	IC	N	4392 1891
80105		2-6-4T	OC	Bton	1955
(D9524) 144-8		0-6-0DH		Sdn	1964
1861	ELLESMERE	0-4-0WT	OC	H(L)	244 1861
No.13 NCB 13	KELTON FELL	0-4-0ST	OC	N	2203 1876
	CITY OF ABERDEEN	0-4-0ST	OC	BH	912 1887
No.1		0-6-0T	IC	NR	5710 1902
No.3		0-6-0ST	OC	AB	1458 1916
THE WEMYSS COAL CO LTD					
No.20		0-6-0T	IC	AB	2068 1939
No.6		0-4-0CT	OC	AB	2127 1942
	4wWE			BTH	1908
	4wBE			EE	701 1926
F.82	FAIRFIELD	4wBE		EE	1131 1940
D 180/001		0-6-0DM		AB	343 1941
D 88/001		4wDM		RH	262998 1949
ROTTERDAM MIDDEN No.1					
		0-4-0DE		RH	423658 1958
No.1		0-4-0DE		RH	421439 1958
No.3		0-4-0DE		RH	423662 1958
No.6		0-4-0DE		RH	434781 1960
D 88/003		4wDM		RH	506500 1965
970213		2w-2PMR		Wkm	6049

Gauge : 3'0".

	BORROWSTOUNNESS	0-4-0T	OC	AB	840 1899

Gauge : 2'6".

No.2		4wWE		BLW	20587 1902

DUMFRIES & GALLOWAY

BRITISH NUCLEAR FUELS LTD, CHAPELCROSS WORKS, ANNAN.
Gauge : 5'4".   (NY 216695)

No.1	4301/G/0001	4wDM		RH	411320	1958
No.2	4301/G/0002	4wDM		RH	411321	1958

IMPERIAL CHEMICAL INDUSTRIES LTD, NOBEL DIVISION, DUMFRIES FACTORY,
DRUNGANS, near MAXWELLTOWN.
Gauge : 4'8½".   (NX 947751)

No.1	0-6-0DM	AB	385	1952
No.2	4wDM	RH	398613	1956 +

          + Carries plate 398163 in error.

MINISTRY OF DEFENCE, ROYAL ORDNANCE FACTORY, EASTRIGGS.

          For full details see Section Five.

NOBELS EXPLOSIVES CO LTD, POWFOOT, ANNAN. (Subsidiary of I.C.I.Ltd.)
Gauge : 4'8½".   (NY 163663)

No.1	0-4-0DM	AB	347	1941
220	0-4-0DM	AB	359	1941

Gauge : 2'6".   (NY   )

7329	4wDM	HE	7329	1973
7330	4wDM	HE	7330	1973

RICHARDSONS MOSS LITTER CO LTD, NUTBERRY WORKS, EASTRIGGS.
Gauge : 2'0".   (NY 250668)

		4wDM		RH	174532	1936	
No.1	331185	4wDM		RH	222089	1943	
No.2	LOD 758086	4wDM		RH	235641	1945	OOU
		4wDM		FH	3756	1955	
		4wDM		RH	452280	1960	OOU

SHIPBREAKERS (QUEENBOROUGH) LTD, CAIRNRYAN.
Yard (NX 062690) with locos for scrap & resale occasionally present.

FIFE

BRITISH ALUMINIUM CO LTD, BURNTISLAND.
Gauge : 4'8½".   (NT 225862)

	B.A.Co Ltd No.3	0-4-0ST OC	AB	2046	1937	OOU
No.1		0-4-0DM	JF	4210004	1949	
No.2		0-4-0DM	JF	4210045	1951	

CALEDONIAN PEAT PRODUCTS LTD, MOSS MORRAN PEAT WORKS, COWDENBEATH.
Gauge : 2'0".   (NT 179899)

	DRUMBOW EXPRESS	4wDM		L	5114	1933
		4wDM		MR	7512	1938
		4wDM		RH	198228	1940

CARNEGIE DUNFERMLINE TRUST, PITTENCRIEFF PARK, DUNFERMLINE.
Gauge : 4'8½".   (NT 086872)

No.11		0-4-0ST	OC	GH		1898

LOCHTY PRIVATE RAILWAY CO, ANSTRUTHER.
Gauge : 4'8½".   (NO 522080)

60009	UNION OF SOUTH AFRICA	4-6-2	3C	Don	1853	1937 +
		0-4-0ST	OC	P	1376	1915
10	FORTH	0-4-0ST	OC	AB	1890	1926
No.16		0-6-0ST	IC	WB	2759	1944
No.21		0-4-0ST	OC	AB	2292	1951
No.4	NORTH BRITISH	4wDM		RH	421415	1958

+ Loco usually stored at Markinch Goods Shed (NO 299013) between 'Special' workings.

MINISTRY OF DEFENCE, AIR FORCE DEPARTMENT, LEUCHARS.
Gauge : 4'8½".   (NO 453212, 456209)

	A.M.W. No.217	0-4-0DM		JF	22964	1941
	A.M.W. No.218	0-4-0DM		JF	22965	1941

Also uses M.O.D.,A.D. locos. For full details see Section Five.

MINISTRY OF DEFENCE, NAVY DEPARTMENT.
Royal Naval Armament Depot, Crombie.
Gauge : 4'8½".   (NT 04x84x)

3	YD No.94	0-4-0DM	HE	2641	1941	OOU
	YD No.107	0-4-0DM	HE	3132	1944	
	YD No.1100	4wDH	CE	B1844	1979	

Gauge : 2'6".   (NT   )

3	YARD No.3	0-4-0DM	HE	2242	1940	
4	YARD No.4	0-4-0DM	HE	2243	1940	
	YARD No.873	4wBE	GB	3537		OOU
	YARD No.874	4wBE	GB	3538		
PM 11	YARD No.875	4wBE	GB	3539		
P 9303	YARD No.1018	4wBE	VE	7667		OOU
B 2	YARD No.1066	4wBE	CE	B0483	1976	
	YARD No.1067	4wBE	CE	B0483	1976	
	YARD No.1068	4wBE	CE	B0483	1976	
	YARD No.1072	4wBE	CE	B0943	1976	
	YARD No.1073	4wDM	HE	7450	1976	
	YARD No.1074	4wDM	HE	7451	1976	
	YARD No.1075	4wDM	HE	7447	1976	
	YARD No.1076	4wDM	HE	7448	1976	

Royal Naval Dockyard, Rosyth.
Gauge : 4'8½".   (NT 108821)

	YARD No.2868	4wDM	FH	3739	1955
	YARD No.3080	4wDM	FH	3777	1956
YSM	YARD No.3081				
	JINGLING GEORDIE	4wDM	FH	3778	1956
	YARD No.3082	4wDM	FH	3779	1956

T.MUIR, SCRAP MERCHANTS, EASTER BALBEGGIE, near THORNTON.
Gauge : 4'8½".   (NT 291962)

No.61			0-4-0ST	OC		GR	272	1894	OOU
No.3			0-4-0ST	OC		AB	946	1902	OOU
No.22			0-4-0ST	OC		AB	1069	1906	OOU
No.10			0-6-0T	OC		AB	1245	1911	OOU
No.53			0-4-0ST	OC		AB	1807	1923	OOU
No.47			0-4-0ST	OC		AB	2157	1943	OOU
No.15			0-6-0ST	IC		AB	2183	1945	OOU
No.6			0-4-0ST	OC		AB	2261	1949	OOU
No.7			0-4-0ST	OC		AB	2262	1949	OOU
No.18			0-6-0ST	IC		HE	3809	1954	OOU
			4wDM			RH	235521	1945	OOU

Gauge : 2'8".

			4wDM			RH	195854	1940	Dsm

Gauge : 2'0".

554	H 643		0-4-0DM			HE			Dsm

NATIONAL COAL BOARD.

   For full details see Section Four.

NORTH EAST FIFE DISTRICT COUNCIL, CRAIGTOUN MINIATURE RAILWAY, near ST. ANDREWS.
Gauge : 1'3".   (NO 482141)

278	RIO GRANDE		2-8-0PH	S/O		SL		R8	1976

SCOTTISH GRAIN DISTILLERS LTD, CAMERON BRIDGE DISTILLERY, WINYGATES, LEVEN.
Gauge : 4'8½".   (NO 348001)

			4wDM			RH	321733	1952	

SOUTH OF SCOTLAND ELECTRICITY BOARD.
Kincardine Power Station.
Gauge : 4'8½".   (NS 923887)

5	8010	H/P 54	0-4-0DE			RH	402801	1956	OOU
9	8010	H/P 95	0-4-0DE			RH	431764	1960	
	8010		0-4-0DE			RH	449753	1961	
			0-4-0DH			AB	517	1966	OOU

Methil Power Station.
Gauge : 4'8½".   (NO 382001)

8		HP/55	0-4-0DE			RH	418597	1957	
		HP 134	0-4-0DH			RH	506399	1964	

THOS. W. WARD LTD, SHIPBREAKING YARD, INVERKEITHING.
Gauge : 4'8½".   (NT 127823)

			0-4-0DH			JF	4220003	1959	

GRAMPIAN

ABERDEEN CITY COUNCIL, DUTHIE PARK, ABERDEEN.
Gauge : 4'8½".   (NJ 940044)

| | MR. THERM | 0-4-0ST | OC | | AB | 2239 1947 |

ALFORD VALLEY RAILWAY CO LTD, ALFORD.
Gauge : 2'0".   (NJ 579159)

	SACCHARINE	0-4-2T	OC		JF	13355 1912
A.V.R. No.1	HAMEWITH	4wDM			L	3198 c1930
		4wDM			MR	5342 1931
		4wDM			MR	9215 1946

FRASERBURGH TOWN COUNCIL, FRASERBURGH MINI-RAILWAY.
Gauge : 2'0".   (NK 001659)

| 677 | KESSOCK KNIGHT | 4wDM | S/O | | LB | 53541 1963 |

CAPTAIN J. HAY, DELGATIE CASTLE, TURRIF, ABERDEEN.
Gauge : 2'0".   (   )

| | | 4wDM | | | LB | 52610 1961 |

SCOTTISH AGRICULTURAL INDUSTRIES LTD, SANDILANDS CHEMICAL WORKS, MILLER STREET, ABERDEEN.
Gauge : 4'8½".   (NJ 945063)

| 2/188 | | 4wDM | | | RH | 275880 1949 |

GEORGE WATSON & SONS, PEAT WORKS, MIDDLEMUIR WORKS, NEW PITSLIGO.
Gauge : 2'0".   (NJ 901573)

2563		4wPM			OK	2563 1929	OOU
		4wDM			LB	53162 1962	
		4wDM			LB	54781 1965	

HIGHLANDS

BRITISH ALUMINIUM CO LTD.
Invergordon Works.
Gauge : 4'8½".   (NH 713704)

| | | 4wDM | R/R | | S&H | 7505 1967 |
| | | 4wDH | R/R | | S&H | 7516 1969 |

Lochaber Works. Fort William.
Gauge : 4'8½".   (NN 126751)

RHEOLA	0-4-0DM	JF	22893	1940

Gauge : 3'0".   (NN 130753)   RTC.

W6/2-2

4wDM	RH	418770	1957

HOWARD-DORIS LTD, STROMEFERRY. (Closed)
Gauge : 4'8½".   (NG 866347)

D2007	HD-LM1	TRIBRUIT	0-6-0DM	HC	D917	1956	OOU
D2008	HD-LM2	GUINNION	0-6-0DM	HC	D918	1956	OOU

KEY & KRAMER LTD, COATING DIVISION, SALTBURN, near INVERGORDON.
Gauge : 900mm.   (NH 724702)

3

4wDH	CS	3848	1973
4wDH	CS	3849	1973
4wDH	CS	3850	1973

STRATHSPEY RAILWAY CO LTD.
Gauge : 4'8½".   Locos are kept at :-

Aviemore.   (NH 898131)
Boat Of Garten.   (NH 943189)

(45025)	5025			4-6-0	OC	VF	4570	1934
46428				2-6-0	OC	Crewe		1948
46464				2-6-0	OC	Crewe		1950
(57566)	828			0-6-0	IC	St Rollox		1899
44008				2-6w+6w-2DE		Derby		1959
Sc 79979				4wDMR		A.C.Cars		1958
No.17				0-6-0T	IC	AB	2017	1935
No.4		BALMENACH		0-4-0ST	OC	AB	2020	1936
No.1				0-4-0ST	OC	AB	2047	1937
No.9	9114/9	DAILUAINE		0-4-0ST	OC	AB	2073	1939
48	9103/48	CAIRNGORM		0-6-0ST	IC	RSH	7097	1943
No.60				0-6-0ST	IC	HE	2864	1943
No.3		CLYDE		0-6-0ST	IC	HE	3686	1948
A.M. No.147	INVERKEITHING			0-4-0ST	OC	AB	2315	1951
	CLAN	010		0-4-0DM		HC	D613	1939
	QUEEN ANNE			4wDM		RH	265618	1948
010	INVERESK PAPER	CLAN		0-4-0DE		RH	260756	1950
No.39	MADALENE			0-4-0DM		RH	304471	1951
No.2D				0-4-0DM		RH	304473	1951
3	8020			0-4-0DH		NB	27549	1956
	INTER VILLAGE 2-5			4wDM		MR	5763	1957
	SHARA			2w-2PMR		Wkm	1288	1933 +

+ Currently under renovation elsewhere.

WIGGINS TEAPE & CO LTD, SCOTTISH PULP & PAPER MILLS, CORPACH, FORT WILLIAM.
Gauge : 4'8½".   (NN 083766)

08077		0-6-0DE	Derby		1955
V 38	MARGARET	0-4-0DH	EEV	D1126	1966

LOTHIAN

D.ALLISON & M.LISTON, EDINBURGH AREA.
Gauge : 2'0".   (   )

	4wDM	HE	4440	1952

BLUE CIRCLE INDUSTRIES LTD, OXWELLMAINS CEMENT WORKS, DUNBAR.
Gauge : 4'8½".   (NT 708768)

ADAM	4wDH	S	10022	1959
No.2	4wDH	RR	10266	1967

EAST LOTHIAN DISTRICT COUNCIL, DANDERHILL CHILDRENS RECREATION PARK, DANDERHALL.
Gauge : 4'8½".   (NT 308698)

No.29	0-4-0ST OC	AB	1142	1908

LEYLAND VEHICLES LTD, MEDIUM/LIGHT VEHICLE DIVISION, BATHGATE PLANT.
Gauge : 4'8½".   (NS 978676)

248 PAUL	0-4-0DH	JF	4220011	1961

NATIONAL COAL BOARD.
> For full details see Section Four.

PRESTONGRANGE MINING MUSEUM, PRESTONGRANGE.
Gauge : 4'8½".   (NT 374737)

No.7 PRESTONGRANGE	0-4-2ST OC	GR	536	1914
17	0-4-0ST OC	AB	2219	1946
No.21	0-4-0ST OC	AB	2284	1949
	4wDM	MR	9925	1963

D.RITCHIE & C.YOUNG, EDINBURGH.
Gauge : 3'0".   (   )

	4wDM	RH	466591	1961

Gauge : 2'6".

	4wDM	RH	242916	1946
	4wDM	RH	273843	1949
10553	0-4-0DM	RH	338429	1955

Gauge : 2'0".

TERRAS	4wDM	MR	7189	1937
	4wDM	HE	2654	1942
	4wDM	RH	249530	1947

ROYAL SCOTTISH MUSEUM, CHAMBERS STREET, EDINBURGH.
Gauge : 5'0".   (NT 258734)

"WYLAM DILLY"	4w VCG	WmHedley	1813

Gauge : 1'7".

WYLAM DILLY	4w VCG	Royal Scottish Museum	

STRATHSPEY RAILWAY ASSOCIATION, c/o S.A.I. LTD, LEITH DOCKS.
Gauge : 4'8½".   (NT 276764)

No.20

                                    0-6-0ST OC   AB   1833 1924

TEXACO OIL CO LTD, GRANTON, EDINBURGH.
Gauge : 4'8½".   (NT 230773)

                M.F.P. No.3   0-4-0DM   JF   4210140 1958
                Loco on hire from the Department Of Energy.

R.A.WATSON LTD, WHITE MOSS PEAT WORKS, BOGSBANK, WEST LINTON.
Gauge : 2'0".   (NT 148597)

                                    2wPM   R.A.Watson   OOU

YOUNGS PARAFFIN, LIGHT & MINERAL OIL CO LTD, PUMPHERSTON. (Subsidiary of B.P.Ltd)
Gauge : 4'8½".   (NT 073696)   RTC.

                                    0-4-0DM   HC   D697 1950 OOU

STRATHCLYDE

JOHN R. ADAM & SONS LTD, SCRAP MERCHANT, LONDON ROAD SCRAPYARD, PARKHEAD.
Gauge : 4'8½".   (NS 617638)

   26
                                    4wDM   NB   27546 1960 OOU

ALLIS-CHALMERS GREAT BRITAIN LTD, WAGON WORKS, WISHAW.
Gauge : 4'8½".   (NS 788542)

No.5   2589   0-4-0DM   AB   472 1961

ARNOTT YOUNG & CO (SHIPBREAKERS) LTD, DALMUIR YARD.
Gauge : 4'8½".   (NS 483709)

   (D2866)   AY 1021   0-4-0DH   YE   2849 1961
                                 0-6-0DM   HE   2064 1940

ASSOCIATED BRITISH MALSTERS LTD, THE MALTINGS, STEPENDS ROAD, AIRDRIE.
Gauge : 4'8½".   (NS 792649)

   (08046)
                                 0-6-0DE   Derby   1954

AYRSHIRE RAILWAY PRESERVATION GROUP, MINNIVEY COLLIERY, DALMELLINGTON.
Gauge : 4'8½".   (NS 475073)

No.19		0-4-0ST	OC	AB	1614	1918
		0-4-0ST	OC	AB	1889	1926 +
No.10		0-4-0ST	OC	AB	2244	1947
No.25		0-6-0ST	OC	AB	2358	1954
No.1		0-4-0ST	OC	AB	2368	1955
		0-4-0DM		JF	22888	1939
		0-4-0DM		JF	4200028	1948
M/C 324	BLINKIN' BESS	4wDM		RH	284839	1950 +
144-7		0-4-0DM		AB	399	1956
68002	DB 965330	2w-2PMR		Wkm	10180	1968 +

Gauge : 2'6".

No.1		4wDM	RH	211681	1942 +
2		4wDM	RH	183749	1937 +
No.3		4wDM	RH	210959	1941 +

+ Currently stored at a secret location.

BABCOCK POWER LTD, CONSTRUCTION DIVISION, RENFREW.
Gauge : 4'8½".   (NS 493668, 495666)   RTC.

P 6686		0-4-0DE	RH	312983	1951 OOU

ANDREW BARCLAY, SONS & CO LTD, LOCOMOTIVE BUILDERS, CALEDONIA WORKS, KILMARNOCK.
Gauge : 4'8½".   (NS 425382)

1483	4wDM	HE	5306	1958
	0-4-0DM	RH	421697	1959
57	0-4-0DH	S	10057	1961
93	0-6-0DH	S	10093	1962
94	0-6-0DH	S	10094	1962
95	0-6-0DH	S	10095	1962
96	0-4-0DH	S	10096	1962
100	0-4-0DH	S	10100	1962
103	0-4-0DH	S	10103	1962
104	0-4-0DH	S	10104	1962
130	0-4-0DH	S	10130	1963
132	0-4-0DH	S	10132	1963
134	0-4-0DH	S	10134	1963
	4wDH	TH	185V	1967
	4wDH	TH	214V	1969

Locos under construction & repair usually present.

BRITISH STEEL CORPORATION, BRITISH STEEL SERVICES CENTRES, REINFORCEMENT STEEL
SERVICES DIVISION, CLYDE WORKS, MOSSEND, BELLSHILL.
Gauge : 4'8½".   (NS 746608)

4764	0-4-0DM	JF	4210126	1957

BRITISH STEEL CORPORATION, SCOTTISH DIVISION.
Clydebridge Works, Cambuslang, Glasgow.
Gauge : 4'8½".   (NS 632620)

No.7		0-4-0DH	HE	7045	1971 OOU
	DH 2803/260	0-4-0DH	AB	591	1974
	DH 2803-270	0-4-0DH	HE	7044	1971
	DH 2803-280	0-4-0DH	HE	7048	1971
	DH 2803-290	0-4-0DH	HE	7051	1971

Craigneuk Mills, Motherwell.
Gauge : 4'8½".   (NS 774551)

	7222/70/01	0-4-0DE	RH	323599	1953	
No.2	7222/70/02	0-4-0DE	RH	323605	1954	

Dalzell Works, Motherwell.
Gauge : 4'8½".   (NS 759569)

DH 8	0-4-0DH	HE	7187	1971	
DH 9	0-4-0DH	HE	7343	1973	OOU
DH 9	0-4-0DH	HE	7047	1971	Dsm
DH 11	0-4-0DH	HE	7324	1973	OOU
DH 12	0-4-0DH	HE	7345	1973	
DH 14	0-4-0DH	AB	605	1976	
DH 15	4wDH	TH	267V	1976	
DH 16	4wDH	TH	266V	1976	

Gartcosh Works, Gartcosh, Glasgow.
Gauge : 4'8½".   (NS 702678, 712678)

47/03	0-4-0DE	RH	458642	1961	OOU
47/04	0-4-0DE	RH	384140	1955	OOU
DH 11	0-4-0DH	HE	7423	1976	

Glengarnock Works, Glengarnock.
Gauge : 4'8½".   (NS 319533, 325532)

No.4	0-4-0DE	RH	423663	1958	OOU
	0-4-0DE	RH	449747	1960	OOU
No.8	0-4-0DE	RH	449748	1960	
DH 9	0-4-0DH	HE	7046	1971	
DH 10	0-4-0DH	HE	7050	1971	OOU
DH 11	0-4-0DH	AB	602	1975	

Hallside Works, Hallside, Cambuslang.   (Closed)
Gauge : 4'8½".   (NS 664604, 666604)

1	72/21/41	0-4-0DE	RH	349088	1954	Dsm
2	72-21-42	0-4-0DE	RH	349086	1954	OOU

Lanarkshire Works, Flemington, Motherwell.   (Closed)
Gauge : 4'8½".   (NS 768563)

DE 1	0-4-0DE	RH	395295	1956	Dsm
DH 7	0-4-0DH	HE	7184	1970	Dsm

Mossend Engineering Works, Mossend, Bellshill.
Gauge : 4'8½".   (NS 745608)   RTC.

	4wDM	RH	299097	1950	OOU
DE 5	0-4-0DE	RH	449751	1960	OOU

Ravenscraig Works, Motherwell.
Gauge : 4'8½".   (NS 778562)

	0-4-0DE	RH	395297 1956	+ Dsm
	0-4-0DE	RH	458645 1961	+ Dsm
	0-4-0DE	RH	477813 1962	+ Dsm
32	0-6-0DH	RR	10286 1969	
33	0-6-0DH	RR	10287 1969	
39	0-6-0DH	RR	10289 1970	
42	0-6-0DH	S	10082 1961	
46	0-6-0DH	S	10071 1961	
47	0-6-0DH	S	10050 1961	
	0-6-0DH	RR	10213 1964	
	0-6-0DH	RR	10216 1965	
242	0-6-0DH	TH	242V 1972	
243	0-6-0DH	TH	243V 1972	
244	0-6-0DH	TH	244V 1972	
245	0-6-0DH	TH	245V 1973	
246	0-6-0DH	TH	255V 1975	
247	0-6-0DH	TH	253V 1975	
248	0-6-0DH	TH	254V 1975	
249	0-6-0DH	TH	256V 1975	
250	0-6-0DH	TH	252V 1974	
DH 8	0-4-0DH	HE	7049 1971	
DH 401	0-4-0DH	HE	7262 1972	OOU
DH 402	0-4-0DH	HE	7263 1972	Dsm
DH 403	0-4-0DH	HE	7322 1972	
DH 404	0-4-0DH	HE	7323 1972	
DH 405	0-4-0DH	AB	586 1973	OOU
DH 406	0-4-0DH	AB	587 1973	
DH 407	0-4-0DH	AB	588 1973	
DH 408	0-4-0DH	AB	589 1973	OOU
DH 409	0-4-0DH	AB	590 1974	OOU
DH 410	0-4-0DH	AB	595 1975	
DH 411	0-4-0DH	AB	610 1976	
DH 412	0-4-0DH	AB	611 1976	
DH 601	0-6-0DH	HE	6974 1969	
602	0-6-0DH	HE	7190 1970	
	Rebuilt	AB	6808 1981	
DH 63	0-6-0DH	HE	7188 1970	OOU
DH 64	0-6-0DH	HE	7061 1971	
DH 65	0-6-0DH	HE	7280 1972	
DH 66	0-6-0DH	HE	7306 1972	
DH 607	0-6-0DH	HE	7353 1973	OOU
DH 608	0-6-0DH	HE	7354 1974	
DH 609	0-6-0DH	AB	596 1975	
DH 610	0-6-0DH	AB	597 1975	
DH 611	0-6-0DH	AB	598 1975	
DH 612	0-6-0DH	AB	599 1975	
	2w-2DMR	Wkm	8856 1961	
	2w-2PMR	Wkm	10482 1970	
	4wDMR	Donelli Spa	163 1979	
3	0-4-0WE	RSH	7852 1957	
4	0-4-0WE	RSH	8207 1961	
	0-4-0WE	RSH	8298 1962	
	0-4-0WE	(GECT	5370 1973	
		(BD	3684 1973	
	0-4-0WE	GECT	5574 1979	

+ Frames in use as Barrier Wagons.

Gauge : 6'9½".   Used in Strip Mill.

	4wBE	GB	6064 1962

BRITISH STEEL CORPORATION, SHEFFIELD DIVISION, PROFIT CENTRES, B.S.C. FORGES,
FOUNDRIES & ENGINEERING, FULLWOOD FOUNDRIES, NEW STEVENSTON WORKS,
NEW STEVENSTON, MOTHERWELL.
Gauge : 4'8½".   (NS 758600)

	MOSSEND No.1	0-4-0DE		RH	408302	1957
		4wBE		CE	B1571	1978

BRITISH STEEL CORPORATION, TUBES DIVISION, GENERAL TUBES WORKS GROUP.
Clydesdale Works, Bellshill.
Gauge : 4'8½".   (NS 753596, 753598)

E	0-4-0DE		RH	412717	1958
F	0-4-0DE		RH	461961	1965
G	0-4-0DE		RH	461963	1967
No.5	0-4-0DE		RH	434775	1960	OOU

Imperial Works, Martyn Street, Airdrie.
Gauge : 4'8½".   (NS 752648)

No.5	0-4-0DM	JF	4210088	1953
	0-4-0DH	EEV	3991	1970

BUTLINS LTD, HEADS OF AYR HOLIDAY CAMP.
Gauge : 2'0".   (NS 299187)

145	C.P.HUNTINGTON	4-2-4DH S/O	Chance	

CALEDONIAN PEAT PRODUCTS LTD, RYEFLAT MOSS, CARSTAIRS.
Gauge : 3'0".   (NS 953478)

4	4wDM	MR	60S362	1968
5	4wDM	MR	60S382	1969

GEORGE H. CAMPBELL & CO (GLASGOW) LTD, IRON & STEEL MERCHANTS, ATLAS WORKS, AIRDRIE.
Gauge : 4'8½".   (NS 774653)

	4wDM	FH	3700	1955	Dsm

Also other locos for scrap occasionally present.

CITY OF GLASGOW DISTRICT COUNCIL, COPLAWHILL TRANSPORT MUSEUM, 25, ALBERT DRIVE,
GLASGOW.
Gauge : 4'8½".   (NS 581632)

		0-4-0VBT	VCG	Chaplin	2368	1885
123		4-2-2	IC	N	3553	1886
103		4-6-0	OC	SS	4022	1894
256	(62469) GLEN DOUGLAS	4-4-0	IC	Cowlairs		1913
9		0-6-0T	OC	NB	21521	1917
No.1		0-6-0F	OC	AB	1571	1917
49	(62277) GORDON HIGHLANDER	4-4-0	IC	NB	22563	1920

Gauge : 4'0".

	SUBWAY CAR No.1	4w-4wRER	Oldbury	1896

CLYDE CEMENT CO LTD, (RIBBLESDALE GROUP OF COMPANIES), GARTSHERRIE WORKS.
Gauge : 4'8½".   (NS 730661)

38	0-6-0DH	RR	10217	1965
No.5	4wDH	RR	10230	1965

CLYDESIDE CONSTRUCTIONAL CO LTD. PLANT DEPOT. 11. RANFURLY CASTLE TERRACE.
BRIDGE OF WEIR.
Gauge : 2'0".   (NS 396648)

	4wDM	RH	179005	1936	
CCC 51	4wDM	MR	7330	1938	

COSTAIN CONCRETE CO LTD. COLTNESS FACTORY. NEWMAINS.
Gauge : 4'8½".   (NS 823553)

COSTAIN 1	4wDM	RH	408494	1957	OOU
COSTAIN 2	4wDM	RH	326065	1952	OOU
	4wDH	HE	7430	1977	

ESSO PETROLEUM CO LTD. BOWLING, near DUMBARTON.
Gauge : 4'8½".   (NS 436738)

ESSO	4wDH	RR	10199	1964

P.FIELDING, c/o WILLIAM DeVENNY & SON LTD. CAMSAIL WORKS. ROSNEATH.
Gauge : 4'8½".   (NS 258826)

	4wVBT VCG	S	9593	1954	Dsm

ALEX FINDLAY & CO LTD. BRIDGE BUILDERS. PARKNEUK WORKS. MOTHERWELL.
Gauge : 4'8½".   (NS 749578)

	0-4-0DM	JF	4210005	1949
	0-4-0DM	AB	444	1958

FIRTH BROWN LTD. PARKHEAD FORGE.
Gauge : 4'8½".   (NS 625645)   RTC.

A5	0-4-0DH	RH	504547	1963	OOU
No.6	0-4-0DH	JF	4220026	1963	OOU

HEWDEN (CONTRACTS) LTD. MOODIESBURN PEAT WORKS.
Gauge : 2'0".   (NS 705708)

	4wDM	MR	9846	1952

IMPERIAL CHEMICAL INDUSTRIES LTD. ORGANICS DIVISION. STEVENSTON WORKS.
Gauge : 4'8½".   (NS   )

1	4wDH	S	10049	1961
2	4wDH	S	10052	1961

IMPERIAL CHEMICAL INDUSTRIES LTD. PETROCHEMICALS DIVISION. ARDEER WORKS.
Gauge : 4'8½".   (NS 278414)

	0-4-0DH	AB	551	1968

IRELAND FERROUS LTD. SCRAP DEALERS. LANGLOAN WORKS. COATBRIDGE.
Gauge : 4'8½".   (NS 725643)   RTC.

	0-4-0DM	HE	3125	1946	Dsm

LILLEY/WADDINGTON LTD. PLANT DEPOT, CHARLES STREET, SPRINGBURN, GLASGOW.
Yard (NS 609667) with locos present between contracts.

See entry under Essex for loco fleet details.

SIR WILLIAM LITHGOW, HARDRIDGE RAILWAY, KILMACOLM.
Gauge : 2'0".   (NS 313675)

		4wPM	MR	2097	1922
		4wPM	MR	2171	1922
No.2		4wDM	MR	8700	1941

R.G.McCARROL LTD. CROWHILL ROAD, BISHOPBRIGGS, GLASGOW.
Gauge : 2'0".   (NS 609697)

| | | 4wDM | MR | 8564 | 1940 |

Here for overhaul only.

MINISTRY OF DEFENCE, ARMY DEPARTMENT, INCHTERF GUN RANGE, near KIRKINTILLOCH.

For full details see Section Five.

MINISTRY OF DEFENCE, NAVY DEPARTMENT, ROYAL NAVAL ARMAMENT DEPOT, GIFFEN, BEITH.
Gauge : 4'8½".   (NS 345510)

BE 1		4wDM	RH	221561	1943
BE 3		4wDM	RH	221564	1943
	YARD No.AK 1	4wDM	RH	224352	1943
M 3571	YARD No.AC/118	0-4-0DM	AB	366	1943

MINISTRY OF DEFENCE, ROYAL ORDNANCE FACTORY, BISHOPTON.
Gauge : 4'8½".   (NS 438703)

868		0-4-0DM	AB	338	1939	OOU
		4wDM	FH	3894	1958	
		0-4-0DH	RSHD/WB	8364	1962	
	LOCO No.9	4wDH	TH	277V	1977	

Gauge : 2'6".   (NS   )

(9)		4wBE	WR	2334	1942	+ Dsm
25		4wBE	WR	6214	1959	
29		4wBE	WR	1614	1940	OOU
(36)	36/1335	4wBE	WR	6215	1959	
38		4wBE	WR	5766	1957	OOU
35		4wBE	GB	898	1939	OOU
36		4wBE	GB	899	1939	Dsm
37		4wBE	GB	988	1940	Dsm
39		4wBE	GB	990	1940	Dsm
41		4wBE	GB	992	1940	Dsm
(44)		4wBE	GB			Dsm
46		4wBE	GB	3166		Dsm
48		4wBE	GB	3168		Dsm
53		4wBE	GB	3815		
54		4wBE	GB	3816		
55		4wBE	GB	3817		
1		4wBE	BV	562	1970	
2		4wBE	BV	563	1970	
3		4wBE	BV	307	1968	
4		4wBE	BV	308	1968	
5		4wBE	BV	690	1974	
6		4wBE	BV	694	1974	

7		4wBE	BV	611	1972
8		4wBE	BV	695	1974
9		4wBE	BV	696	1974
10		4wBE	BV	697	1974
11		4wBE	BV	564	1970
12		4wBE	BV	565	1970
13		4wBE	BV	691	1974
15		4wBE	BV	309	1968
16	38/540	4wBE	BV	1142	1976
17		4wBE	BV	614	1972
18		4wBE	BV	1143	1976
19		4wBE	BV	613	1972
20		4wBE	BV	612	1972
25		4wBE	BV	692	1974
26		4wBE	BV	693	1974
28		4wBE	BV	306	1968
38	38/518	4wBE	BV	698	1974
39		4wBE	BV	608	1971
40		4wBE	BV	699	1974
42		4wBE	BV	700	1974
43		4wBE	BV	701	1974
45		4wBE	BV	703	1974
46		4wBE	BV	609	1971
47	38/516	4wBE	BV	702	1974
48		4wBE	BV	610	1971
1		4wDM	HE	7513	1976
2		4wDM	HE	7512	1976
3		4wDM	HE	8827	1979
4		4wDM	HE	8828	1979
5		4wDM	HE	8829	1979
6		4wDM	HE	8964	1979
7		4wDM	HE	8965	1979
8		4wDM	HE	8830	1979
9		4wDM	HE	8967	1980
10		4wDM	HE	8966	1980
11		4wDM	HE	8968	1980
12		4wDM	HE	8969	1980

+ Frame used as a battery stand.

MOTHERWELL BRIDGE & ENGINEERING CO LTD, MOTHERWELL.
Gauge : 4'8½".   (NS 747575)

802		0-4-0DH	RH	457299	1962

MOTHERWELL MACHINERY & SCRAP CO LTD.
Glen Yard, Motherwell.
Gauge : 4'8½".   (NS 771559)

FGM 88D		0-4-0DM	JF	4200009	1947	OOU

Inshaw Works, Motherwell.
Gauge : 4'8½".   (NS 776549)

		0-4-0DM	JF	23010	1945
DH 10		0-4-0DH	HE	7043	1971

Also other locos for scrap occasionally present.

NATIONAL COAL BOARD.

For full details see Section Four.

NOBELS EXPLOSIVES CO LTD, ARDEER WORKS. (Susidiary of I.C.I.Ltd.)
Gauge : 2'6".   (NS 303383)

1		4wDM	RH	422569	1959	
3		4wDM	RH	191655	1938	
5		4wDM	RH	191656	1938	
8		4wDM	RH	211689	1942	Dsm
11		4wDM	RH	323583	1951	
12		4wDM	RH	310042	1951	
13		4wDM	RH	323595	1951	
18		4wDM	RH	381717	1955	Dsm
20		4wDM	RH	412420	1958	
21		4wDM	AB	554	1970	
22		4wDM	AB	555	1970	
23		4wDM	AB	556	1970	
24		4wDM	AB	557	1970	
25		4wDM	AB	562	1971	
26		4wDM	AB	560	1971	
27		4wDM	AB	561	1971	
		4wDM	RH	174139	1935	+
		4wDM	RH	221602	1943	+
		4wDM	RH	354043	1953	+
31		4wDM	MR	101GA022	1969	Dsm
32		4wDM	MR	101GA023	1969	
33		4wDM	MR	101GA024	1969	OOU
34		4wDM	MR	101GA025	1969	Dsm

    + These are numbered 28, 29 & 30; which is which is unknown.

EDMUND NUTTALL LTD, CIVIL ENGINEERS, PLANT DEPOT, KILSYTH.
Gauge : 3'0".   (NS 702776)

	BETTY		4wDM	MR	60S393	1970
EN 62			4wBE	GB	420245/1	1970
EN 63			4wBE	GB	420245/2	1970
EN 86			4wDH	MR	110U081	1970
EN 87			4wDH	MR	110U080	1970
			4wDH	MR	110U082	1970
EN 89	No.2		4wDH	MR	110U083	1970

Gauge : 2'6".

EN 78	286		4wBE	SIG	706-716	1976
EN 79	287		4wBE	SIG	706-717	1976

Gauge : 2'0".

EN 52		4wBE	CE	5590	1969
EN 53		4wBE	CE	5590	1969
54		4wBE	CE	5590	1969
EN 55		4wBE	CE	5590	1969
EN 56		4wBE	CE	5590	1969
EN 57		4wBE	CE	5590	1969
EN 58		4wBE	CE	5590	1969
EN 59		4wBE	CE	5590/8	1969
EN 60		4wBE	CE	5590/9	1969
EN 43		4wBE	CE	5943	1972
EN 61		4wBE	CE	5949	1972
EN 70		4wBE	CE	5949A	1972
65		4wBE	CE	5949B	1972
67		4wBE	CE	5949C	1972
EN 71		4wBE	CE	5949D	1972
		4wBE	CE	5949E	1972
		4wBE	CE	5949F	1972
		4wBE	CE	5949G	1972
		4wBE	CE	5949H	1972

70	4wBE		CE	BO129	1973
	4wBE		CE	BO951A	1976
75	4wBE		CE	BO951B	1976
	4wBE		CE	BO951C	1976
77	4wBE		CE	BO951D	1976
EN 81	4wDM	RH	7002/0467/2	1966
	4wDM	RH	7002/0467/6	1966
EN 83	4wDM	RH	7002/0467/8	1966
	4wDH	MR	110U133	1973
EN 92	4wDH	MR	110U134	1973

Locos present in yard between contracts.

PEAT DEVELOPMENT LTD, KIRKFIELDBANK ROAD, DOUGLAS WATER.
Gauge : 2'0".   (NS 875382)

	4wDM		MR	7066 1938
	4wDM		MR	8863 1944

REDPATH ENGINEERING LTD, STRUCTURAL DIVISION, WESTBURN WORKS, CAMBUSLANG, GLASGOW.
(Subsidiary of British Steel Corporation.)
Gauge : 4'8½".   (NS 658608)

6					
		0-4-0DE		YE	2529 1953
	TEES-SIDE No.8	4wDH		TH	115V 1962

RICHARDSONS MOSS LITTER CO LTD, FANNYSIDE WORKS, LONGRIGGEND.
Gauge : 2'6".   (NS 801739)

4wDM		LB	55870 1968

J.P.ROBERTSON, SCRAP METAL MERCHANT, GATEHEAD, near KILMARNOCK.
Gauge : 2'6".   (NS 392364)

4wDM		HE	6057 1962	OOU

ROYAL SCOTTISH MUSEUM, BIGGAR GASWORKS MUSEUM, BIGGAR.
Gauge : 2'0".   (NT 039376)

5				
	0-4-0T	OC	AB	988 1903

SCOTTS ENGINEERING CO LTD, CARTSBURN DOCKYARD, GREENOCK.
Gauge : 4'8½".   (NS 285756)

4wBE		CE	2556 1954
4w-4wBE		CE	3338 1954

SHELL U.K. OIL, NORTH CRESCENT, ARDROSSAN.
Gauge : 4'8½".   (NS 230428)   R.S.W.

0-4-0F	OC	AB	1952 1928

SHIPBREAKING INDUSTRIES LTD, FASLANE.
Gauge : 4'8½".   (NS 243897)

5/2				
	0-4-0DM		AB	350 1941
	4wDM		RH	312430 1951

SIR WILLIAM ARROL, N.E.I. CLARKE CHAPMAN CRANES LTD, DALMARNOCK IRONWORKS, PARKHEAD.
Gauge : 4'8½".   (NS 617637)

201515	ARROL	4wDM	RH	458958 1961

SOUTH OF SCOTLAND ELECTRICITY BOARD, BARONY POWER STATION, AUCHINLECK.
Gauge : 4'8½".   (NS 525217)

7	8030	0-4-0DH	AB	515 1966	OOU
	8030	0-4-0DH	AB	516 1966	

STRATHCLYDE PASSENGER TRANSPORT EXECUTIVE, GOVAN WORKSHOPS, GOVAN ROAD.
Gauge : 4'0".   (NS 555655)

L 3	4wBE	WR	583 1927
L 2	4wBE	CE	B0965A 1977
	4wBE	CE	B0965B 1977
	4wBE	NNM	78101E 1979

STRATHCLYDE SAWMILLS LTD, CASTLE ROAD, DUMBARTON.
Gauge : 4'8½".   (NS 403748)

4wDM	MR	5762 1956

TALBOT SCOTLAND LTD, LINWOOD, PAISLEY.   (Closed)
Gauge : 4'8½".   (NS 457640)

1	0-4-0DM	JF	4210070 1952	OOU
2	0-4-0DM	JF	4210134 1957	OOU

R.B.TENNENT LTD, WHIFFLET FOUNDRY, COATBRIDGE.
Gauge : 4'8½".   (NS 738643)   R.S.W.

JOHN	4wVBT VCG	S	9561 1953	Dsm
RANALD	4wVBT VCG	S	9627 1957	
ROBIN	4wVBT VCG	S	9628 1957	
DENIS	4wVBT VCG	S	9631 1958	

WAGON REPAIRS LTD, REPARCO WORKS, DOUGLAS PARK LANE, HAMILTON.
Gauge : 4'8½".   (NS 712559)

4wDM	RH	284835 1950	OOU
4wDM	RH	321731 1952	

JOHN WALKER & SONS LTD, WHISKY BLENDERS, HURLFORD, near KILMARNOCK.
Gauge : 4'8½".   (NS 456359)

JOHNNIE WALKER	4wDM	RH	417890 1959

JOHN WILLIAMS (WISHAW) LTD, EXCELSIOR IRON & STEEL WORKS, WISHAW.
Gauge : 4'8½".   (NS 777552)

0-4-0DM	JF	4210138 1958

WILSON KINMOND & MARR LTD, PLANT DEPOT, BURNFIELD ROAD, THORNLIEBANK, GLASGOW.
Gauge : c2'0".   (NS 555599)

0-4-0BE	WR	7662 1975
0-4-0BE	WR	7663 1975

Locos present in yard between contracts.

JOHN WOODROW (BUILDERS) LTD, PLANT DEPOT, MAIN STREET, BRIDGE OF WEIR.
Gauge : 2'0".   (NS 390656)

		4wDM		MR	9982	1954
		4wDM		MR	22012	1958

TAYSIDE

BRECHIN RAILWAY PRESERVATION SOCIETY, BRECHIN, near MONTROSE.
Gauge : 4'8½".   (NO 603603)

		0-4-0ST	OC	AB	807	1897
17		0-4-0ST	OC	AB	1863	1926
		0-6-0ST	OC	HE	2879	1943
N.C.B. 6		0-6-0ST	IC	WB	2749	1944
Y.S.M. YARD No.DY 326		4wDM		FH	3743	1955
144-6		0-4-0DM		RH	421700	1959

JOHN DEWAR & SONS LTD, WHISKEY BLENDERS, INVERALMOND, PERTH.
Gauge : 4'8½".   (NO 097258)

		4wDM		RH	458957	1961

WESTERN ISLES

STORNOWAY WATERWORKS, ISLE OF LEWIS.
Gauge : 2'0".   (NB 410375)

		4wPM		(MR?)		Dsm

SECTION 3 WALES
ADRAN 3 CYMRU

CLWYD	209
DYFED	213
GWENT	216
GWYNEDD	219
MID GLAMORGAN	227
POWYS	228
SOUTH GLAMORGAN	228
WEST GLAMORGAN	232

CLWYD

L.S.BECKETT LTD, WHIXALL MOSS PEAT WORKS, near BETTISFIELD.
Gauge : 2'0".   (SJ 478367, 503368)   RTC.

	4wPM	MR	1934	1919	OOU
	4wDM	MR	4023	1926	OOU
	4wDM	RH	171901	1934	Dsm
	4wDM	RH	191679	1938	OOU

BRITISH STEEL CORPORATION, SHOTTON WORKS SPORTS CLUB, SHOTTON, DEESIDE.
Gauge : 4'8½".   (SJ 309692)

41	0-4-0DH	HC	D1020	1956	Pvd

BRITISH STEEL CORPORATION, WELSH DIVISION, SHOTTON WORKS, DEESIDE.
Gauge : 4'8½".   (SJ 305719, 306703, 311695, 314699)

2	0-6-0DH		NB	27649	1956	+ Dsm
No.4	0-6-0DH		NB	27411	1954	@ OOU
	0-6-0DH		NB	27412	1954	+ Dsm
	0-6-0DH		NB	27749	1957	+ Dsm
	0-6-0DE		YE	2527	1953	@ Dsm
No.5	0-6-0DH		NB	28034	1960	+ Dsm
10	0-6-0DH		EEV	D3985	1970	@ OOU
12	0-6-0DH		EEV	5352	1971	
13	0-6-0DH		EEV	D3998	1970	@ OOU
14	0-6-0DH		EEV	D3999	1970	@ OOU
15	0-6-0DH		EEV	5353	1971	
16	0-6-0DH		GECT	5391	1973	
17	0-6-0DH		GECT	5392	1973	
18	0-6-0DH		GECT	5396	1975	@ OOU
19	0-6-0DH		GECT	5397	1975	@ OOU
20	0-6-0DH		GECT	5398	1975	
21	0-6-0DH		GECT	5399	1975	
22	0-6-0DH		GECT	5400	1975	
23	0-6-0DH		GECT	5401	1975	@ OOU
24	0-6-0DH		GECT	5402	1975	
27	0-6-0DM		HE	5671	1960	@ OOU
29	0-6-0DM		HE	5132	1957	@ OOU
37	0-4-0DH		HC	D762	1952	@ OOU
38	0-4-0DH		HC	D763	1952	@ OOU
39	0-4-0DH		HC	D764	1952	
		Rebuilt	HB		1972	@ OOU
40	0-4-0DH		HC	D927	1955	
		Rebuilt	HB		1972	@ OOU
42	0-4-0DH		HC	D1409	1969	@ OOU
43	0-4-0DH		GECT	5575	1979	@ OOU
44	0-4-0DH		GECT	5576	1979	@ OOU
45	0-4-0DH		GECT	5577	1979	OOU
106	0-6-0DH		NB	27768	1959	@ OOU
109	0-6-0DH		NB	27870	1960	@ OOU
111	0-6-0DH		NB	28045	1961	@ OOU
112	0-6-0DH		NB	28046	1962	@ OOU
115	0-6-0DH		AB	489	1964	@ OOU
117	0-6-0DH		AB	507	1966	@ OOU
118	0-6-0DH		AB	508	1966	@ OOU
	0-6-0DH		NB	27402	1954	+ Dsm
(M 51692)	4w-4wDMR		Cravens		1960	
(M 51691)	4w-4wDMR		Cravens		1960	

1		0-4-0WE	EE	5360	1971	OOU
2		0-4-0WE	EE	5361	1971	OOU

+ Converted to slag bogies, etc.
@ Stored, awaiting disposal, at Hot Strip Mill. (SJ 310703)

Gauge : 2'6".   (SJ 302704, 305705)   Cold Strip Mill.

A		4wBE	GB	2187	1949	Dsm
(B)		4wBE	GB	2186	1949	
C		4wBE	GB	2974	1959	Dsm
D		4wBE	GB	2973	1959	OOU
E		4wBE	GB	2643	1955	OOU
F		4wBE	GB	2188	1949	Dsm
(G)		4wBE	GB	6114	1965	Dsm
H	6	4wBE	GB	420155	1968	Dsm
No.1	A	4wBE	GB	420330/1	1972	
No.2		4wBE	GB	420330/2	1972	Dsm
No.3		4wBE	WR	Q7628	1976	
4		4wBE	WR	Q7807	1976	
No.5		4wBE	WR	Q7808	1976	
6		4wBE	WR	Q7809	1976	
48		0-6-0DM	HB	D1417	1971	
49		0-6-0DM	HB	D1418	1971	
50		0-6-0DM	HB	D1419	1971	

CENTRAL ELECTRICITY GENERATING BOARD, CONNAH'S QUAY POWER STATION.
Gauge : 4'8½".   (SJ 287704)

10	D.ARTIS	0-4-0DH	JF	4210001	1949	
1		0-4-0DH	JF	4210077	1952	
2		0-4-0DM	JF	4210090	1954	
3		0-4-0DM	JF	4210069	1952	OOU

COURTAULDS LTD, GREENFIELD WORKS, HOLYWELL JUNCTION.
Gauge : 4'8½".   (SJ 202775)

	4wDH	RR	10251	1966
	4wDH	RR	10252	1966

C.C.CRUMP & CO, DENTITH WAGON REPAIR WORKS, CONNAH'S QUAY.
Gauge : 4'8½".   (SJ 296697)

MARIE	0-4-0DH	JF	22882	1939
M.O.P. No.7	0-4-0DM	JF	4210144	1958

P.DOBBINS (CHESTER) LTD, SCRAP METAL PROCESSORS, SALTNEY FERRY WORKS.
Gauge : 4'8½".   (SJ 367652)

3	0-4-0ST OC	P	2084	1948	OOU

GLYN QUARRY ENTERPRISES, WYNNE QUARRY, GLYN CERIOG.
Gauge : 2'0".   (SJ 199379)

	4wDM	MR	8720	1941
BEAR	4wDM	RH	339209	1952

GUEST KEEN & NETTLEFOLDS LTD. BRYMBO STEELWORKS.
Gauge : 4'8½".   (SJ 296536)

	ESMOND	0-6-ODE	YE	2604	1955		
	HOPE	0-6-ODE	YE	2632	1957		
	JOHN	0-6-ODE	YE	2658	1957		
	SPENCER	0-6-ODE	YE	2659	1957		
No.38		0-6-ODE	YE	2792	1961		
	WILLIAM	0-6-ODE	YE	2800	1962		
	NEVILLE	0-4-ODE	YE	2853	1961		
	AUSTIN	0-4-ODE	YE	2855	1961		
	CHARLES	0-4-ODE	YE	2858	1961		
	EMRYS	0-6-ODE	YE	2867	1962		
No.8		0-6-ODE	YE	2884	1962		
	WINSTON	0-6-ODE	YE	2942	1965		

HOLYWELL-HALKYN MINING & TUNNEL CO LTD. OLWYN GOCH MINE. RHYDYMWYN.
(Subsidiary of Courtaulds Ltd.)
Gauge : 1'10½".   (SJ 203677)   (Underground)

	4wDM	RH	182138	1936		
	4wDM	RH	183727	1937		
	4wDM	RH	221593	1943		Dsm
	4wDM	RH	226309	1943	+	Dsm
	4wDM	RH	331250	1952		OOU
	4wDM	RH	354029	1953		OOU
744	4wBE	WR	744	1929		OOU
773	4wBE	WR	773	1930		OOU
898	4wBE	WR	898	1935		
899	4wBE	WR	899	1935		
	0-4-OBE	WR	1080	1937		
	0-4-OBE	WR	5311	1955		

          + Dumped at Pen Y Bryn shaft, Halkyn.

I.B.JOLLY, 1, LLEWELLYN DRIVE, BRYN-Y-BAAL, near MOLD.
Gauge : 4'8½".   (SJ 261647)

3	(9036)	2w-2PMR	Wkm	8196	1958	+	

Gauge : 60cm.

LR 2718	4wPM	MR	997	1918	+	Dsm
LR 2832	4wPM	MR	1111	1918	+	
	4wPM	MR	c1916			
No.21	4wPM	MR	c1920			
	4wDM	MR	5852	1933	+	Dsm
	4wDM	MR	8723	1941	+	Dsm
	2w-2PM	Wkm	3030	1941	+	

Gauge : 1'11½".

	4wPM	MR	6013	1931		
No.12/14	4wPM	L	30233	1946		
	4wDM	MR	20558	1955	+	

Gauge : 2'8".

4wPM	MR	c1918		
	Rebuilt	P.Finnie		+ Dsm

          + These locos are elsewhere for renovation.

LLANGOLLEN STEAM RAILWAY SOCIETY, LLANGOLLEN STATION.
Gauge : 4'8½".   (SJ 211423, 214422)

7754		0-6-0PT	IC	NB	24042	1930
1243	RICHBORO	0-6-0T	OC	HC	1243	1917
3		0-6-0ST	IC	P	1567	1920
5459 1	BURTONWOOD BREWER	0-6-0ST	IC	K	5459	1932
D 2	ELISEG	0-4-0DM		JF	22753	1939
		0-4-0DM		JF	4000007	1947
D 4		0-4-0DM		HC	D1012	1956
D 1		0-4-0DH		NB	27734	1958

LLOYDS SPAR QUARRIES (MOLD) LTD, HENDRE MINE, EFAIL PARCY, near MOLD.
Gauge : 2'0".   (SJ 185676)   (Underground)

0-4-0BE	WR	N7661	1974

MOSTYN DOCK & TRADING LTD, MOSTYN DOCK.
Gauge : 4'8½".   (SJ 156811)

1	0-4-0DE	YE	2627	1956
2	0-4-0DE	YE	2819	1960

NATIONAL COAL BOARD.
For full details see Section Four.

THE RHYL MINIATURE RAILWAY, MARINE LAKE LEISURE PARK, RHYL.
Gauge : 1'3".   (SJ 002806)

101		JOAN	4-4-2	OC	A.Barnes	101	1920
		RAILWAY QUEEN	4-4-2	OC	A.Barnes	102	1922 +
104	No.1	BILLIE	4-4-2	OC	A.Barnes	104	1928 +
No.106		BILLY	4-4-2	OC	A.Barnes	106	1934
			4w-4wPM				1964

+ Currently stored elsewhere.

SYNTHITE LTD, CHEMICAL MANUFACTURERS, MOLD.
Gauge : 4'8½".   (SJ 233649)

4wDM	MR	1944	1919

TILCON LTD, DYSERTH QUARRY.
Gauge : 2'7".   (SJ 063790)   RTC.

4	4wDM	MR	5025	1929	Dsm
	4wDM	RH	296091	1949	OOU

TUNNEL CEMENT LTD, PADESWOOD HALL CEMENT WORKS, BUCKLEY.
Gauge : 4'8½".   (SJ 290622)

	0-4-0DE	YE	2854	1961
No.6	4wDH	RR	10235	1965
No.7	4wDH	RR	10276	1967

DYFED

AMOCO (U.K.) LTD, HERBRANDSTON, MILFORD HAVEN.
Gauge : 4'8½".   (SM 888085)

	0-4-ODH		YE	2808	1960
	4wDM	R/R	Unilok	1802	1973
	4wDH		TH	286V	1979

BATCHELOR ROBINSON METALS & CHEMICALS LTD, NEVILL'S DOCK, LLANELLI.
Gauge : 4'8½".   (SS 505990)

	4wDM	RH	338420	1954
D2	0-4-ODH	EEV	D1204	1967
	0-4-ODE	RH	544998	1969

BRITISH RAILWAYS, LONDON MIDLAND REGION, VALE OF RHEIDOL LIGHT RAILWAY, ABERYSTWYTH.
Gauge : 1'11½".   (SN 587812)   R.S.W.

7	OWAIN GLYNDWR	2-6-2T	OC	Sdn		1923
8	LLYWELYN	2-6-2T	OC	Sdn		1923
9	PRINCE OF WALES	2-6-2T	OC	DM	2	1902

See also Section Seven.

BRITISH STEEL CORPORATION, WELSH DIVISION, PROFIT CENTRES, B.S.C.TINPLATE,
TROSTRE WORKS, LLANELLI.
Gauge : 4'8½".   (SS 531994)

6	0-6-ODE	BT/WB	3021	1951
3	0-4-ODE	BT/WB	3096	1956
3	0-6-ODH	NB	28044	1961

CENTRAL ELECTRICITY GENERATING BOARD, CARMARTHEN BAY POWER STATION, BURRY PORT.
Gauge : 4'8½".   (SN 449004)

No.1	0-4-ODM	AB	392	1954
No.2	0-4-ODM	AB	393	1954

G.DAVIES, STEPHENS GREEN FARM, MILTON, near PEMBROKE.
Gauge : 4'8½".   (SN 041019)

	0-4-OST	OC	WB	2565	1936

DEPARTMENT OF THE ENVIRONMENT, TRECWN.
Gauge : 2'6".   (SM 971325)

MW 12270	4wDM	HE	6340	1970	Dsm

DUPORT STEELS LTD, DUPORT STEELWORKS, LLANELLI.   (Closed)
Gauge : 4'8½".   (SN 496003)

104	0-6-ODH	NB	27593	1957	OOU
113	0-6-ODH	NB	28050	1962	
116	0-6-ODH	AB	490	1964	OOU
D4	0-6-ODH	EEV	D924	1966	OOU
1	4wDH	CE	5632	1969	OOU
No.2	4wDH	CE	5632	1969	OOU
3	4wDH	CE	5862	1971	OOU

DYFED COUNTY COUNCIL, SCOLTON MANOR, near HAVERFORDWEST.
Gauge : 4'8½".   (SM 991222)

No.1378	GWEANDRAETH RAILWAY CO No.2,				
	KIDWELLY. MARGARET	0-6-0ST IC		FW	410 1878
A 123 W	PWM 1922	2w-2PMR		Wkm	3361 1942

DYFED RAILWAY CO LTD, VALE OF TEIFI NARROW GAUGE RAILWAY, HENLLAN, LLANDYSSUL.
Gauge : 2'0".   (SN 357406)

35	4wDM	MR	7126 1936
1	4wDM	MR	8683 1941

ESSO PETROLEUM CO LTD, HERBRANDSTON, MILFORD HAVEN.
Gauge : 4'8½".   (SM 878068)

	0-6-0DH	EES	8424 1963
1	0-6-0DH	HC	D1372 1965
	0-6-0DH	HE	8998 1981

GULF OIL REFINING LTD, WATERSTON, near MILFORD HAVEN.
Gauge : 4'8½".   (SM 935055)

(D2046)	2	0-6-0DM		Don	1958
			Rebuilt	HE	6644 1967
(03113)		0-6-0DM		Don	1960
		4wDH		TH	193V 1968

GWILI RAILWAY CO LTD.
Gauge : 4'8½".   Locos are kept at :-

	Bronwydd Arms.	(SN 417239)
	c/o Iron Rails Ltd, (Transport Contractors),	
	Carmarthen.	(SN 416197)
	Penybont Halt.	(SN 408259)

7820	DINMORE MANOR	4-6-0 OC		Sdn	1950
	LITTLE LADY	0-4-0ST OC		P	1903 1936
	MERLIN//MYRDDIN	0-4-0ST OC		P	1967 1939
	OLWEN	0-4-0ST OC		RSHN	7058 1942
71516		0-6-0ST IC		RSH	7170 1944
No.1		0-6-0ST OC		HC	1885 1955
		4wDM		RH	207103 1941
B 154 W	PWM 2222	2w-2PMR		Wkm	4139 1947

KILN PARK RAILWAY, TENBY.
Gauge : 2'0".   (SN 124000)

	ROCKET	0-2-2+4wPH S/0		Group 4,
				Birmingham 1970

LIME FIRMS LTD, PANTYRODIN LIMEWORKS, LLANDYBIE.
Gauge : 2'11".   (SN616168)   RTC.

	4wDM	RH	404976 1957 OOU

MILFORD DOCKS RAILWAY CO, MILFORD HAVEN.
Gauge : 4'8½".   (SM 899061)

CHARLES NEWBON	0-4-0DM	JF	4200015 1947 OOU
F.B.ROBJENT	0-4-0DM	JF	4200016 1947
MARGARET BRISTOWE	0-4-0DM	HC	D850 1954

MINISTRY OF DEFENCE, NAVY DEPARTMENT, ROYAL NAVAL ARMAMENT DEPOT.
Milford Haven.
Gauge : 4'8½".   (SM 917051)

ND 3815			0-4-0DM	HE	2389	1941	OOU
ND 3644	YARD No.343		0-4-0DM	HE	3282	1945	OOU
	YARD No.8690		4wDM	FH	3857	1957	
ND 10022	YARD No.9677		4wDM	FH	3968	1961	

Gauge : Metre. (SM 917051)

ND 3646	YARD No.173		4wDM	RH	210961	1941	OOU
ND 3645	YARD No.174	No.2	4wDM	RH	211679	1941	
ND 3647			4wDM	MR	22144	1962	
ND 3824	YARD No.A497		4wDM	HE	6647	1967	
	YARD No.A498		4wDM	HE	6648	1967	OOU

Trecwn.
Gauge : 4'8½".   (SM 970324)

ND 3066	B 16		0-4-0DM	HE	2390	1941	
ND 3067	B 17		0-4-0DM	HE	2391	1941	
ND 3068	B 18		0-4-0DM	HE	2392	1941	
			4wDM	FH	3811	1956	OOU
	YARD No.9034		4wDM	FH	3901	1959	
	YARD No.9354		4wDM	FH	3950	1961	
	YARD No.9355		4wDM	FH	3951	1961	

Gauge : 2'6".   (SM 970325)

ND 3051	YARD No.B 2		0-4-0DM	HE	2022	1939	
ND 3052	YARD No.B 3		0-4-0DM	HE	2263	1940	
ND 3053	YARD No.B 4		0-4-0DM	HE	2264	1940	
ND 3054	YARD No.B 5		0-4-0DM	HE	2265	1940	
ND 3055	YARD No.B 6		0-4-0DM	HE	2266	1940	
ND 3056	YARD No.B 7		0-4-0DM	HE	2267	1940	
ND 3057	YARD No.B 8		0-4-0DM	HE	2268	1940	OOU
ND 3058	YARD No.B 9		0-4-0DM	HE	2269	1940	OOU
ND 3059	YARD No.B 10		0-4-0DM	HE	2270	1940	
ND 3060	YARD No.B 19		0-4-0DM	HE	2398	1941	OOU
ND 3061	YARD No.B 20		0-4-0DM	HE	2399	1941	
ND 3062	YARD No.B 21		0-4-0DM	HE	2400	1941	
ND 3063	YARD No.B 22		0-4-0DM	HE	2401	1941	OOU
ND 3064	YARD No.B 23		0-4-0DM	HE	2402	1941	
ND 3065	YARD No.B 24		0-4-0DM	HE	2403	1941	OOU
ND 3301	B 25		4wBE	WR	1367	1939	
	B 26		4wBE	WR	1368	1939	
ND 3302	YARD No.B 27		4wBE	WR	1770	1941	
ND 3304	YARD No.B 28		4wBE	WR	1771	1941	
ND 3305	B 40		4wBE	WR	3805	1948	
ND 3306	YARD No.B 48		4wBE	GB	3545	1948	
ND 3307	YARD No.B 49		4wBE	GB	3546	1948	OOU
ND 3308	YARD No.B 50		4wBE	GB	3547	1948	
ND 2965	YARD No. 54		4wDMR	FH	2196	1940	
ND 10261	A 1		4wDM	BD	3751	1980	
ND 10262	A 2		4wDM	BD	3752	1980	
ND 10260	A 3		4wDM	BD	3753	1980	
ND 10363	MAINT No.17.200	722	4wDM	HE	6646	1967	

NATIONAL COAL BOARD.

For full details see Section Four.

PENDINE WILD LIFE & LEISURE PARK, PENDINE.
Gauge : 2'0".   (SN 244083)

20483		4wDH	S/O	HU	LX1001	1968

REES INDUSTRIES LTD, SCRAP METAL MERCHANTS, SARON WORKS, BELLEVUE ROAD,
LLWYNHENDY, LLANELLI.
Gauge : 4'8½".   (SH 545993)

	4wDM		RH	476143	1963	OOU

THYSSEN (GREAT BRITAIN) LTD, PLANT DEPOTS, BYNEA and OLD CASTLE, LLANELLI.
Gauge : 2'0".   (SS 554986, 499999)

010206	4wBE	WR	D6686	1964
	4wBE	WR	C6697	1963
	4wBE	WR	C6698	1963
	4wBE	WR	C6791	1963
4	4wBE	WR	D6841	1964
010219	4wBE	WR	H7196	1968
	4wBE	GB	420140	
	4wBE	GB	420221	

Locos present in yards between contracts.

Gauge : 4'8½".

(D2950)	D4	0-4-0DM		HE	4625	1954	Pvd

GWENT

A.R.ADAMS LTD, ROBERT STREET, NEWPORT DOCKS.
Gauge : 4'8½".   (ST 321869)

12054	0-6-0DE	Derby		1949
	0-6-0DM	(VF	D78	1948
		(DC	2252	1948

Locos are hired out to various concerns, but they are stored at Rowecord
Engineering Ltd, West Side, Old Town Dock, Newport between hirings.

ALPHA STEEL LTD, CORPORATION ROAD, NEWPORT.
Gauge : 4'8½".   (ST 334846)   RTC.

	0-4-0DM		RH	252686	1949	OOU

BLAENAVON MUSEUM TRUST, BLAENAVON.
Gauge : 4'8½".   (SO 237093)

	0-4-0ST	OC	AB	1619	1919
THE BLAENAVON TOTO No.6	0-4-0ST	OC	AB	1680	1920

BRITISH STEEL CORPORATION, SHEFFIELD DIVISION, PROFIT CENTRES, B.S.C. STAINLESS,
PANTEG WORKS, GRIFFITHSTOWN, PONTYPOOL.
Gauge : 4'8½".   (ST 296988, 297983)

1		4wDH	S	10083	1961
2	840177	0-4-0DH	S	10165	1964
3	MAUD	0-4-0DH	S	10142	1962

BRITISH STEEL CORPORATION, WELSH DIVISION, LLANWERN WORKS, NEWPORT.
Gauge : 4'8½".   (ST 385863)

31	153/2501	0-8-0DH		YE	2893	1962	Dsm
101		0-6-0DH		EEV	D1246	1968	
102	153/1522	0-6-0DH		EEV	D1247	1968	
103	153/1523	0-6-0DH		EEV	D1248	1968	
104	153/1524	0-6-0DH		EEV	D1249	1968	
			Rebuilt	GECT		1975	
105	153/1525	0-6-0DH		EEV	D1253	1968	
106	153/1526	0-6-0DH		EEV	D1226	1967	
107	153/1527	0-6-0DH		EEV	D1251	1968	
108	153/1534	0-6-0DH		EEV	D1254	1968	
109	153/1535	0-6-0DH		EEV	D1252	1968	
			Rebuilt	GECT		1975	
301	153/1528	0-6-0DH		GECT	5378	1972	
302	153/1529	0-6-0DH		GECT	5379	1972	
303	153/1530	0-6-0DH		GECT	5380	1972	
304	153/1531	0-6-0DH		GECT	5381	1972	
305	153/1532	0-6-0DH		GECT	5382	1973	
306	153/1533	0-6-0DH		GECT	5383	1973	
DE1	153/2503	6wDE		GECT	5409	1976	
DE2	153/2504	6wDE		GECT	5410	1976	
DE3	153/2505	6wDE		GECT	5411	1976	
DE4	153/2506	6wDE		GECT	5412	1976	
DE5	153/2507	6wDE		GECT	5413	1976	
		4wBE		S&H		c1961	OOU
1		0-4-0WE		GB	2519	1954	Dsm
		0-4-0WE		GB	2998	1961	
		0-4-0WE		GB	2999	1961	
		0-4-0WE		GECT	5389	1973	
2		0-4-0WE		GECT	5390	1973	
		0-4-0WE		Llanwern		1977	

BRITISH STEEL CORPORATION, WELSH DIVISION, PROFIT CENTRES, B.S.C. ASSOCIATED PRODUCTS GROUP.
Orb Works, Newport.
Gauge : 4'8½".   (ST 323862)   RTC.

No.36	0-4-0DE		(BT	91	1958	
	0-6-0DM		(BP	7856	1958	OOU
			HC	D1037	1958	OOU

Whitehead Works, Newport.
Gauge : 4'8½".   (ST 308868)

	0-4-0DE		YE	2779	1960

BRITISH STEEL CORPORATION, WELSH DIVISION, PROFIT CENTRES, B.S.C. TINPLATE.
Abercarn Works, Abercarn.
Gauge : 4'8½".   (ST 214949)

No.1	0-6-0DM		JF	22497	1938

Ebbw Vale Works, Ebbw Vale.
Gauge : 4'8½".   (SO 173094, 175073)

No.	Name	Type		Builder	Works No.	Date	Status
120		0-4-0DH		YE	2822	1961	OOU
124		0-4-0DH		YE	2842	1963	OOU
125		0-4-0DH		YE	2824	1963	OOU
150		0-8-0DH		S	10060	1962	OOU
151		0-8-0DH		S	10061	1962	OOU
153		0-8-0DH		S	10063	1962	OOU
154		0-8-0DH		S	10135	1962	OOU
157		0-8-0DH		S	10191	1964	OOU
158		0-8-0DH		S	10192	1964	OOU
170		0-8-0DH		HE	7063	1971	
171		0-8-0DH		HE	7064	1971	
172		0-8-0DH		HE	7065	1971	
173		0-8-0DH		HE	7200	1972	
		2w-2PMR		Wkm	9359	1963	
No.17		0-6-0DH		NB			+

+ Frame rebuilt in 1972 & used for carrying metal ingots.

JOHN CASHMORE LTD, NEWPORT DOCKS.
Yard (ST 319874) with locos for scrap occasionally present.

CENTRAL ELECTRICITY GENERATING BOARD.
Rogerstone Power Station.
Gauge : 4'8½".   (ST 261892)

		Type		Builder	Works No.	Date	
		0-4-0DM		JF	4210127	1957	
		0-4-0DM		AB	446	1959	

Uskmouth Power Station, West Nash, Newport.
Gauge : 4'8½".   (ST 327837)

		Type		Builder	Works No.	Date	
		0-4-0DM		AB	445	1959	
	USKMOUTH 3	0-4-0DH		HE	5622	1960	

Gauge : 2'0".   (ST 331840)   Pleasure Line.

		4wDM		HE	6013	1961	

R.L.DEAN, "GLEN USK", PLAS LLECHA, near CAERLEON.
Gauge : 4'8½".   (ST 379935)

No.	Name	Type		Builder	Works No.	Date	
111	ALDWYTH	0-6-0ST	IC	MW	865	1882	

FAIRFIELD-MABEY LTD, ENGINEERS, CHEPSTOW.
Gauge : 4'8½".   (ST 538938)

No.	Name	Type		Builder	Works No.	Date	Status
		0-4-0ST	OC	N	2119	1876	OOU
D 249		0-4-0WT	OC	KS	3063	1918	OOU
		0-4-0DM		JF	22897	1940	
		4wDM		FH	3783	1955	OOU
15	ROSEDALE	4wDH		S	10070	1961	
128		0-4-0DH		S	10128	1963	

GWENT COAL DISTRIBUTION CENTRE, GWENT, NEWPORT.
Gauge : 4'8½".   (ST 317875)

No.	Name	Type		Builder	Works No.	Date	
(D2181)	PRIDE OF GWENT	0-6-0DM		Sdn		1962	

<u>W.J.HARRIS & SONS, SCRAP DEALERS, PONTYPOOL ROAD.</u>
Yard (SO 297000) with locos for scrap occasionally present.

<u>HAYWOOD PLANT SALES LTD, PONTYPOOL.ROAD.</u>
Yard (SO 295001) with locos for resale occasionally present.

<u>MINISTRY OF DEFENCE, AIR FORCE DEPARTMENT, CAERWENT.</u>

        Uses M.O.D.,A.D. locos as required. For full details see Section Five.

<u>MINISTRY OF DEFENCE, ROYAL ORDNANCE FACTORY, GLASCOED.</u>
Gauge : 4'8½".   (SO 345016)

31469	R.O.F. No.1	0-4-0DH		TH	132C 1963
		Rebuild of 0-4-0DM	JF	22982 1942
19091	R.O.F. GLASCOED No.2	0-4-0DH	RSHD/WB	8366 1962
19092	R.O.F. GLASCOED No.3	0-4-0DH	RSHD/WB	8365 1962
	R.O.F. No.4	4wDH		TH	292V 1980

<u>MONSANTO CHEMICALS LTD, NEWPORT.</u>
Gauge : 4'8½".   (ST 335855)   R.S.W.

52-1		0-4-0F	OC	AB	1966 1929

<u>NATIONAL COAL BOARD.</u>

        For full details see Section Four.

<u>VAYNOR QUARRIES LTD, MACHEN QUARRY, near NEWPORT.</u>
Gauge : 4'8½".   (ST 223886)

PONTYPOOL No.2	4wDM	RH	252844 1948
	4wDM	FH	3832 1957

<u>WELSH NATIONAL WATER DEVELOPMENT AUTHORITY, USK SEWAGE DIVISION,</u>
<u>NASH TREATMENT WORKS, USKMOUTH.</u>
Gauge : Metre.   (ST 334837)   RTC.

59.01.083	4wDH	MR	121UA117 1974	OOU

GWYNEDD

<u>ABERLLEFENI SLATE QUARRIES LTD, ABERLLEFENI.</u>
Gauge : 2'3".   (SH 769102)

4wBE	BE		OOU
4wBE	LMM		OOU
4wBE	CE	BO457 1974	

<u>ANGLESEY ALUMINIUM METAL LTD, PENRHOS WORKS, HOLYHEAD.</u>
Gauge : 4'8½".   (SH 264807)

52-060	0-4-0DH	HE	7183 1970

ASSOCIATED OCTEL CO LTD, AMLWCH.
Gauge : 4'8½".   (SH 446936)

1		4wDM		RH	321727 1952
		0-4-0DH		HE	7460 1977

BALA LAKE RAILWAY LTD // RHEILFFORDD LLYN TEGID CYF, LLANUWCHLLYN.
Gauge : 1'11½".   (SH 881300)

No.10	JONATHAN	0-4-0ST	OC	HE	678 1898	
No.3	HOLY WAR	0-4-0ST	OC	HE	779 1902	
	ALICE	0-4-0ST	OC	HE	780 1902	Dsm
No.4	MAID MARIAN	0-4-0ST	OC	HE	822 1903	
3114	ASHOVER	0-4-0ST	OC	KS	3114 1918	
727/69		4wDM		HU	38384 c1930	
		4wDM		MR	5821 1934	
		4wDM		RH	189972 1938	
	CHILMARK	4wDM		RH	194771 1939	
		4wDM		HE	1974 1939	
No.11	CERNYW	4wDM		RH	200744 1940	
1		4wDM		RH	200748 1940	
		4wDM		FH	2544 1942	
U 190	T.R.A. 15	4wDM		RH	283512 1949	
		4wDM		L	34025 1949	
2 44052	ALISTER	4wDM		L	44052 c1958	
	MEIRIONNEDD	4w-4wDH		SL	22 1973	

D.W.BATES, c/o NORTH WALES TRAMWAY MUSEUM, TAL-Y-CAFN.
Gauge : 4'8½".   (SH 787716)

(DE 900338)		2w-2PMR		Wkm	509 1932

BRITISH RAILWAYS, HOLYHEAD BREAKWATER. (Closed)
Gauge : 4'8½".   (SH 235836)

01001		0-4-0DM		AB	396 1956 OOU
01002		0-4-0DM		AB	397 1956 OOU

BUTLINS LTD, PWLLHELI HOLIDAY CAMP, PENYCHAIN.
Gauge : 2'0".   (SH 434363)

157	C.P.HUNTINGTON	4-2-4DH	S/O	Chance	

CONWAY VALLEY RAILWAY MUSEUM, BETWS-Y-COED.
Gauge : 60cm.   (SH 796565)

	SGT. MURPHY	0-6-0T	OC	KS	3117 1918

CORRIS RAILWAY SOCIETY // CYMDEITHAS RHEILFFORDD CORRIS, MAESPOETH.
Gauge : 2'3".   (SH 753069)

		4wDM		MR	22258 1965

FAIRBOURNE RAILWAY LTD, FAIRBOURNE.
Gauge : 1'3".   (SH 616128)

	COUNT LOUIS	4-4-2	OC	BL	32	1923
	ERNEST W. TWINING	4-6-2	OC	G&S	10	1949
	KATIE	2-4-2	OC	Guest	14	1954
	SIAN	2-4-2	OC	Guest	18	1963
	GWRIL	4wPM		L	20886	1941
	RACHEL	0-6-ODM		G&S	15	1959
	SYLVIA	4w-4wPM		G&S	14	1961

FESTINIOG RAILWAY CO // Y RHEILFFORDD FFESTINIOG.
Gauge : 1'11½".   Locos are kept at :-

Boston Lodge Shed & Works.	(SH 584379,	585378)
Llyn Ystodu.	(SH 677443)	
Minffordd P.W. Depot.	(SH 599386)	
Moelwyn Tunnel (South Entrance).	(SH 679428)	
Tan-Y-Bwlch.	(SH 650415)	

No.2	PRINCE	0-4-0STT	OC	GE		1863	
No.5	WELSH PONY	0-4-0STT	OC	GE	(234?)	1867	OOU
No.10	MERDDIN EMRYS	0-4-4-OT	4C	Boston Lodge		1879	
No.11	LIVINGSTONE THOMPSON	0-4-4-OT	4C	Boston Lodge		1885	Dsm
	BLANCHE	2-4-0STT	OC	HE	589	1893	
	LINDA	2-4-0STT	OC	HE	590	1893	
1	BRITOMART	0-4-0ST	OC	HE	707	1899	
	MOUNTAINEER	2-6-2T	OC	AL	57156	1916	
	"VOLUNTEER"	0-6-0ST	OC	P	2050	1944	OOU
	EARL OF MERIONETH/						
	IARLL MEIRIONNYDD	0-4-4-OT	4C	Boston Lodge		1979	
	MARY ANN	4wDM		MR	(?590	1917)	+
	MOELWYN	2-4-ODM		BLW	49604	1918	
		4wDM		MR	3694	1924	
			Rebuild of	MR	1895	1919	Dsm
	ANDREW	4wDM		RH	193984	1939	Dsm
8565	JANE JGF 1	4wDM		MR	8565	1940	
	TYKE	4wDM		HE	2290	1941	Dsm
	ALISTAIR	4wDM		RH	201970	1940	
		4wDM		MR	8788	1943	
		4wDM		FH	3307	1948	
	UPNOR CASTLE	4wDM		FH	3687	1954	
	MOEL HEBOG	0-4-ODM		HE	4113	1955	
	SLUDGE JGF 4	4wDM		L	41545	1955	
JGF 2	THE LADY DIANA	4wDM		MR	21579	1957	
	BALFOUR BEATTY No.4362	4wDM		MR	21615	1957	Dsm
JGF 3	SANDRA	4wDM		MR	22119	1961	
		2w-2PMR		Wkm	1543	1934	

+ Carries plate 507/1917.

Gauge : 2'6".

	YD No.1016	4wDM	FH	3831	1958

FFESTINIOG RAILWAY MUSEUM, PORTHMADOG GOODS SHED.
Gauge : 1'11½".   (   )

No.1	PRINCESS	0-4-0STT	OC	GE		1863

J.W.GREAVES & SONS LTD.
Llechwedd Slate Quarries, Blaenau Ffestiniog.
Gauge : 2'0".　　　(SH 702468, 705466)　　RTC.

4wDM	RH	174542	1935	OOU
4wPM	Greaves	c1936	Dsm

Maen Offeren Slate Quarry, Blaenau Ffestiniog.
Gauge : 2'0".　　　(SH 713467, 716465)　　(Some locos work underground.)

4wDM	RH	175127	1935
4wDM	RH	174536	1936
4wBE	WR	918	1936
4wBE	GB?		

GWYNFYNYDD GOLD MINE, near DOLGELLAU.
Gauge : 1'11½".　　(SJ 736261)

4wBE	WR	5537	1956 +

　　　　　+ Currently in store elsewhere.

KINGSTON MINERALS LTD, PENMAENMAWR GRANITE QUARRIES.
Gauge : 3'0".　　　(SH 701758, 715765)　　RTC.

(PENMAEN)	0-4-0VBT VC	DeW		1878	Dsm
	4wDM	RH	202987	1941	OOU

THE NARROW GAUGE RAILWAY CENTRE OF NORTH WALES.
Blaenau Ffestiniog B.R. Station.
Gauge : 2'0".　　　(　　　　)

2207	BLAENAU FFESTINIOG	4wDM	HE	2207	1941

Dolwyddelan B.R. Station.
Gauge : 2'6".　　　(SH 738522)

2209	DOLWYDDELAN	4wDM	HE	2209	1941

Gloddfa Ganol, Blaenau Ffestiniog.
Gauge : 4'8½".　　　(SH 693470)　　(Not all locomotives are on public display.)

		0-4-0DM	JF	22900	1941
TR 27	PWM 2215	2w-2PMR	Wkm	4132	1947
TR 8		2w-2PMR	Wkm	4151	1948
TR 10		2w-2PMR	Wkm	4162	1948
TR 12	(PWM 2188) A156W	2w-2PMR	Wkm	4165	1948
TR 2	(PWM 2779)	2w-2PMR	Wkm	6878	1954
TR 7	A33M	2w-2PMR	Wkm	8503	1960

Gauge : Metre.

50823 1915	4wPM	RP	50823	1915

Gauge : 3'0".

No.3	LLANFAIR	0-4-0VBT VC	DeW		1895
1082		0-4-0DM	Ruhrthaler	1082	c1936
4	3930044	4wDM	JF	3930044	1950
105H006		4wDH	MR	105H006	1969
(c 13)		2w-2PMR	Wkm	2449	1938	Dsm
6/501		2w-2PMR	Wkm	4091	1946	Dsm

W6/504		2w–2PMR		Wkm	4092	1946	Dsm
C 18	4808	2w–2PMR		Wkm	4808	1948	
C 20	4810	2w–2PMR		Wkm	4810	1948	
C 23		2w–2PMR		Wkm	4813	1948	Dsm
(C 26)		2w–2PMR		Wkm	4816	1948	Dsm
C 37		2w–2PMR		Locospoor	B7281E		Dsm

Gauge : 2'6".

984	"WEE PUG"	0–4–0T	OC	AB	984	1903	
45913		4wPM		HU	45913	1932	

Gauge : 2'4".

398102		4wDM	RH	398102	1956	

Gauge : 2'0".

1568	DOROTHY	0–4–0ST	OC	WB	1568	1899	Dsm
14005	STEAM TRAM	4wVBT	G	L	14005	1940	
		Rebuilt from 4wPM in				1969	
		4wBE		BE	16306	c1917	Dsm
No.760		0–4–0PM		BgC	760	1918	
774	OAKELEY	0–4–0PM		BgC	774	1919	
		4wPM		HU	39924	1924	Dsm
		4wPM		MH	A110	1925	Dsm
640	WELSH PONY	4wWE		WR	640	1926	
1	SPONDON	4wBE		Spondon		1926	
No.1568		4wPM		FH	1568	1927	
36863		4wDM		HU	36863	1929	
D564		4wDM		HC	D564	1930	
		4wPM		H	982	1930	Dsm
		4wPM		FH	1747	1931	Dsm
3916		4wDM		L	3916	1931	
257081	DELTA	0–4–0DM		Dtz	257081	c1931	
164346		4wDM		RH	164346	1932	
		4wDM		RH	166028	1932	Dsm
164350		4wDM		RH	164350	1933	
1881		4wPM		FH	1881	1934	
6299		4wPM		L	6299	1935	
		4wDM		FH	2025	1937	
1298	LITTLE GEORGE	0–4–0BE		WR	1298	1938	
2201		4wDM		FH	2201	1939	
		4wDM		RH	198297	1939	
2666		4wDM		HE	2666	1942	
No.235	1840	4wBE		GB	1840	1942	
235711	PEN-YR-ORSEDD	4wDM		RH	.235711	1945	
	THAKEHAM	4wPM		Thakeham		c1946	
3424		4wPM		FH	3424	1949	
No.3205	BREDBURY	4wPM		Bredbury		c1954	
		4wDM		RH	393327	1956	Dsm
ZM 32	416214	4wDM		RH	416214	1957	
		4wDM		RH	432664	1959	
6018		4wDM		HE	6018	1961	
	TRENT WATER AUTHORITY	4wDM		MR	22128	1962	
	RAIL TAXI	4–2–0PMR		R.P.Morris		1967	

Gauge : 60cm.

		0–6–0T	OC	KS	2442	1915
		0–6–0T	OC	KS	2451	1915
3010		0–6–0T	OC	KS	3010	1916
3014		0–6–0T	OC	KS	3014	1916
461	WD 2182	4wPM		MR	461	1917
No.646		0–4–0PM		BgC	646	1918
No.736		0–4–0PM		BgC	736	1918

4470		4wPM		OK	4470 1930	
1835		4wDM		HE	1835 1937	Dsm
		4wDM		HE	2024 1940	Dsm
211647		4wDM		RH	211647 1941	

Gauge : 1'11½".

No.3	FESTINIOG	4wDM		Festiniog Rly	1974

Gauge : 1'10¼".

	KATHLEEN	0-4-0VBT VC		DeW	1877

Gauge : 1'6".

551		4wBE		WR	551 1924

Gauge : 1'3".

6502	WHIPPIT QUICK	4wPMR		L	6502 1935
			Rebuilt	Fairbourne	

NARROW GAUGE RAILWAY MUSEUM TRUST, WHARF STATION, TYWYN.
Gauge : 2'6". (SH 586004)

5	NUTTY	4wVBT ICG		S	7701 1929 +

+ Currently not on public display.

Gauge : 2'0".

		0-4-0WT OC		KS	721 1901

Gauge : 1'10¼".

	GEORGE HENRY	0-4-0VBT VC		DeW	1877
	ROUGH PUP	0-4-0ST OC		HE	541 1891
	JUBILEE 1897	0-4-0ST OC		MW	1382 1897

Gauge : 1'10".

13		0-4-0T IC		Spence	1895

Gauge : 1'6".

	PET	0-4-0ST IC		Crewe	1865
	DOT	0-4-0WT OC		BP	2817 1887

Also here are the side frames of 1'3" gauge KATIE 0-4-0T OC DB 4 1896.

NATIONAL MUSEUM OF WALES, NORTH WALES QUARRY MUSEUM, GILFACHDDU, LLANBERIS.
Gauge : 1'11½". (SH 586603)

	UNA	0-4-0ST OC		HE	873 1905

NATIONAL TRUST, INDUSTRIAL RAILWAY MUSEUM, PENRHYN CASTLE // CASTELL PENRHYN,
LLANDEGAI, near BANGOR.
Gauge : 4'8½". (SH 603720)

No.1		0-4-0WT OC		N	1561 1870
	HAYDOCK	0-6-0T IC		RS	2309 1879
	HAWARDEN	0-4-0ST OC		HC	526 1899
	VESTA	0-6-0T IC		HC	1223 1916

Gauge : 4'0".

	FIRE QUEEN	0-4-0 OC		AH	1848

Gauge : 3'0".

	KETTERING FURNACES					
	No.3	0-4-0ST	OC	BH	859	1885
	(WATKIN)	0-4-0VBT	VC	DeW		1893

Gauge : 1'10¾".

	CHARLES	0-4-0ST	OC	HE	283	1882
	HUGH NAPIER	0-4-0ST	OC	HE	855	1904

OAKELEY SLATE QUARRIES CO LTD, NEW MANOD QUARRY, near LLAN FFESTINIOG.
Gauge : 2'0".   (SH 731454)

		4wBE	BE	16303	c1917	OOU

D.G.OWEN, BUILDER STREET WEST, LLANDUDNO.
Gauge : 4'8½".   (SH 781816)

3211	0-4-0DM	AB	349	1941
	4wDM	FH	3147	1947

QUARRY TOURS LTD, LLECHWEDD SLATE MINE, BLAENAU FFESTINIOG.
Gauge : 2'0".   (SH 699468)

		4wBE	WR	308	1921	Pvd
		4wBE	WR	323	1921	Pvd
No.4	THE ECLIPSE	0-4-0WE	Greaves		1927	
		Rebuilt from 0-4-0ST/OC WB 1445 1895 Pvd				
	THE COALITION	0-4-0WE	Greaves		1930	
		Rebuilt from 0-4-0ST/OC WB 1278 1890 Pvd				
MBS 387		4wBE	LMM	1053	1950	Pvd
		4wBE	LMM	1066	1950	Pvd
LE/12/65		4wBE	WR	C6766	1963	
LE/12/64		4wBE	WR	E6807	1965	
		4wBE	WR			
		4wBE	WR			
		4wBE	WR			

RHEILFFORDD LLYN PADARN, CYFYNGEDIG, LLANBERIS LAKE RAILWAY LTD, GILFACHDDU, LLANBERIS.
Gauge : 1'11½".   (SH 586603)

No.1		ELIDIR	0-4-0ST	OC	HE	493	1889	
No.2		WILD ASTER	0-4-0ST	OC	HE	849	1904	
No.3		DOLBADARN	0-4-0ST	OC	HE	1430	1922	
No.5		HELEN KATHRYN	0-4-0WT	OC	Hen	28035	1948	
No.8		TWLL COED						
		YD No.AD689	4wDM		RH	268878	1952	
No.9		DOLGARROG	4wDM		MR	22154	1962	
No.12		MOR LEIDR	4wDM		RH		1954	+
15	5		4wDM		MR	5861	1934	@
No.17		GARRET	4wDM		MR	7902	1939	
No.18		"BRAICH"	4wDM		MR	7927	1941	Dsm
No.19		LLANELLI	4wDM		RH	451901	1961	
		A.M.W. No.196	4wDM		RH	198286	1940	
		A.M.W. No.231	4wDM		RH	203031	1941	
2			4wDM		RH	425796	1958	
No.3		164/137	4wDM		RH	441427	1961	

+ Either RH 375315 or 375316.
@ Converted into a brake van.

SNOWDON MOUNTAIN RAILWAY LTD, LLANBERIS.
Gauge : 2'7½".   (SH 582597)   R.S.W.

2	ENID	0-4-2T	OC	SLM	924	1895
3	WYDDFA	0-4-2T	OC	SLM	925	1895
4	SNOWDON	0-4-2T	OC	SLM	988	1896
5	MOEL SIABOD	0-4-2T	OC	SLM	989	1896
6	PADARN	0-4-2T	OC	SLM	2838	1922
7	RALPH SADLER	0-4-2T	OC	SLM	2869	1923
8	ERYRI	0-4-2T	OC	SLM	2870	1923

TALYLLYN RAILWAY CO, TYWYN.
Gauge : 2'3".   (SH 590008)   R.S.W.   (Loco shed at Pendre Station.)

No.1	TALYLLYN	0-4-2ST	OC	FJ	42	1865
No.2	DOLGOCH	0-4-0WT	OC	FJ	63	1866
No.3	SIR HAYDN	0-4-2ST	OC	HLT	323	1878
No.4	EDWARD THOMAS	0-4-2ST	OC	KS	4047	1921
DIESEL No.5	MIDLANDER	4wDM		RH	200792	1940
No.6	DOUGLAS	0-4-0WT	OC	AB	1431	1918
No.7	IRISH PETE	0-4-2T	OC	Pendre		1974
DIESEL No.8	MERSEYSIDER	4wDH		RH	476108	1964
No.8A		4wDH		RH	476109	1964 Dsm
No.9	ALF	0-4-0DM		HE	4136	1950
No.9A		0-4-0DM		HE	4135	1950 Dsm
6292		4wDH		HE	6292	1967 @ Dsm
		2w-2PMR		Towyn		1954
19		2w-2PMR		Towyn		1952 + Dsm

@ Frames utilised in a hydraulic press.
+ Converted into a flat wagon.

WELSH HIGHLAND LIGHT RAILWAY (1964) LTD, GELERT FARM, PORTHMADOG.
Gauge : 2'0".   (SH 571393)

	RUSSELL	2-6-2T	OC	HE	901	1906
	NANTMOR	0-6-0WT	OC	OK	9239	1921
	PEDEMOURA	0-6-0WT	OC	OK	10808	1924 +
No.7	KAREN	0-4-2T	OC	P	2024	1942
36	CNICHT	4wDM		MR	8703	1941
No.9		4wDM		MR	9547	1950
1	GLASLYN	4wDM		RH	297030	1952
	KINNERLEY	4wDM		RH	354068	1953
		4wDM		HE	6285	1968

+ Currently under renovation at Richards Engineering Ltd,
  Rhosemar, Clwyd. (SJ 209685)

DAVID WOODS, DINORWIC VILLAGE.
Gauge : 1'11½".   (SH   )

	4wPMR		Oakeley

MID GLAMORGAN

AMALGAMATED ROADSTONE CORPORATION LTD, WESTERN DIVISION, PENDERYN QUARRY, HIRWAUN.
Gauge : 4'8½".   (SN 950088)

		0-4-0DM	JF	4210081	1953	OOU
2371		0-4-0DM	JF	4210120	1956	OOU
		4wDH	RR	10222	1965	

BRITISH BENZOLE & COAL DISTILLATION LTD, BEDWAS COKING & BY-PRODUCTS PLANT.
Gauge : 4'8½".   (ST 183893)

	0-4-0DM	JF	4160010	1955
	0-6-0DH	NB	27935	1962
	0-4-0WE	WSO	2097	1935
	0-4-0WE	WSO	4496	1946

BRITISH STEEL CORPORATION, SHEFFIELD DIVISION, PROFIT CENTRES, B.S.C. FORGES,
FOUNDRIES & ENGINEERING, DOWLAIS FOUNDRY, DOWLAIS, MERTHYR TYDFIL.
Gauge : 4'8½".   (SO 069081)   RTC.

No.1		0-4-0DM	HC	D984	1955	OOU
No.3		0-4-0DM	HC	D1246	1961	OOU

CAERPHILLY RAILWAY SOCIETY LTD, HAROLD WILSON INDUSTRIAL ESTATE, VAN ROAD,
CAERPHILLY.
Gauge : 4'8½".   (ST 163865)

41312		2-6-2T	OC		Crewe		1952
		0-6-2T	IC		Cdf		1897
	DESMOND	0-4-0ST	OC		AE	1498	1906
	FORESTER	0-4-0ST	OC		AB	1260	1911
	VICTORY	0-4-0ST	OC		AB	2201	1945
No.2	HAULWEN	0-6-0ST	IC		VF	5272	1945
				Rebuilt	HE	3879	1961
	DEIGHTON	0-4-0ST	OC		RSHN	7705	1952
		0-4-0DE			YE	2731	1959

CONEY BEACH MINIATURE RAILWAY, CONEY BEACH, PORTHCAWL.
Gauge : 1'3".   (SS 821768)

1935	SILVER JUBILEE	4-6-4PE	S/O	1935
1936	CONEY QUEEN	4-6-4PE	S/O	1936

FORD MOTOR CO LTD, BRIDGEND FACTORY.
Gauge : 4'8½".   (SS 935783)

	0-6-0DH		HC	D1377	1966
		Rebuilt	HE	9039	1979

NATIONAL COAL BOARD.

For full details see Section Four.

POWYS

BRECON MOUNTAIN RAILWAY CO LTD // RHEILFFORDD MYNYDD BRYCHEINIOG,
PONTSTICILL STATION.
Gauge : 1'11½". (SO 063120)

	SYBIL	0-4-0ST	OC		HE	827	1903
		0-4-0VBT	VC		Redstone		1905
	GRAF SCHWERIN-LÖWITZ	0-6-2WTT	OC		Jung	1261	1908
2		4-6-2	OC		BLW	61269	1930
No.9		0-4-0WT	OC		OK	12722	1936
		0-4-0VBT	VC		Hill & Bailey		
No.2		4wDM			RH	425798	1958
		4wDM			RH	444207	1961
	RHYDYCHEN	4wDM			MR	11177	1961

WELSHPOOL & LLANFAIR LIGHT RAILWAY PRESERVATION CO LTD, LLANFAIR CAEREINION.
Gauge : 2'6". (SJ 107069) R.S.W.

No.2	(823)	THE COUNTESS	0-6-0T	OC		BP	3497	1902
No.6		MONARCH	0-4-4-0T	4C		WB	3024	1953
7		CHATTENDEN						
		YARD No.690	0-6-0DM			Bg/DC	2263	1949
No.8		DOUGAL	0-4-0T	OC		AB	2207	1946
10	699.01	SIR DREFALDWYN	0-8-0T	OC		FB	2855	1944
No.11		FERRET YARD No.86	0-4-0DM			HE	2251	1940
No.12		JOAN	0-6-2T	OC		KS	4404	1927
No.14	85		2-6-2T	OC		HE	3815	1954
			4wDM			RH	191680	1938 + Dsm
6		YARD No.6	0-4-0DM			HE	2245	1940
			2w-2PMR			Wkm	2904	1940

                    + Stored at Heniarth. (SJ 123032)

J.WOODRUFFE, RHIW VALLEY LIGHT RAILWAY, LOWER HOUSE, MANAFON, near WELSHPOOL.
Gauge : 1'3". (SJ 143028)

	POWYS	0-6-2T	OC		SL	20 1973

SOUTH GLAMORGAN

ABERTHAW & BRISTOL CHANNEL PORTLAND CEMENT CO LTD.
Aberthaw Cement Works.
Gauge : 4'8½". (ST 033675)

		0-4-0DM	JF	4210084 1953	Dsm
	BLUEBELL	0-4-0DM	JF	4210100 1955	
	IRIS	0-4-0DM	JF	4210114 1956	

Rhoose Cement Works.
Gauge : 4'8½". (ST 063661)

	HYACINTH	0-4-0DM	JF	4210092 1954	OOU
	PRIMROSE	0-4-0DM	JF	4210121 1956	

ALLIED STEEL & WIRE LTD, CASTLE & TREMORFA WORKS, CARDIFF.
Gauge : 4'8½".   (ST 198754, 207763)

370	CARLISLE	0-6-0DE		YE	2755	1959	
371	KIN-KENADON	0-6-0DE		YE	2763	1959	
372	BAMBOROUGH	0-6-0DE		YE	2760	1959	
378	TINTAGEL	0-6-0DH		RR	10290	1970	
379	CAERLEON	0-6-0DE		YE	2620	1956	
380	CAMELOT	0-6-0DH		RR	10288	1969	
390	AMESBURY	0-6-0DE		YE	2756	1959	
391	ASTOLAT	0-6-0DE		YE	2630	1956	
392	SARUM	0-6-0DE		YE	2619	1956	
394	CUNETIO	0-6-0DE		YE	2770	1959	
396	LEMANIS	0-6-0DE		YE	2761	1959	
398	CALLEVA	0-6-0DE		YE	2769	1959	
		0-6-0DE		YE	2633	1957	Dsm
		0-6-0DE		YE	2640	1957	Dsm

B.P. CHEMICALS (U.K.) LTD, BARRY WORKS, SULLY.
Gauge : 4'8½".   (ST 143683)   R.S.W.

		0-4-0F	OC	AB	2238	1948

BUTETOWN HISTORIC RAILWAY SOCIETY, BUTE ROAD STATION, CARDIFF.
Gauge : 4'8½".   (ST 192749)

	SIR GOMER	0-6-0ST	OC	P	1859	1932
107		0-6-0DH		NB	27932	1959

CENTRAL ELECTRICITY GENERATING BOARD, ABERTHAW POWER STATION.
Gauge : 4'8½".   (ST 024658)

	0-6-0DM	P	5014	1959
2	0-6-0DH	EES	8199	1963

DOW CORNING LTD, CARDIFF ROAD, BARRY.
Gauge : 4'8½".   (ST 142685)

4wDM	FH	3947	1960

A.E.KNILL & CO LTD, No.1 DOCK, BARRY.
Gauge : 4'8½".   (ST 118676)

No.3229/25

4wDM	FH	2601	OOU

NATIONAL MUSEUM OF WALES, INDUSTRIAL AND MARITIME MUSEUM, BUTE CRESCENT, BUTETOWN, CARDIFF.
Gauge : 4'8½".   (ST 192745)

No.10

0-6-0ST	IC	HC	544	1900

Gauge : 4'4".

4wG	Nat.Museum Wales	1981

Gauge : 3'0".

4wDM	RH	187100	1937 +

    + Not yet on public display.

POWELL DUFFRYN WAGON CO LTD.
Maindy Works.
Gauge : 4'8½".   (ST 172782)

| | | | 4wDM | | FH | 3814 | 1956 |
| 49 | | | 4wDM | | FH | 3959 | 1961 |

Radyr Works.
Gauge : 4'8½".   (ST 139798)

| | | | 4wDM | | RH | 312433 | 1951 |
| | | | 0-4-0DH | | RH | 512462 | 1965 |

ROATH & WELLFIELD COAL CO LTD, 5, WELLFIELD ROAD, ROATH, CARDIFF.
Gauge : 4'8½".   (ST 205776)

| 99 | | | 0-6-0DH | | NB | 27931 | 1959 + |

+ Loco currently under overhaul at B.R. Canton Depot.

SOUTH WALES WAREHOUSES LTD, PENARTH TIDAL HARBOUR, CARDIFF.
Gauge : 4'8½".   (ST 176729)

| | | | 4wDH | | FH | 3890 | 1959 | OOU |

WIGGINS TEAPE LTD, ELY PAPER WORKS, CARDIFF.
Gauge : 4'8½".   (ST 150767)   RTC.

| ELY | | | 4wDH | | FH | 3967 | 1961 | OOU |

WOODHAM BROS, SCRAP DEALERS & SHIPBREAKERS, BARRY.
Gauge : 4'8½".   (ST 111670)   (Ex B.R. locos for disposal, all OOU.)

2859		2-8-0	OC	Sdn		1918
2861		2-8-0	OC	Sdn		1918
2873		2-8-0	OC	Sdn		1918
2874		2-8-0	OC	Sdn		1918
3802		2-8-0	OC	Sdn		1939
3803		2-8-0	OC	Sdn		1939
3814		2-8-0	OC	Sdn		1940
3845		2-8-0	OC	Sdn		1942
3850		2-8-0	OC	Sdn		1942
3855		2-8-0	OC	Sdn		1942
3862		2-8-0	OC	Sdn		1942
4115		2-6-2T	OC	Sdn		1936
4247		2-8-0T	OC	Sdn		1916
4248		2-8-0T	OC	Sdn		1916
4253		2-8-0T	OC	Sdn		1917
4270		2-8-0T	OC	Sdn		1919
4277		2-8-0T	OC	Sdn		1920
4953	PITCHFORD HALL	4-6-0	OC	Sdn		1929
4979	WOOTTON HALL	4-6-0	OC	Sdn		1930
5199		2-6-2T	OC	Sdn		1934
5227		2-8-0T	OC	Sdn		1924
5526		2-6-2T	OC	Sdn		1928
5538		2-6-2T	OC	Sdn		1928
5539		2-6-2T	OC	Sdn		1928
5552		2-6-2T	OC	Sdn		1928
5553		2-6-2T	OC	Sdn		1928
5668		0-6-2T	IC	Sdn		1926
5967	BICKMARSH HALL	4-6-0	OC	Sdn		1937
6023	KING EDWARD II	4-6-0	4C	Sdn		1930
6686		0-6-2T	IC	AW	974	1928

No.	Name	Type	Cyl	Builder	Works No.	Year
6984	OWSDEN HALL	4-6-0	OC	Sdn		1948
7229		2-8-2T	OC	Sdn		1935
7927	WILLINGTON HALL	4-6-0	OC	Sdn		1950
9682		0-6-0PT	IC	Sdn		1949
30499		4-6-0	OC	Elh		1920
30825		4-6-0	OC	Elh		1927
30830		4-6-0	OC	Elh		1928
34010	SIDMOUTH	4-6-2	3C	Bton		1945
34028	EDDYSTONE	4-6-2	3C	Bton		1946
34046	BRAUNTON	4-6-2	3C	Bton		1946
34053	SIR KEITH PARK	4-6-2	3C	Bton		1947
34058	SIR FREDERICK PILE	4-6-2	3C	Bton		1947
34070	MANSTON	4-6-2	3C	Bton		1947
34072	257 SQUADRON	4-6-2	3C	Bton		1948
34073	249 SQUADRON	4-6-2	3C	Bton		1948
35006	PENINSULAR & ORIENTAL S.N. CO	4-6-2	3C	Elh		1941
35009	SHAW SAVILL	4-6-2	3C	Elh		1942
35010	BLUE STAR	4-6-2	3C	Elh		1942
35011	GENERAL STEAM NAVIGATION	4-6-2	3C	Elh		1944
35022	HOLLAND-AMERICA LINE	4-6-2	3C	Elh		1948
35025	BROCKLEBANK LINE	4-6-2	3C	Elh		1948
35027	PORT LINE	4-6-2	3C	Elh		1948
42859		2-6-0	OC	Crewe	5981	1930
44123		0-6-0	IC	Crewe	5658	1925
44901		4-6-0	OC	Crewe		1945
45163		4-6-0	OC	AW	1204	1935
45293		4-6-0	OC	AW	1348	1936
45337		4-6-0	OC	AW	1392	1937
47406		0-6-0T	IC	VF	3977	1926
48173		2-8-0	OC	Crewe		1943
48305		2-8-0	OC	Crewe		1943
48518		2-8-0	OC	Don	1966	1944
73096		4-6-0	OC	Derby		1955
73156		4-6-0	OC	Don		1956
75079		4-6-0	OC	Sdn		1956
76077		2-6-0	OC	Hor		1956
76084		2-6-0	OC	Hor		1957
78059		2-6-0	OC	Dar		1956
80072		2-6-4T	OC	Bton		1953
80097		2-6-4T	OC	Bton		1954
80098		2-6-4T	OC	Bton		1954
80104		2-6-4T	OC	Bton		1955
80150		2-6-4T	OC	Bton		1956
92207		2-10-0	OC	Sdn		1959
92219		2-10-0	OC	Sdn		1960
92245		2-10-0	OC	Crewe		1958

WEST GLAMORGAN

B.P. CHEMICALS (U.K.) LTD, BAGLAN BAY WORKS.
Gauge : 4'8½".   (SS 744924)

ZZ44	0-6-0DH	EEV	D3989	1970
ZZ67	0-6-0DH	EEV	D4003	1972

B.P. OIL LLANDARCY REFINERY LTD. LLANDARCY.
Gauge : 4'8½".   (SS 718960)

5	0-6-0DH	TH	157V	1965
6	0-6-0DH	TH	194V	1968
7	0-6-0DH	TH	230V	1971
8	0-6-0DH	TH	246V	1973

BRITISH STEEL CORPORATION, SHEFFIELD DIVISION, PROFIT CENTRES, B.S.C. FORGES,
FOUNDRIES & ENGINEERING, LANDORE FOUNDRY, MORRISTON, SWANSEA. (Closed)
Gauge : 4'8½".   (SS 670955)

9/7354	E.S.C. No.42	0-4-0DH	NB	27939	1959	OOU
		0-4-0DH	NB	27941	1961	

BRITISH STEEL CORPORATION, WELSH DIVISION, PORT TALBOT WORKS, PORT TALBOT.
Gauge : 4'8½".   (SS 773885, 775861, 781871)

501	0-4-0DE	BT/WB	3066	1954	
502	0-4-0DE	BT/WB	3067	1954	+
503	0-4-0DE	BT/WB	3068	1954	
504	0-4-0DE	BT/WB	3069	1954	
505	0-4-0DE	BT/WB	3070	1954	
506	0-4-0DE	BT/WB	3071	1954	
507	0-4-0DE	BT/WB	3072	1954	
508	0-4-0DE	BT/WB	3098	1956	
509	0-4-0DE	BT	3099	1956	
510	0-4-0DE	BT	3100	1956	
511	0-4-0DE	BT	3101	1956	
512	0-4-0DE	BT	3102	1956	+
513	0-4-0DE	BT	3103	1957	+
514	0-4-0DE	BT	3120	1957	+
801	4w-4wDE	AL	77120	1950	
802	4w-4wDE	AL	77776	1950	
803	4w-4wDE	AL	77777	1950	
804	4w-4wDE	AL	77778	1950	
805	4w-4wDE	AL	77779	1950	
901	4w-4wDE	PT/WB	3063	1955	
902	4w-4wDE	BT/WB	3064	1955	
903	4w-4wDE	BT/WB	3065	1955	
904	4w-4wDE	BT/WB	3137	1957	
905	4w-4wDE	BT/WB	3138	1957	
906	4w-4wDE	BT/WB	3139	1957	
907	4w-4wDE	BT/WB	3140	1957	
908	4w-4wDE	BT/WB	3141	1957	
909	4w-4wDE	BT	3142	1957	
910	4w-4wDE	BT	3143	1957	
951	4w-4wDE	BT	3111	1957	
952	4w-4wDE	BT	3112	1957	
953	4w-4wDE	BT	3113	1957	
19	4wDE	WB	3003	1951	@ Dsm
20	4wDE	WB	2972	1953	@ Dsm

A		0-4-0WE	GB	1661	1940	
C		0-4-0WE	GB	2283	1949	
D		0-4-0WE	GB	2282	1949	OOU
E		0-4-0WE	GB	2307	1951	
F		0-4-0WE	GB	2801	1958	
G		0-4-0WE	GB	2802	1958	
		0-4-0WE	GB			+1
		4wWE	(BD	3748	1979	
			(GECT	5476	1979	
		4wWE	(BD	3749	1979	
			(GECT	5477	1979	

+ Converted to slave units for use with locos 501/8/9/10/11.
@ Converted to a Brake Tender Runner.
+1 Either GB 2591/1955 or 2737/1956.

BRITISH STEEL CORPORATION, WELSH DIVISION, PROFIT CENTRES, B.S.C. TINPLATE
VELINDRE WORKS, LLANGYFELACH, SWANSEA.
Gauge : 4'8½".   (SS 642997)

712		4wDE	BT/WB	2974	1953
LD1		0-4-0DE	BT/WB	3097	1956
114		0-4-0DH	NB	27878	1962

BUILDWELL CONCRETE PRODUCTS, CLYDACH-ON-TAWE.
Gauge : 4'8½".   (SN 688009)

	0-4-0ST OC	AB	1081	1909	OOU

GEORGE COHEN, SONS & CO LTD, SCRAP DEALERS, BEAUFORT WORKS, MORRISTON.
Gauge : 4'8½".   (SS 672970)

ND 5953		0-4-0DM	JF	22876	1939	OOU
600	A.M.W. No.169	0-4-0DM	JF	22878	1939

Also other locos for scrap occasionally present.

FORD MOTOR CO LTD, CRYMLYN BURROWS, SWANSEA.
Gauge : 4'8½".   (SS 704935)

4wDH	TH	163V	1966

INCO EUROPE LTD, CLYDACH.
Gauge : 4'8½".   (SN 696013)

P 0026	M.N.C. No.1	0-4-0ST OC	P	1345	1914
C 0120		0-4-0DH	EEV	D1205	1967

LLEWELLYN & WILKINS LTD, PARC COLLIERY, PEN RHIWFAWR, YSTALYFERA.
Gauge : 2'0".   (SN 729102)

	JANET	4wDM	RH	432647	1959	
	PEARL	4wDM	RH	432648	1959	
S 141		4wDM	RH	462361	1960	Dsm
	WENDY	4wDM	RH	504546	1963	

METAL BOX INDUSTRIAL COMPONENTS LTD, CANAL SIDE, NEATH.
Gauge : 4'8½".   (SS 743964)

2	OA 229	4wDM	RH	394014	1956	OOU
		0-4-0DM	RSHD/WB	7910	1963	OOU

<u>NATIONAL COAL BOARD.</u>

For full details see Section Four.

<u>SWANSEA INDUSTRIAL & MARITIME MUSEUM, COAST LINES WAREHOUSE, SOUTH DOCKS, SWANSEA.</u>
Gauge : 4'8½".   (SS 659927)

ROSYTH No.1	0-4-0ST OC	AB	1385	1914
SIR CHARLES	0-4-0F OC	AB	1473	1916
SWANSEA VALE No.1	4wVBT VCG	S	9622	1958
DYLAN	4wDM	RH	393302	1955
LANDORE FOUNDRY No.1	0-4-0DH	NB	27654	1956

<u>WAGON REPAIRS LTD, PORT TENNANT WORKS.</u>
Gauge : 4'8½".   (SS 681933)

D 1	4wDM	RH	224353	1945
	0-4-0DM	Bg	3590	1962

<u>THOS. W. WARD LTD, GIANTS GRAVE, BRITON FERRY.</u>
Gauge : 4'8½".   (SS 735948)

75	0-4-0DH	NB	28037	1961
76	0-4-0DH	NB	28038	1961

<u>WEST GLAMORGAN COUNTY COUNCIL, CEFN COED STEAM CENTRE, CRYNANT.</u>
Gauge : 4'8½".   (SN 786034)

0-6-0ST IC	HE	3846	1956

SECTION 4

NATIONAL COAL BOARD

The current surface locomotive stock of the N.C.B. is listed here under Area headings. Each of these lists has been arranged so that locos are listed in chronological order of construction.

Locomotives working entirely underground have been omitted as have those temporarily on the surface for repair, transfers, etc. Those working partly underground and partly on the surface (drift mines, etc) have been included with the surface locos.

PAGE	AREA	REFERS TO COUNTIES

NATIONAL COAL BOARD

PAGE	AREA	REFERS TO COUNTIES
237	SCOTTISH	Central, Fife, Lothian, Strathclyde.
238	NORTH EAST	Durham, Northumberland, Tyne & Wear.
242	NORTH YORKSHIRE	West Yorkshire.
244	DONCASTER	South Yorkshire.
245	BARNSLEY	South Yorkshire, West Yorkshire.
246	SOUTH YORKSHIRE	Nottinghamshire, South Yorkshire.
248	WESTERN	Clwyd, Cumbria, Greater Manchester, Merseyside, Staffordshire.
250	NORTH DERBYSHIRE	Derbyshire.
251	NORTH NOTTINGHAMSHIRE	Nottinghamshire.
252	SOUTH NOTTINGHAMSHIRE	Nottinghamshire.
253	SOUTH MIDLANDS	Derbyshire, Kent, Leicestershire, Warwickshire, West Midlands.
255	SOUTH WALES	Dyfed, Gwent, Mid Glamorgan, West Glamorgan.

COAL PRODUCTS DIVISION

NATIONAL SMOKELESS FUELS LTD.

PAGE	AREA	REFERS TO COUNTIES
258	DURHAM MANAGEMENT UNIT	Durham, Tyne & Wear.
259	MIDLANDS MANAGEMENT UNIT	Derbyshire, South Yorkshire, West Midlands.
259	WALES MANAGEMENT UNIT	Mid Glamorgan.
260	THOMAS NESS LTD	Mid Glamorgan, Tyne & Wear.

OPENCAST EXECUTIVE

PAGE	AREA	REFERS TO COUNTIES
260	SCOTTISH REGION	Fife, Strathclyde.
260	NORTHERN REGION	Northumberland, Tyne & Wear.
261	CENTRAL (EAST) REGION	Derbyshire, Leicestershire, Nottinghamshire, West Yorkshire.
262	CENTRAL (WEST) REGION	Clwyd, Greater Manchester.
262	SOUTH WESTERN REGION	Dyfed, Mid Glamorgan, West Glamorgan.

MARKETING DEPARTMENT

NATIONAL FUEL DISTRIBUTORS LTD - COAL CONCENTRATION DEPOTS.

See under County headings.

MINING RESEARCH & DEVELOPMENT ESTABLISHMENT

| 263 | | Derbyshire. |

NATIONAL COAL BOARD
SCOTTISH AREA

(Headquarters : Green Park, Greenend, Liberton, Edinburgh, Lothian.)

Locomotives are kept at the following locations :-

CENTRAL .

Kn	(NS 986812)	Kinneil Colliery, Bo'ness. (Coal Winding ceased.)
Pm	(NS 838914)	Polmaise Colliery, Fallin.

FIFE

Bw	(NT 213957)	Bowhill Coal Preparation Plant, Cardenden.
CCW	(NT 169916)	Central Workshops, Cowdenbeath.
Cm	(NT 007910)	Comrie Colliery, Saline.
Fs	(NT 310939)	Frances Colliery, Dysart.
Sf	(NT 277895)	Seafield Colliery, Kirkcaldy.

LOTHIAN

BG	(NT 274652)	Bilston Glen Colliery, Loanhead.
NCW	(NT 334636)	Central Workshops, Newbattle, Newtongrange.
Pk	(NS 936639)	Polkemmet Colliery, Whitburn.
AMS	(NT 336620)	Withdrawn Machinery Stores, Arniston, Gorebridge.
WTC	(NT 311701)	Woolmet Training Centre, Millerhill.

STRATHCLYDE

By	(NS 528217)	Barony Colliery, Auchinleck.
Bd	(NS 720703)	Bedlay Colliery, Glenboig. (Closed)
Cd	(NS 665685)	Cardowan Colliery, Stepps.
Kl	(NS 478205)	Killoch Colliery, Ochiltree.

Gauge : 4'8½".

No.8	DARDANELLES	0-6-0ST	OC		+	AB	1175	1909	Pk	Pvd
No.9		0-6-0T	IC			HC	895	1909	Bd	OOU
No.16		0-4-0ST	OC			AB	1116	1910	By	
No.8		0-6-0T	OC			AB	1296	1912	Pk	OOU
No.12		0-6-0ST	OC			AB	1829	1924	Pk	OOU
		(Rebuilt with parts of				GR	539	1917)		
No.29		0-4-0ST	OC			AB	1996	1934	BG	OOU
No.6		0-4-0ST	OC			AB	2043	1937	Bd	OOU
No.17		0-6-0ST	IC			HE	2880	1943	Pk	OOU
No.33		4wDM				RH	221647	1943	Pm	
No.7		0-6-0ST	IC			WB	2777	1945	Cm	OOU
No.30		0-4-0ST	OC			AB	2259	1949	Fs	OOU
No.23		0-4-0ST	OC			AB	2260	1949	Cd	
No.17		0-4-0ST	OC			AB	2296	1952	Bd	OOU
No.19		0-6-0ST	IC			HE	3818	1954	Cm	
No.4		0-4-0DE				RH	349089	1954	Kl	OOU
No.20		0-4-0ST	OC			AB	2357	1955	CCW	Dsm
No.8		0-4-0ST	OC			AB	2369	1955	By	OOU
No.5		0-6-0ST	IC			HE	3837	1955	Cm	OOU
No.10		0-4-0DH				AB	414	1957	BG	
No.6		0-4-0DM				AB	417	1957	Kl	
No.10		0-6-0DH				NB	27591	1957	Bw	
No.12	H 662	0-4-0DH				NB	27732	1957	Cm	
No.2		0-4-0DM				AB	418	1958	Kl	

	BL 98	0–4–0DM	AB	421	1958	BG	
No.32		0–4–0DH	AB	447	1959	BG	
No.3		0–4–0DM	AB	450	1959	Kl	
		4wDM	RH	458960	1962	Pm	
		0–6–0DH	EEV	D1119	1966	Cd	
No.31		4wDH	RR	10231	1966	By	

+ Carries plates AB 1296 1912 in error.

Gauge : 3'6".

	TRAINING LOCO No.2	0–6–0DM	HE	4074	1955	WTC	
	TRAINING LOCO No.1	0–6–0DM	HE	4075	1955	WTC	
3		0–6–0DM	HE	4820	1955	WTC	Dsm

Gauge : 3'0".

		4wBE	GB	2009	1946	Fs	
No.1		4wDM	RH	256273	1948	Kl	
C/H 1116	THE FOX	0–4–0DM	HE	4634	1954	BG	
4		4wBE	GB	2915	1958	Sf	
		4wBE	GB	6045	1962	Sf	
		0–4–0DM	HC	DM1405	1968	BG	

Gauge : 2'8".

H 881		0–4–0DM	HE	3200	1945	Cm	
		4wDM	RH	506415	1964	Cm	

Gauge : 2'6".

31		4wDM	RH	252860	1948	Bd	OOU
No.1	C/H 1120	4wBE	WR	5291	1954	Bd	OOU
	MS 12727	4wBE	(EE	2395	1957	Kn	
			(Bg	3493	1957		
		0–4–0D	RH	476133	1962	Cd	
		4wBE	CE	5871A	1971	Pm	

NORTH EAST AREA

(Headquarters : Team Valley, Gateshead, Tyne & Wear.)

Locomotives are kept at the following locations :-

DURHAM

STC	(NZ 412496)	Area Training Centre, Seaham.
Bl	(NZ 462394)	Blackhall Colliery, Blackhall.
Da	(NZ 436480)	Dawdon Colliery, Seaham Harbour.
E	(NZ 437443)	Easington Colliery, Easington.
EH	(NZ 346370)	East Hetton Colliery, Quarrington Hill.
H	(NZ 442421)	Horden Colliery, Horden.

Mu	(NZ 400470)	Murton Colliery, Murton. (Part of Hawthorn Colliery.)
Se	(NZ 410495)	Seaham Colliery, New Seaham.
SWW	(NZ 427491)	Seaham Wagon Works, Seaham Harbour.
SH	(NZ 384453)	South Hetton Colliery, South Hetton. (Works traffic for Hawthorn Colliery & Coking Plant.)
VT	(NZ 425500)	Vane Tempest Colliery, Seaham.

NORTHUMBERLAND

ATC	(NZ 266879)	Area Training Centre, Ashington.
As	(NZ 264881)	Ashington Colliery, Ashington. (Supplies locos to Lynemouth Colliery.)
Bt	(NZ 304823, 309823)	Bates Colliery, Blyth.
El	(NZ 283917)	Ellington Colliery, Ellington.
Ly	(NZ 299925)	Lynemouth Colliery, Bewick Drift Mine, Lynemouth.
Sb	(NU 215080)	Shilbottle Colliery, Shilbottle. (RTC)
Wt	(NU 176064)	Whittle Colliery, Newton-on-the-Moor.

TYNE & WEAR

CD	(NZ 187605)	Clockburn Drift Mine, Winlaton Mill.
D	(NZ 204631)	Derwenthaugh Loco Shed, Blaydon. (Works traffic for Clockburn Drift Mine & Derwenthaugh Coking Plant.)
Ep	(NZ 364483)	Eppleton Colliery, Hetton-le-Hole. (Part of Hawthorn Colliery.)
Hn	(NZ 342533)	Herrington Colliery, New Herrington.
LEW	(NZ 336525)	Lambton Engine Works, (Central Workshops), Philadelphia.
P	(NZ 335523)	Philadelphia Loco Shed. Philadelphia. (Lambton Railway.)
Wa	(NZ 313607)	Wardley Loco Shed, Wardley. (Bowes Railway.)
Wm	(NZ 393580)	Wearmouth Colliery, Sunderland.
We	(NZ 373667)	Westoe Colliery, South Shields. (Harton Railway and branch to Boldon Colliery.)

Gauge : 4'8½".

2			4wWE	Siemens		1908	We	
4			4w–4wWE	Siemens		1909	We	
9			4w–4wWE	AEG	1565	1913	We	
E10			4wWE	Siemens	862	1913	We	
11			4w–4wWE	(EE	1795	1951	We	
				(Bg	3351	1951		
12			4w–4wWE	(EE	1794	1951	We	
				(Bg	3350	1951		
13			4w–4wWE	(EE	2308	1957	We	
				(Bg	3469	1957		
14			4w–4wWE	(EE	2599	1959	We	
				(Bg	3519	1959		
15			4w–4wWE	(EE	2600	1959	We	
				(Bg	3520	1959		
		2502/7	0-6-0ST IC	WB	2779	1945	VT	OOU
512	(12060)		0-6-0DE	Derby		1949	P	
514	(12084)		0-6-0DE	Derby		1950	Bt	
513	(12098)		0-6-0DE	Derby		1952	P	
509	(12119)		0-6-0DE	Dar		1952	P	
511	(12133)		0-6-0DE	Dar		1952	P	
	(D3088)		0-6-0DE	Derby		1954	LEW	
		2100/526	0-6-0DH	NB	27588	1957	SH	
		20/110/701	0-6-0DH	NB	27717	1957	SH	
No.7D		2446/43	0-6-0DM	HE	5177	1958	EH	
No.3D		2404/70	0-6-0DM	HE	5302	1958	Wm	
41		9207/5	0-6-0DE	RH	421438	1958	VT	
No.10D		2309/57	0-6-0DM	HE	5304	1959	EH	
64		2235/64	0-6-0DH	NB	27763	1959	SH	
65		2235/65	0-6-0DH	NB	27764	1959	SH	

66		2235/66	0-6-0DH	NB	27765	1959	SH	
	D4056		0-6-0DE	Dar		1961	Sb	OOU
No.56	(D4068)		0-6-0DE	Dar		1961	Wt	
No.51	(D4069)		0-6-0DE	Dar		1961	Wt	
No.52	(D4070)		0-6-0DE	Dar		1961	LEW	
No.53	(D4072)		0-6-0DE	Dar		1961	LEW	
			2w-2DMR	+ Bg	3585	1962	We	
			0-4-0DH	AB	478	1963	We	
		05200/100	4wDM	MR	5766	1963	SWW	
			0-6-0DH	S	10157	1963	Wa	
			0-6-0DH	S	10158	1963	Wa	
			0-6-0DH	AB	488	1964	E	
		2303/63	0-6-0DH	AB	491	1964	H	
		8411/03	0-4-0DH	RR	10201	1964	D	
No.1	(D9500)	(9312/92)	0-6-0DH	Sdn		1964	As	
	D9502	(9312/97)	0-6-0DH	Sdn		1964	As	
506	(D9504)		0-6-0DH	Sdn		1964	As	
No.9	(D9508)	9312/99	0-6-0DH	Sdn		1964	As	
38	(D9513)	D1/9513	0-6-0DH	Sdn		1964	As	
No.4	(D9514)	9312/96	0-6-0DH	Sdn		1964	As	
No.8	(D9517)	9312/93	0-6-0DH	Sdn		1965	As	
No.7	(D9518)	9312/95	0-6-0DH	Sdn		1964	As	
No.3	(D9521)	9312/90	0-6-0DH	Sdn		1964	As	
507	(D9525)		0-6-0DH	Sdn		1965	As	
No.6	(D9527)	9312/94	0-6-0DH	Sdn		1965	As	
No.2	(D9528)	9312/100	0-6-0DH	Sdn		1965	As	OOU
	(D9531)	D2/9531	0-6-0DH	Sdn		1965	As	
37	(D9535)	9312/59	0-6-0DH	Sdn		1965	As	
No.5	(D9536)	9312/91	0-6-0DH	Sdn		1965	As	
No.36	(D9540)	2233/508	0-6-0DH	Sdn		1965	As	
	D9555	(9107/57)	0-6-0DH	Sdn		1965	As	
			0-6-0DH	AB	498	1965	Bt	
No.500			0-6-0DH	HE	6611	1965	Wa	
No.501			0-6-0DH	HE	6612	1965	Wm	
No.502			0-6-0DH	HE	6613	1965	Wa	
No.503			0-6-0DH	HE	6614	1965	Wa	
No.504			0-6-0DH	HE	6615	1965	Wa	
		21201/208	0-6-0DH	HE	6616	1965	D	
No.209			0-6-0DH	HE	6617	1965	D	
		2120/210	0-6-0DH	HE	6618	1965	D	
		9101/66	0-6-0DH	HE	6662	1966	D	
		2120/211	0-6-0DH	AB	514	1966	Wa	
			0-4-0DH	AB	523	1967	Se	
		2233/242	0-4-0DH	AB	524	1967	Se	
			0-4-0DH	AB	547	1967	VT	
			0-4-0DH	AB	548	1967	Se	
No.67		9101/0067	0-6-0DH	AB	549	1967	We	
No.157			0-4-0DH	HE	6676	1967	D	
		05200/92	0-4-0DH	AB	550	1968	Se	
No.61			0-6-0DH	AB	582	1973	LEW	
No.72			0-6-0DH	AB	583	1973	LEW	
No.69		2446/83	0-6-0DH	AB	584	1973	SH	
No.71			0-6-0DH	AB	585	1973	P	
No.594		20.104.997	0-6-0DH	AB	594	1974	Bt	
		20-110-704	0-6-0DH	AB	603	1976	H	
			0-6-0DH	AB	604	1976	EH	
			0-6-0DH	AB	608	1976	E	
		20/110/708	0-6-0DH	AB	609	1976	Da	
			0-6-0DH	AB	612	1976	SH	
			0-6-0DH	AB	613	1977	As	
			0-6-0DH	AB	614	1977	E	
			0-6-0DH	AB	615	1977	Bt	
			0-6-0DH	AB	616	1977	Bt	
			0-6-0DH	AB	623	1978	As	

20/110/739		0-6-0DH	AB	624	1978	Da
		0-6-0DH	AB	646	1979	We
		0-6-0DH	AB	647	1979	P

+ Self-propelled Tower Wagon.

Gauge : 3'6".   (Underground locos working out on surface.)

		0-6-0DM	HC	DM632	1947	CD
		0-6-0DM	HC	DM639	1947	CD
709		0-6-0DM	HC	DM709	1955	CD
993		0-6-0DM	HC	DM993	1956	CD
		0-6-0DM	HC	DM1063	1957	CD
1428	20/280/21	0-6-0DM	(HC	DM1428	1977	CD
			(HE	8525	1977	

Gauge : 3'0".

	C/H 1117	0-4-0DM	HE	4635	1954	El
		4wBE	CE	5921	1972	Ly

Gauge : 2'8¼".

		0-4-0DM	HE	6980	1968	ATC

Gauge : 2'6".

No.2		4wDM	RH	256314	1949	LEW
No.1		4wDM	RH	256323	1949	We
		4wBE	(EE	1809	1952	Wm
			(Bg	3363	1952	
		4wDH	CE	B1819	1978	Wm

Gauge : 2'0".

		0-6-0DM	HC	DM804	1951	Da
		0-4-0DM	HE	4109	1952	Hn
		0-4-0DM	HE	4110	1953	Bl
		0-4-0DM	HE	4387	1953	ACW
		0-6-0DM	HC	DM842	1954	STC
		4wBE	(EE	2519	1958	STC
			(Bg	3500	1958	
		0-4-0DM	HE	5596	1961	STC
		0-4-0DM	HE	6619	1966	STC
		4wDM	HE	6628	1966	Bl
		4wDM	HE	7080	1971	EH
	9125/62	4wBE	CE	B0112	1973	EH
		4wDM	HE	6347	1975	Ep
		4wDM	HE	6348	1975	Mu
		4wBE	CE	B0451	1975	Wt
		4wBE	CE	B1539	1977	Wt

NORTH YORKSHIRE AREA

(Headquarters : P.O. Box 13, Allerton Bywater, Castleford, West Yorkshire.)

Locomotives are kept at the following locations :-

WEST YORKSHIRE

AH	(SE 423205)	Ackton Hall Colliery, Featherstone.
AB	(SE 421279)	Allerton Bywater Colliery, Allerton Bywater.
ACW	(SE 429277)	Central Workshops, Allerton Bywater.
Fs	(SE 455269)	Fryston Colliery, Fryston.
Kl	(SE 527233)	Kellingley Colliery, Knottingley.
KTC	(SE 529231)	Kellingley Training Centre, Knottingley.
LL	(SE 430306)	Ledston Luck Colliery, Kippax.
Nm	(SE 365253)	Newmarket Colliery, Stanley.
Ns	(SE 399170)	Nostell Colliery, Nostell.
Pf	(SE 439326)	Peckfield Coal Preparation Plant, Micklefield.
PW	(SE 451226)	Prince of Wales Colliery, Pontefract.
Sv	(SE 392273)	Savile Colliery, Methley.
Sh	(SE 383202)	Sharlston Colliery, New Sharlston.
Wt	(SE 359181)	Walton Colliery, Crofton. (Closed)
Wd	(SE 441262)	Wheldale Colliery, Castleford.

Gauge : 4'8½".

7		0-6-0ST	IC	HE	3168	1944	Wd	
		0-6-0ST	IC	WB	2746	1944	AH	OOU
		0-4-0DH		NB	27095	1953	Ns	OOU
No.1	ROSE LOUISE	0-6-0DM		HC	D972	1956	Fs	OOU
	HENDON	4wDM		FH	3865	1957	Ns	OOU
No.5		0-6-0DM		HC	D1070	1958	Sv	
	G.H. 1	0-6-0DM		HC	D1071	1958	Sv	
No.8		0-6-0DM		HC	D1073	1958	AB	
No.10		0-6-0DM		HC	D1134	1959	Nm	
45		0-4-0DH		YE	2674	1959	AB	
No.17	2.11.27	0-6-0DM		HC	D1137	1960	Nm	
108		0-4-0DH		S	10089	1962	Kl	
109		0-4-0DH		S	10118	1962	Kl	
110		0-4-0DH		S	10119	1962	Kl	
111		0-4-0DH		S	10120	1963	Kl	
3		0-4-0DH		S	10155	1963	Ns	
6		0-6-0DH		S	10182	1964	Nm	
7		4wDH		S	10183	1964	Wd	
No.9	L 40947	0-4-0DH		TH/S	151C	1965	Pf	
No.8	L 40885	0-6-0DH		TH/S	154C	1965	Nm	
No.10		0-6-0DH		RR	10256	1966	Ns	
No.14	40995	0-6-0DH		RR	10267	1967	Ns	
44		0-6-0DH		HE	6684	1968	AB	
		0-6-0DH		HE	6685	1968	Wd	
		0-6-0DH		HE	7062	1971	AH	
		0-4-0DH		HE	7259	1971	Fs	
		0-4-0DH		HE	7260	1971	Sv	
	HUNTER	0-6-0DH		HE	7276	1972	Fs	
38		0-6-0DH		HE	7277	1972	Wd	
39		0-6-0DH		HE	7278	1972	AB	
		0-6-0DH		HE	7279	1972	AH	
40		0-6-0DH		HE	7307	1973	Wd	
No.41		0-6-0DH		HE	7352	1973	Wd	
42		0-6-0DH		AB	592	1974	AB	
43		0-6-0DH		AB	593	1974	AB	
No.44		0-6-0DH		TH	248V	1974	Pf	
No.45		0-6-0DH		TH	249V	1974	Pf	

| No.46 | | 0-6-0DH | | HE | 7408 | 1975 | Fs | |
| 47 | | 0-6-0DH | | HE | 7540 | 1977 | AH | |

Gauge : 3'6".

| | | 4wBE | | GB | 6081 | 1963 | KTC | |
| | | 0-6-0DM | | HB | DM1421 | 1972 | KTC | |

Gauge : 3'0".

ML 18		0-6-0DM		HE	4033	1948	Ns	
4053		0-6-0DM		HE	4053	1950	Wt	
		0-6-0DM		HE	5434	1958	KTC	

Gauge : 2'6".

728		0-6-0DM		HC	DM728	1949	LL	OOU
No.1		0-6-0DM		HC	DM730	1950	Wd	OOU
		0-6-0DM		HC	DM670	1953	LL	
985		0-6-0DM		HC	DM985	1955	LL	
		4wDM		HE	5694	1960		
			Rebuilt	HE	6638	1967	K1	
		4wDM		HE	5695	1960		
			Rebuilt	HE	6639	1968	K1	
18		4wDH		SMH	101T018	1979	LL	
19		4wDH		SMH	101T019	1979	LL	
20		4wDH		SMH	101T020	1979	LL	
		4wDM		SMH	60SD752	1979	Wd	
		4wDM		SMH	60SD753	1979	PW	
		4wDM		SMH	60SD754	1980	PW	
		4wDM		SMH	60SD755	1980	Wd	
	KIRSTIN	4w-4wDH		GMT		1981	LL	

Gauge : 2'0".

| | | 4wDM | | HE | 8840 | 1980 | Sh | |

DONCASTER AREA

(Headquarters : St. George's, Thorne Road, Doncaster, South Yorkshire.)

Locomotives are kept at the following locations :-

SOUTH YORKSHIRE

AM	(SE 557139)	Askern Colliery, Askern.
Bn	(SE 569074)	Bentley Colliery, Bentley.
BTC	(SE	Bentley Training Centre, Bentley.
Bd	(SE 528077)	Brodsworth Colliery, Woodlands.
Fk	(SE 464097)	Frickley Colliery, South Elmsall.
Hf	(SE 654113)	Hatfield Colliery, Stainforth.
Hk	(SE 464053)	Hickleton Colliery, Thurnscoe.
MkM	(SE 616046)	Markham Main Colliery, Armthorpe.
Rs	(SK 601985)	Rossington Colliery, Rossington.
YM	(SK 543993)	Yorkshire Main Colliery, Edlington.

Gauge : 4'8½".

No.43	BRM 21		0-6-0DM		HE	4513	1955	Bd	
46005			0-6-0DM		HE	5240	1957	YM	OOU
No.47			0-6-0DM		HC	D1068	1958	Bd	
3			0-6-0DM		HC	D1189	1960		
				Rebuilt	HE	8901	1978	Bd	
D2519			0-6-0DM		HC	D1210	1962	Hf	OOU
	HERBERT		0-6-0DH		HE	5590	1964	MkM	
	F/SE/353		0-4-0DH		HC	D1340	1966	Hk	
	TOMMY		0-4-0DH		HC	D1342	1966	Hk	
	ROBERT		0-4-0DH		HC	D1386	1966	Hf	
			0-4-0DH		HE	7405	1974	MkM	
			0-4-0DH		HE	7422	1976	YM	

Gauge : 3'0".

No.0			0-4-0DM	HE	3426	1946	Bd
			0-4-0DM	HE	3427	1946	YM
	BRM 301		4wDM	RH	249550	1947	Bd
			0-6-0DM	HC	DM1120	1958	BTC
No.10			0-6-0DM	HC			BTC

Gauge : 2'6".

| | | 4wDM | RH | 331263 | 1952 | Rs |

Gauge : 2'3".

| | | 0-4-0DM | HE | 3573 | 1948 | Bn |

Gauge : 2'1½".

| | | 4wDM | RH | 379659 | 1955 | Fk |

Gauge : 2'0".

3		4wDM	RH	222066	1946	Rs
		4wDM	RH	252863	1947	Rs
No.681		0-4-0DM	HC	DM689	1948	Hf
No.682		0-4-0DM	HC	DM690	1948	Hf
		0-4-0DM	HC	DM749	1949	AM

BARNSLEY AREA

(Headquarters : Grimethorpe, Barnsley, South Yorkshire.)

Locomotives are kept at the following locations :-

SOUTH YORKSHIRE

Bw	(SE 359025)	Barrow Colliery, Worsborough Bridge.
DM	(SE 401039)	Darfield Main Colliery, Wombwell.
Dw	(SE 312058)	Dodworth Colliery, Dodworth.
EM	(SE 230112)	Emley Moor Colliery, Skelmanthorpe.
Gt	(SE 406083)	Grimethorpe Colliery, Grimethorpe.
HM	(SE 419060)	Houghton Main Colliery, Little Houghton.
NG	(SE 330098)	North Gawber Colliery, Mapplewell.
Wl	(SE 313108)	Woolley Colliery, Darton.

WEST YORKSHIRE

Pm	(SE 260115)	Parkmill Colliery, Clayton West. (Drift mine – includes u/g locos.)

Gauge : 4'8½".

8	TL 60	WR 28	0-6-0ST	IC		HE	3183	1944	Wl
	(D2239)		0-6-0DM			(VF	D289	1956	Dw
						(DC	2563	1956	
			0-4-0DM			HC	D1094	1959	Pm
	08679		0-6-0DE			Hor		1959	NG
No.1	(D2057)		0-6-0DM			Don		1959	
				Rebuilt	HE	6645		Gt
No.2	(D2093)		0-6-0DM			Don		1960	
				Rebuilt	HE	6643		Gt
	D2284		0-6-0DM			(RSH	8102	1960	Wl
						(DC	2661	1960	
(D2199)	ROCKINGHAM COLLIERY 1		0-6-0DM			Sdn		1961	Bw
			4wDH			S	10058	1961	Dw
			0-4-0DM			HC	D1259	1962	Wl
	DH 17		4wDH			S	10176	1964	Bw
	TL 13		4wDH			TH/S	142C	1964	NG
			0-4-0DH			RR	10203	1965	DM
	TL 40		4wDH			TH/S	156C	1965	Bw
	TL 39		4wDH			TH/S	158C	1965	EM
	TL 10		4wDH			TH/S	171C	1966	DM
	WOOLLEY No.1		0-6-0DH			AB	553	1969	NG

Gauge : 2'6".

746		0-4-0DM		HC	DM746	1951	Pm
747		0-4-0DM		HC	DM747	1951	Pm
		0-4-0DM		HC	DM748	1951	Pm
890		0-4-0DM		HC	DM890	1955	Pm
1356		0-4-0DM		HC	DM1356	1965	Pm

Gauge : 2'2".

TL 42		4wDM		HE	6273	1965	HM
		4wDH		HE	7530	1977	HM

Gauge : 2'0".

| | 4wDM | HE | 7274 1973 | DM |

Gauge : 1'9".

| | 4wDM | HE | 6631 1965 | Dw |

SOUTH YORKSHIRE AREA

(Headquarters : Wath-on-Dearne, Rotherham, South Yorkshire.)

Locomotives are kept at the following locations :-

NOTTINGHAMSHIRE

Mn	(SK 609783)	Manton Colliery, Manton.
So	(SK 558809)	Shireoaks Colliery, Shireoaks.
St	(SK 552784)	Steetley Colliery, Steetley.

SOUTH YORKSHIRE

Bb	(SE 475033)	Barnburgh Colliery, Barnburgh.
Bh	(SK 454843)	Brookhouse Colliery, Beighton.
Cy	(SK 512997)	Cadeby Colliery, Conisbrough.
Cw	(SE 407015)	Cortonwood Colliery, Wombwell.
Dn	(SK 517867)	Dinnington Colliery, Dinnington.
Ec	(SE 390003)	Elsecar Colliery, Elsecar.
KP	(SK 494826)	Kiveton Park Colliery, Kiveton Park.
Mb	(SK 550925)	Maltby Colliery, Maltby.
Mv	(SE 449015)	Manvers Coal Preparation Plant & Colliery (SE 448012), Wath-on-Dearne.
MTC	(SE 454014)	Manvers Training Centre, Wath-on-Dearne.
Sw	(SK 475941)	Silverwood Colliery, Thrybergh.
Tc	(SK 498897)	Thurcroft Colliery, Thurcroft.
Tt	(SK 437877)	Treeton Colliery, Treeton.
Wh	(SE 437020)	Wath Colliery, Wath-on-Dearne.

Gauge : 4'8½".

No.16	(D2209)	TRACEY		0-6-0DM	(VF (DC	D210 1953 2484 1953	KP	
	D2225	DEBRA		0-6-0DM	(VF (DC	D274 1955 2548 1955	Wh	OOU
No.5	D2229			0-6-0DM	(VF (DC	D278 1955 2552 1955	Bh	
	(D2238)	CAROL		0-6-0DM	(VF (DC	D288 1955 2562 1955	Mv	
No.18	(D2248)	SUE	2243	0-6-0DM	(RSH (DC	7867 1957 2580 1957	Mb	
No.23		CLARRIE No.54		0-6-0DM	HC	D1090 1958	Dn	
		TERRY No.56		0-6-0DM	HC	D1091 1958	Sw	OOU
No.11		DINNINGTON		0-6-0DM	HC	D1113 1958	Dn	

No.	D No.	Name	Type		Builder	Works No.	Date	Loc
No.8		DICK No.55	0-6-0DM		HC	D1115	1958	Cy
No.20		MANTON	0-6-0DM		HC	D1121	1958	Mn
No.22		DAVID No.58	0-6-0DM		HC	D1128	1958	Cy
No.19			0-6-0DM		HC	D1133	1958	Mn
		ALEX No.59	0-6-0DM		HC	D1138	1958	Sw
		KERRY No.57	0-6-0DM		RH	347748	1958	Mv
		GEOFFERY No.60	0-6-0DM		RH	347749	1959	Mv
No.7			0-6-0DM		HC	D1139	1959	Cy
No.4		DL 4	0-6-0DM		HC	D1152	1959	Bh
No.21		CARL No.61	0-6-0DM		HC	D1154	1959	MV
No.15		DL 5	0-6-0DM		HC	D1174	1959	KP
	D2607		0-6-0DM		HE	5656	1960	St
(No.30)	D2300		0-6-0DM		(RSH	8159	1960	Mn
					(DC	2681	1960	
No.10	D2317		0-6-0DM		(RSH	8176	1960	Cw
					(DC	2698	1960	
No.24	D2322		0-6-0DM		(RSH	8181	1961	KP
					(DC	2703	1961	
No.12	(D2327)	521/12	0-6-0DM		(RSH	8186	1961	Dn
					(DC	2708	1961	
No.31	(D2328)	DINNINGTON No.2	0-6-0DM		(RSH	8187	1961	St
					(DC	2709	1961	
	D2332	LLOYD	0-6-0DM		(RSH	8191	1961	So
					(DC	2713	1961	
No.33	D2334		0-6-0DM		(RSH	8193	1961	Tc
					(DC	2715	1961	
No.3	(D2337)	DOROTHY	0-6-0DM		(RSH	8196	1961	Mv
					(DC	2718	1961	
No.1	(D2373)	DAWN	0-6-0DM		Sdn		1961	Mv
		KEN No.67	0-6-0DH	+	S	10180	1964	Cy
		LESLIE No.68	0-6-0DH		S	10181	1964	Sw
No.13		WALTER No.69	0-6-0DH		HE	6230	1964	Sw
No.26		521/61	0-4-0DH		HC	D1343	1965	So
No.34			0-6-0DH		YE	2913	1965	Tt
No.25			0-6-0DH		YE	2939	1965	Tt
		FRANK No.70	0-6-0DH		RR	10223	1965	Mv
No.35		No.71	0-6-0DH		HE	6286	1965	Mv
No.14		WILF No.72	0-6-0DH		HE	6287	1965	Ec
No.2		HARRY No.73	0-6-0DH		HE	6661	1966	Bb
		RAYMOND No.74	0-6-0DH		RR	10261	1966	Mv
		ERNEST	0-6-0DH		TH	250V	1974	Mv
			0-4-0DH		HE	7426		St

+ Carries plate 10181 in error.

Gauge : 3'0".

			4wDM	+	MR			Cy
			0-6-0DM		HE	3517	1948	Cy
		KATY	4wDM		MR	40S280	1968	Sw

+ Either 7406 or 8814.

Gauge : 2'2".

			4wDM	MR	9695	1952	Wh
			4wDM	RH	382808	1955	Wh

Gauge : 2'0".

No.101			0-6-0DM	HC	DM630	1948	MTC
			0-4-0DM	HC	DM808	1953	KP

WESTERN AREA

(Headquarters : Staffordshire House, Berry Hill Road, Stoke-on-Trent, Staffordshire.)

Locomotives are kept at the following locations :-

CLWYD

Bs	(SJ 315483)	Bersham Colliery, Rhostyllen.
PA	(SJ 127837)	Point Of Ayr Colliery, Talacre.

CUMBRIA

Wm	(NX 973189)	William Loco Shed, Whitehaven Harbour. (Serves loading point for Haig Colliery.)

GREATER MANCHESTER

Ac	(SD 798012)	Agecroft Colliery, Pendlebury.
Bi	(SJ 629999)	Bickershaw Colliery, Leigh.
Wk	(SD 733029)	Walkden Central Workshops, Walkden.

MERSEYSIDE

Bo	(SJ 549935)	Bold Colliery, St. Helens.
Ct	(SJ 473892)	Cronton Colliery, Whiston.
OB	(SJ 573976)	Old Boston Training Centre, Haydock.
Pk	(SJ 600947)	Parkside Colliery, Newton-le-Willows.
SM	(SJ 519909)	Sutton Manor Colliery, St. Helens.

STAFFORDSHIRE

SWW	(SJ 882433)	Area Wagon Works, Stafford Colliery, Trentham. (Closed)
Fl	(SJ 914420)	Florence Colliery, Longton.
HH	(SJ 884423)	Hem Heath Colliery, Trentham.
Hd	(SJ 837479)	Holditch Colliery, Chesterton.
KTC	(SJ 887433)	Kemball Training Centre, Trentham.
LH	(SK 060168)	Lea Hall Colliery, Rugeley.
Lt	(SJ 973126)	Littleton Colliery, Huntington.
TMS	(SJ)	Trentham Machinery Stores.
WC	(SK 007141)	West Cannock Colliery, Hednesford.
Ws	(SJ 862480)	Wolstanton Colliery, Wolstanton.

Gauge : 4'8½".

63.000.404	(HORNET)		0-4-0ST	OC	P	1935	1937	Bs
	ROBERT	No.7	0-6-0ST	IC	HC	1752	1943	Bo
	JOSEPH		0-6-0ST	IC	HE	3163	1944	
				Rebuilt	HE	3885	1964	Bo
	WHISTON		0-6-0ST	IC	HE	3694	1950	Bo
63.000.326	BICKERSHAW		0-6-0ST	IC	HE	3776	1952	Bi
63.000.427			4wDM		RH	326068	1953	Bs
63 000 352			4wDM		RH	338413	1953	SWW OOU
63 000 432	WARRIOR		0-6-0ST	IC	HE	3823	1954	Bi
		D 1	0-4-0DH		NB	27650	1956	SM
63/000/311		No.1	0-6-0DM		WB	3117	1956	LH
	HEM HEATH 3D		0-6-0DM		WB	3119	1956	Hd
63/000/325		No.5	0-6-0DM		WB	3123	1957	Ws OOU
	HEM HEATH 4D		0-6-0DM		WB	3134	1957	Hd
63.000.425		D 2	0-4-0DH		NB	27728	1957	Ct
63.000.402		LO 61	0-6-0DE		YE	2660	1957	Ac
63.000.415			0-6-0DE		YE	2712	1958	Pk
			0-6-0DE		YE	2713	1958	SM
63 000 421			0-6-0DE		YE	2717	1958	Pk
63.000.413		D 3	0-4-0DH		NB	27735	1958	Bo

63.000.420		D 5	0-4-0DH	NB	27876	1959	Wm
63/000/365	WOLSTANTON No.1		0-6-0DM	WB	3147	1959	Ws
63/000/366	WOLSTANTON No.3		0-6-0DM	WB	3150	1959	Ws
63/000/320		No.6	0-6-0DE	YE	2748	1959	Lt
		No.7	0-6-0DE	YE	2749	1959	Pk
63.000.401		LO 52	0-6-0DE	YE	2745	1960	Ac
63.000.426		D 6	0-4-0DH	YE	2678	1961	Ct
			4wDH	S	10097	1962	Wm
63/000/362	VICTORIA No.D2						
		LP 61/10	0-6-0DH	RH	512845	1965	Ws
63/000/305		No.2	0-6-0DH	EEV	D1120	1966	LH
63/000/349	HEM HEATH No.10D		6wDH	TH	168V	1966	HH	OOU
63/000/350	HEM HEATH No.11D		6wDH	TH	179V	1967	HH	OOU
63/000/335		No.12D	6wDH	TH	180V	1967	Ws
63/000/336		No.13D	6wDH	TH	181V	1967	Hd
63/000/329		No.1D	0-6-0DH	HE	6663	1969	Bs
63/000/328		No.2D	0-6-0DH	HE	6664	1969	PA
63/000/316		No.3D	0-6-0DH	HE	7181	1970	LH
63/000/342		No.12D	0-6-0DH	HE	7016	1971	HH	OOU
63/000/322		No.6D	0-6-0DH	HE	7017	1971	Wm
63/000/314		No.8D	0-6-0DH	HE	7018	1971	Wk
		No.14D	0-6-0DH	HE	7040	1971	PA
63.000.441	WESTERN ENTERPRISE		6wDE	GECT	5421	1977	Lt
63.000.442	WESTERN PIONEER		6wDE	GECT	5422	1977	Lt
63.000.443	WESTERN PROGRESS		6wDE	GECT	5468	1978	Lt
63.000.444	WESTERN JUBILEE		6wDE	GECT	5478	1979	HH	OOU
63.000.445	WESTERN QUEEN		6wDE	GECT	5479	1979	Bi
63.000.446	WESTERN KING		6wDE	GECT	5480	1979	Bi

Gauge : 2'6".

	A.M.W. No.205	LO 76	4wDM		RH	200803	1941	Ac
	A.M.W. No.232	LO 75	4wDM		RH	203032	1941	Ac
			4wBE		(EE	2416	1957	OB
					(RSH	7935	1957	
63.000.417			0-4-0DM		RH	398116	1958	Pk
63 000 323			4wDM		RH	441945	1959	Fl
63.000.315			4wDM	+	RH	441946	1959	WC
63/000/313		No.1	4wDM		RH	441947	1959	LH
63.000.364			4wDM	a	RH	441948	1959	Ws
63.000.418			4wDM	@	RH	441952	1960	Pk
			4wDM		RH	452293	1960	?
			0-6-0DM		HC	DM1238	1960	OB
63 000 306		7	4wDM		RH	466587	1961	Pk
			4wBE		Bg	3557	1961	KTC
			4wDM		RH	476112	1962	LH
63/000/367			4wDM		RH	487961	1962	HH
63/000/346		No.1	4wDM		RH	497758	1963	HH
			4wDM		RH	497761	1963	HH
63/000/348		LP 61	4wDH		RH	476107	1964	Lt
		LP 61/1	4wDM		RH	506491	1964	Lt
63/000/339		LP 61/47	4wDH		RH	512994	1965	Fl
		5074	4wBE		CE	5074	1965	LH
63/000/309			4wDM	RH	7002/0767/6	1967	LH
63/000/324			4wDM	RH	7002/0867/3	1967	Ws
	IVOR		4wDH		HE	8825	1978	WC
			4wDH		HE	8826	1978	LH
63/000/447			4wDH		HE	8971	1979	Lt
63/000/449			4wDH		HE	8973	1979	LH
	NEWTON		4wDH		HE	8975	1979	Pk

+ Carries plate 441944 in error.
a Known to N.C.B. as 436862 (in error) and carries plate with
  that number.
@ Carries plate 444144 in error.

Gauge : 2'3".

63.000.411		4wDM	+	RH	323587	1952	Bo
63 000 446		4wDM	+	RH	323586	1952	Bo
		4wDH		HE	8970	1979	Bo

+ Each carries the others worksplate.

Gauge : 2'2".

39	DL 5	0-4-0DM		HE	4039	1948	SM	OOU
63.000.424		4wDM		RH	392107	1955	Wk	

Gauge : 2'1".

63.000.347	4wDM		RH	433390	1959	Hd

Gauge : 2'0".

63.000.369	4wDM	RH	497760	1963	TMS
	4wDH	HE	7535	1977	Ps

Gauge : 1'10".

63.000.405	4wDM	RH	497547	1963	Bs

NORTH DERBYSHIRE AREA

(Headquarters : Bolsover, Chesterfield, Derbyshire.)

Locomotives are kept at the following locations :-

DERBYSHIRE

Ak	(SK 428704)	Arkwright Colliery, Duckmanton.
Bs	(SK 460712)	Bolsover Colliery, Bolsover.
Mk	(SK 449722)	Markham Colliery, Duckmanton.
Pl	(SK 498643)	Pleasley Colliery, Pleasley.
RP	(SK 439775)	Renishaw Park Colliery, Staveley.
Sb	(SK 530670)	Shirebrook Colliery, Shirebrook.
Wt	(SK 456793)	Westthorpe Colliery, Killamarsh.
Ww	(SK 533757)	Whitwell Colliery, Whitwell.

Gauge : 4'8½".

		D 5	0-6-0DE	RH	425476	1959	Bs	OOU
DL 24		No.1 N.C.B.	4wDH	TH	114V	1962	Sb	OOU
DL 6		WHITWELL WARRIOR	0-4-0DH	RH	466620	1962	Ww	
DL 8			0-4-0DH	RH	468047	1962	RP	
DL 2			0-6-0DH	YE	2839	1962	Bs	
DL 22			4wDH	S	10141	1962	Ak	
DL 23			4wDH	S	10163	1963	Ak	
DL 10			0-6-0DH	YE	2874	1963	Mk	
DL 9			0-6-0DH	YE	2887	1963	Pl	
DL 11			0-6-0DH	YE	2910	1963	Wt	
DL 12			0-6-0DH	YE	2911	1963	Wt	
DL 13			0-4-0DH	HC	D1279	1963	RP	
DL 20			0-4-0DH	EES	8448	1963	Bs	
DL 14			0-4-0DH	HC	D1292	1964	Ww	
DL 15		10	0-4-0DH	HC	D1344	1965	Sb	
DL 19		No.11	0-4-0DH	EEV	D922	1965	Bs	
DL 21		No.21	0-6-0DH	EEV	D1196	1967	Pl	
DL 16			0-4-0DH	HC	D1387	1967	RP	

NORTH NOTTINGHAMSHIRE AREA

(Headquarters : Edwinstowe, Mansfield, Nottinghamshire.)

Locomotives are kept at the following locations :-

NOTTINGHAMSHIRE

BTC	(SK 696743)	Bevercotes Training Centre, Bevercotes.
Bd	(SK 595567)	Blidworth Colliery, Blidworth.
Cp	(SK 594632)	Clipstone Colliery, Clipstone.
Hw	(SK 627913)	Harworth Colliery, Bircotes.
LHM	(SK 700728)	Lound Hall Mining Museum, Bevercotes.
Mf	(SK 570616)	Mansfield Colliery, Mansfield.
Ot	(SK 661677)	Ollerton Colliery, Ollerton.
Rf	(SK 597604)	Rufford Colliery, Rainworth.
Sw	(SK 536625)	Sherwood Colliery, Mansfield.
Sv	(SK 472617)	Silverhill Colliery, Teversal.
St	(SK 484603)	Sutton Colliery, Sutton-in-Ashfield.
Wb	(SK 582704)	Welbeck Colliery, Welbeck Colliery Village.

Gauge : 4'8½".

		THE WELSHMAN	0-6-0ST	IC	MW	1207	1890	LHM	Pvd
	9		0-6-0ST	OC	YE	2521	1952	LHM	Pvd
D 5		87420	0-4-0DM		HC	D1085	1958	Wb	OOU
		87419	0-6-0DM		HC	D1111	1958	Wb	OOU
D 2		R 1	0-4-0DE		RH	420143	1958	Rf	
D 3			0-4-0DE		RH	425479	1959	Rf	
D 4		R 2	0-4-0DE		RH	431762	1959	Rf	
(D 21)			4wDH		TH	113V	1961	Mf	
			4wDH		S	10073	1961	St	
		LO/62/ML	0-4-0DH		HE	6262	1964	St	

NCB NORTH DERBYSHIRE
NCB NORTH NOTTINGHAMSHIRE

(D 7)	10332		0-4-0DH	S	10189	1964	Wb
D 26	47491	P1	4wDH	RR	10193	1964	Ot
D 23			4wDH	RR	10194	1964	Sw
D 25			4wDH	RR	10195	1964	Sv
	84753	18458	4wDH	TH	140V	1964	Sv
D 22			4wDH	RR	10228	1965	Bd
D 24	No.72241		4wDH	RR	10241	1966	
			Rebuilt	TH	247V	1973	Sw
D 8	47213		0-6-0DH	RR	10257	1966	Ot
			4wDH	TH	169V	1966	Cp
D 10			4wDH	TH	170V	1966	Sw
D 11			4wDH	TH	172V	1966	Mf
D 12	70491		4wDH	TH	174V	1966	Hw
D 13			4wDH	TH	175V	1966	Bd
D 14			4wDH	TH	178V	1967	Hw
D 15	47262		0-6-0DH	RR	10262	1967	Ot
D 21	47453		0-6-0DH	RR	10270	1967	Ot
D 16			4wDH	TH	190V	1967	Cp
D 17			4wDH	TH	191V	1968	Sv
D 18			4wDH	TH	192V	1968	Sv
D 19			4wDH	TH	215V	1970	Cp

Gauge : 2'4".

			4wDM	RH	375347	1954	LHM
5	S20/05979		0-4-0DM	HC	DM1420	1972	BTC

SOUTH NOTTINGHAMSHIRE AREA

(Headquarters : Bestwood, Nottinghamshire.)

Locomotives are kept at the following locations :-

NOTTINGHAMSHIRE

Bb	(SK 534437)	Babbington Colliery, Cinderhill.
Bk	(SK 487549)	Bentinck Colliery, Kirkby-in-Ashfield.
BkTC	(SK 488553)	Bentinck Training Centre, Kirkby-in-Ashfield.
Cv	(SK 603501)	Calverton Colliery, Calverton.
Cg	(SK 653363)	Cotgrave Colliery, Cotgrave.
Gl	(SK 613438)	Gedling Colliery, Gedling.
Hn	(SK 540490)	Hucknall Colliery, Hucknall.
Ly	(SK 535505)	Linby Colliery, Linby.
MG	(SK 481480)	Moor Green Colliery, Newthorpe.
NH	(SK 475585)	New Hucknall Colliery, Huthwaite.
Ns	(SK 521533)	Newstead Colliery, Newstead.
PH	(SK 443522)	Pye Hill Colliery, Jacksdale.

Gauge : 4.8½".

No.3

No.	Name	Type	Builder	Works No.	Year	Loc
		0-6-ODH	RSH	7698	1954	
		Rebuilt from 0-6-ODM by YE			1965	Cv
D 2	I'LL TRY	0-4-ODM	RH	375714	1955	Gl
	PAUL	0-4-ODE	RH	395293	1956	PH
EMFOUR 7	CHARLES BUCHAN	0-6-ODM	RH	326671	1957	Bb
		0-6-ODM	HC	D1018	1957	Cv
D 14 (D3613) DAVID		0-6-ODM	HC	D1114	1958	Bk
(D3618) ROBIN		0-6-ODE	Dar		1958	MG
D 15 (D3619) SIMON		0-6-ODE	Dar		1958	MG
(D2132) LESLEY		0-6-ODE	Dar		1958	MG
D2138		0-6-ODM	Sdn		1960	NH
		0-6-ODM	Sdn		1960	PH
D2299		0-6-ODM	(RSH	8158	1960	Hn
			(DC	2680	1960	
D 8		0-4-ODH	HE	5623	1960	Ns
	FENELLA	0-6-ODH	S	10072	1961	Ly
	JULIAN	0-6-ODH	S	10160	1963	Cg
D 11		0-6-ODH	S	10161	1963	Gl
D 12		0-8-ODH	S	10169	1964	MG
D 13		0-8-ODH	RR	10178	1964	MG
		0-6-ODH	RR	10190	1964	MG
	JAYNE	4wDH	TH/S	137C	1964	PH
	STEPHEN	0-6-ODH	TH/S	141C	1964	Ly
	CHARLES	4wDH	TH	147V	1964	Hn
	ANDREW	0-6-ODH	TH	161V	1965	Bb
	CHARLES	0-6-ODH	YE	2940	1965	Cv
	SUSAN	0-4-ODH	HE	6675	1966	Cv
	GILLIAN	4wDH	TH	176V	1966	Gl
		4wDH	TH	182V	1967	Gl

Gauge : 2'0".

Type	Builder	Works No.	Year	Loc
4wDM	RH	268880	1948	Gl
0-6-ODM	HC	DM812	1953	BkTC
0-6-ODM	HC	DM836	1954	BkTC
4wDH	HE	8909		BkTC

SOUTH MIDLANDS AREA

(Headquarters : Coleorton Hall, Coleorton, Leicestershire.)

Locomotives are kept at the following locations :-

DERBYSHIRE

CH   (SK 279193)   Cadley Hill Colliery, Castle Gresley.

KENT

Sn   (TR 246512)   Snowdown Colliery, Nonington.
Tl   (TR 287503)   Tilmanstone Colliery, Eythorne.

LEICESTERSHIRE

Df	(SK 458069)	Desford Colliery, Bagworth.
Rw	(SK 312162)	Rawdon Colliery, Moira.
Sb	(SK 420144)	Snibston Colliery, Coalville.

WARWICKSHIRE

Bd	(SP 280970)	Baddesley Colliery, Baxterley.
BC	(SK 253000)	Birch Coppice Colliery, Dordon.
DM	(SP 257902)	Daw Mill Colliery, Over Whitacre.

WEST MIDLANDS

Cv	(SP 322843)	Coventry Colliery, Keresley.

Gauge : 4'8½".

	ST. THOMAS	0-6-0ST	OC	AE	1971	1927	Sn	OOU
	ST. DUNSTAN	0-6-0ST	OC	AE	2004	1927	Sn	OOU
	SWIFTSURE	0-6-0ST	IC	HE	2857	1943	CH	
	PROGRESS	0-6-0ST	IC	RSH	7298	1946	CH	
No.1802/B5	15224	0-6-0DE		Afd		1949	Sn	OOU
No.1802/B3	12131	0-6-0DE		Dar		1952	Sn	OOU
		0-4-0DM		JF	4160002	1952	Sn	OOU
	EMPRESS	0-6-0ST	OC	WB	3061	1954	CH	
		0-6-0DH		S	10055	1961	CH	OOU
	CADLEY HILL No.1	0-6-0ST	IC	HE	3851	1962	CH	OOU
	NCB 8 1962	4wDH		TH/S	120C	1962	Sn	
No.1	NCB 9 1963	4wDH		TH/S	126C	1963	Rw	OOU
65		0-6-0ST	IC	HE	3889	1964	CH	OOU
	NCB 13 1964	0-6-0DH		HE	6294	1964	Sb	
	NCB 12 1964	0-6-0DH		HE	6295	1964	Sb	
No.64/1	WARWICKSHIRE 10	0-6-0DH		RR	10187	1964	DM	
No.64/2		0-6-0DH		RR	10188	1964	DM	
		0-6-0DH		RR	10212	1964	Bd	
No.64-37		0-6-0DH		RR	10239	1965	BC	
No.64-38		0-6-0DH		RR	10240	1965	Bd	
		0-6-0DH		RR	10255	1966	Bd	
	NCB 17 1966	0-6-0DH		HE	6288	1966	CH	
	NCB 16 1966	0-6-0DH		HE	6289	1966	Df	
	NCB 18 1967	0-6-0DH		HE	6683	1967	Sb	
		0-6-0DH		EEV	D1197	1967	Rw	
	NCB 20 1967	0-6-0DH		HE	6692	1967	CH	
		0-6-0DH		HE	6693	1967	Df	
	NCB 20 1968	0-6-0DH		EEV	D1200	1967	Sb	
		0-6-0DH		EEV	D1231	1967	Rw	
	NCB 22 1968	0-4-0DH		RH	544875	1968	Df	
	NCB SM AREA No.23 1969	0-6-0DH		HE	6689	1969	Cv	
	NCB SM AREA No.24 1969	0-6-0DH		HE	6690	1969	Cv	
		0-6-0DH		HE	7305	1973	BC	
		0-6-0DH		HE	7396	1974	Cv	
	BADDESLEY No.3	0-6-0DH		TH	257V	1975	DM	
No.4		0-6-0DH		HE	7494	1976	Cv	

Gauge : 2'0".

		0-4-0DM		HC	DM686	1948	Tl	OOU

SOUTH WALES AREA

(Headquarters : Coal House, Llanishen, Cardiff, South Glamorgan.)

Locomotives are kept at the following locations :-

DYFED

Cn	(SN 494077)	Cynheidre Colliery, Five Roads.
Pf	(SN 621114)	Pantyffynnon Colliery Loco Shed, Ammanford. (Serves Wernos Coal Preparation Plant.)

GWENT

CS	(ST 215961)	Celynen South Colliery, Abercarn.
Mr	(SO 189041)	Marine Colliery, Cwm.

MID GLAMORGAN

Ac	(ST 082944)	Abercynon Colliery, Abercynon.
Bg	(ST 154991)	Bargoed Colliery, Bargoed. (Closed)
Cw	(ST 066865)	Cwm Colliery, Llantwit Fardre.
DN	(ST 102972)	Deep Navigation Colliery, Treharris.
Fd	(SS 903914)	Ffaldau Colliery, Blaengarw. (Loco stored only.)
LW	(ST 063943)	Lady Windsor Colliery, Ynysybwl.
Mg	(SS 858918)	Maesteg Coal Preparation Plant, Maesteg. (Serves Coegnant and St. Johns Collieries.)
Md	(SS 963999)	Mardy Colliery, Maerdy.
MV	(ST 074999)	Merthyr Vale Colliery, Aberfan.
MA	(ST 050989)	Mountain Ash Loco Shed & Workshops (ST 051988), Mountain Ash.
Ng	(ST 119856)	Nantgarw Colliery and Training Centre, Treforest.
Pk	(ST 061972)	Penrikyber Colliery, Penrhiwceiber.
TTC	(SS 894841)	Tondu Training Centre, Tondu.
Tw	(SN 943057)	Tower Colliery, Hirwaun.
T4	(SN 926043)	Tower No.4 Colliery, Hirwaun. (Use of loco ceased.)
Ty	(ST 060908)	Tymawr Colliery, Trehafod.

WEST GLAMORGAN

An	(SN 702084)	Abernant Colliery, Cwmgors.
Ap	(SN 863059)	Aberpergwm Colliery, Glyn Neath.
Bt	(SN 785032)	Blaenant Colliery, Crynant.
Bl	(SN 866057)	Blaengwrach Colliery, Glyn Neath.
Bn	(SN 595011)	Brynlliw Colliery, Grovesend.

Gauge : 4'8½".

	SIR JOHN	0-6-0ST	OC		AE	1680	1914	MA	OOU
1426		0-6-0ST	OC		P	1426	1916		
		Incorporates parts of P				1187		Bn	OOU
	LORD CAMROSE	0-6-0ST	OC		AE	2008	1930	MA	OOU
	MENELAUS	0-6-0ST	OC		P	1889	1935	Mr	
	LLANTANAM ABBEY	0-6-0ST	OC		AB	2074	1939	MA	OOU
No.8		0-6-0ST	IC		RSH	7139	1944		
		Rebuilt			HE	3880	1961	MA	OOU
9642		0-6-0PT	IC	+	Sdn		1946	Mg	Pvd
	GWENT	0-6-0DM		++	(VF	D78	1948	Ty	
					(DC	2252	1948		
		0-6-0ST	OC		P	2114	1951	Bn	OOU
D3000		0-6-0DE			Derby		1952	MA	
D3014		0-6-0DE			Derby		1952	MV	

		0-6-0ST	IC	HE	3829	1955	LW	OOU
No.1		0-6-0DE		BT/WB	3073	1955	Cw	
No.2		0-6-0DE		BT/WB	3074	1955	Cw	
D3183		0-6-0DE		Derby		1955	MV	
D3255		0-6-0DE		Derby		1956	Md	
3261		0-6-0DE		Derby		1956	Tw	
	PAMELA	0-6-0ST	IC	HE	3840	1956	Mg	OOU
D 1		0-4-0DH		NB	27655	1956	Lw	
(D2774)		0-4-0DH		NB	28027	1960		
			Rebuilt	AB		1968	CS	
5/521/011		0-4-0DH		NB	28040	1961	CS	
		0-6-0DE		WB	3161	1961	MA	
		0-6-0DE		WB	3191	1961	MA	
No.9 1962		0-4-0DM		AB	475	1962	Pf	
		0-4-0DM		AB	483	1963	Pf	
	ACN No.1	0-4-0DH		EES	8426	1963	MA	Dsm
	ACN No.2	0-4-0DH		EES	8427	1963	Ty	
520/4		0-6-0DH		EES	8428	1963	Cn	
		0-6-0DH		EES	8429	1963	Fd	Dsm
	JAYNE	0-6-0DH		EES	8432	1963	An	
	TOWER No.1	0-6-0DH		EES	8446	1963	Md	
		0-6-0DH		YE	2912	1964	An	
	ABERAMAN No.6	0-6-0DH		AB	496	1964	MA	
		0-6-0DH		EEV	D909	1964	Mg	
		0-6-0DH		EEV	D910	1964	LW	
	TOWER No.2	0-6-0DH		EEV	D911	1964	Tw	
(D9530)		0-6-0DH		Sdn		1965	Md	OOU
		0-6-0DH		EEV	D919	1965	DN	
		0-6-0DH		EEV	D920	1965	LW	
	TAFF MERTHYR No.1	0-6-0DH		EEV	D921	1965	LW	
	MARINE	0-4-0DH		EEV	D923	1965	Mr	
	TAFF MERTHYR No.2	0-6-0DH		EEV	D1116	1966	DN	
	TAFF MERTHYR No.3	0-6-0DH		EEV	D1117	1966	An	
		0-6-0DH		EEV	D1118	1966	Pk	
	APG No.1	0-6-0DH		EEV	D1138	1966	Ap	
	APG No.2	0-6-0DH		EEV	D1139	1966	Ap	
		0-6-0DH		HC	D1374	1966		
			Rebuilt	HE	7485	1976	Mr	
		0-6-0DH		HC	D1375	1966		
			Rebuilt	HE	7487	1976	Mr	
		0-6-0DH		EEV	D1198	1967	Cn	
		0-6-0DH		EEV	D1199	1967	Bn	
No.1		0-6-0DH		GECT	5368	1973	Mg	
No.2		0-6-0DH		GECT	5369	1973	Mg	

+ Privately owned - stored on N.C.B. premises.
++ Property of A.R.Adams Ltd, Gwent.

Gauge : 3'0".

3	4wDM		RH	198267	1939	Bg	
2	4wDM		RH	202988	1940	Bg	
	4wDM	+	RH		1960	Md	
	0-4-0DM		HE	6696	1966	Ng	
	4wDH		HE	8819	1979	Ng	

÷ 444199 or 444200.

Gauge : 2'11".

4wDM		RH	444205	1961	Ac	Dsm

Gauge : 2'10½".

3/521/5093	4wDM	RH	487965	1963	LW

Gauge : 2'10".

521/2004	TRM 207	4wDM	RH	402990	1957	T4
521/2034		4wDM	RH	451904	1962	Ng

Gauge : 2'9½".

3/521/58	4wDM	FH	3734	1955	Cw
3/521/173	4wDM	FH	3735	1955	Cw
1	4wDM	RH	496040	1963	Cw

Gauge : 2'6".

MARY POPINS	4wDM	RH	444196	1960	Ac

Gauge : 2'3".

0—4—ODM	+	HE	6604	1965	An
4wDH		HE	8812	1978	An

+ Incorporates parts of HE 5432.

Gauge : 2'0".

9	0—4—ODM	HE	3358	1945	Bl	
	4wDM	RH	398064	1956	Cn	
12	0—4—ODM	HE	5340	1957	Bl	
4	0—4—ODM	HE	5599	1961	Bl	
2	0—4—ODM	HE	6048	1961	Bl	
3	0—4—ODM	HE	6049	1961	Bl	
520/6	4wDM	RH	487966	1963	Cn	
1	0—4—ODM	HC	DM1312	1963	Bl	OOU
5	0—4—ODM	HC	DM1313	1963	TTC	
6	0—4—ODM	HC	DM1355	1965	Bl	
7	0—4—ODM	HE	6623	1973	Bl	
8	0—4—ODM	HE	7411	1974	Bl	

Trapped Rail Systems :-

Becorit "Roadrailer" (200mm)

3	2 axle drive	BGB	DRL 40/1/501	1971	Ap
4	2 axle drive	BGB			Ap
5	2 axle drive	BGB	DRL 50.200.516	1974	Ap
6	2 axle drive	BGB	DRL 50.200.525	1975	Ap
7	2 axle drive	BGB	DRL 40/3/512	1973	Ap

UMM "Coolie Car"

UMM	40.007	1972	Bt

COAL PRODUCTS DIVISION

(Headquarters : Coal House, Harrow-on-the-Hill, Greater London.)

NATIONAL SMOKELESS FUELS LTD. (Wholly owned subsidiary company.)

DURHAM MANAGEMENT UNIT.

(Headquarters : Coal House, Team Valley, Gateshead 11, Tyne & Wear.)

Locomotives are kept at the following locations :-

COUNTY DURHAM

| FC | (NZ 362317) | Fishburn Coking Plant, Fishburn. |
| HC | (NZ 390460) | Hawthorn Coking Plant, Murton. |

TYNE & WEAR

DC	(NZ 193615)	Derwenthaugh Coking Plant, Blaydon.
LC	(NZ 318511)	Lambton Coking Plant, Fencehouses.
MC	(NZ 316627)	Monkton Coking Plant, Hebburn.
NC	(NZ 238612)	Norwood Coking Plant, Dunston. (Closed)

Gauge : 4'8½".

Shunting Locos.

No.77		0-6-0ST	OC		RSH	7412	1948	NC	OOU
No.2		0-4-0DM			RH	313391	1952	MC	
42		0-4-0DM		+	RH	384141	1955	LC	
No.9D	1	0-6-0DM			HE	5382	1958	FC	
No.11D	2	0-6-0DM			HE	5305	1959	FC	
(03099)		0-6-0DM			Don		1960	MC	
		0-4-0DH			HE	6263	1964	HC	
No.2		0-4-0DH			HE	6677	1967	FC	
No.1		0-4-0DH			HE	6688	1968	NC	

Coke Oven Locos.

	0-4-0WE	GB	2047	1946	NC	
	0-4-0WE	RSH	7692	1953	FC	OOU
	0-4-0WE	RSH	7804	1954	LC	
	0-4-0WE	RSH	7882	1957	DC	
	0-4-0WE	RSH	7886	1958	HC	
	0-4-0WE	RSH	8093	1959	MC	
	0-4-0WE				FC	

Inspection Car.

| | 2w-2DMR | Wkm | 10913 | 1976 | FC |

+ Carries 384161 in error.

MIDLANDS MANAGEMENT UNIT.

(Headquarters : P.O. Box 16, Wingerworth, Chesterfield, Derbyshire.)

Locomotives are kept at the following locations :-

DERBYSHIRE

AvC   (SK 394678)   Avenue Carbonisation Plant, Wingerworth.

SOUTH YORKSHIRE

SwC   (SK 366948)   Smithywood Coking Plant, Chapeltown.

WEST MIDLANDS

CvH   (SP 318845)   Coventry Home Fire Plant, Keresley.

Gauge : 4'8½".

Shunting Locos.

S.W.C.P. No.2	0-6-0ST	IC	HE	3192	1944		
		Rebuilt	HE	3888	1964	SwC	
AVENUE 1	0-6-0DM		HE	4511	1955	AvC	OOU
AVENUE 2	0-6-0DM		HE	4512	1955	AvC	
AVENUE 3	0-6-0DM		HE	4974	1955	AvC	OOU
(AVENUE 4)	0-6-0DM		HE	4514	1955	AvC	OOU
(AVENUE 5)	0-6-0DM		HE	4515	1955	AvC	OOU
	0-4-0DE		YE	2729	1958	SwC	
COVENTRY HOMEFIRE UNIT No.1	0-8-0DH		HE	6657	1965	CvH	
COVENTRY HOMEFIRE UNIT No.2	0-8-0DH		HE	6658	1965	CvH	
	0-6-0DH		EEV	D1250	1968	SwC	
AVENUE 6	4wDH		TH	199V	1968	AvC	
AVENUE 7	4wDH		TH	219V	1970	AvC	
AVENUE 4	0-4-0DH		HC	D1388	1970	AvC	
AVENUE 5	0-4-0DH		HC	D1345	1970	AvC	

2

Coke Oven Locos.

4wWE		GB	2508	1955	AvC
4wWE		GB	2509	1955	AvC
0-4-0WE		GB	2543	1955	SwC

WALES MANAGEMENT UNIT.

(Headquarters : Nantgarw, Cardiff, South Glamorgan.)

Locomotives are kept at the following locations :-

MID GLAMORGAN

AP	(ST 032999)	Aberaman Phurnacite Plant, Abercwmboi.
CEC	(ST 016858)	Coed Ely Coking Plant, Tonyrefail.
CwC	(ST 066865)	Cwm Coking Plant, Llantwit Fardre.
NgC	(ST 116861)	Nantgarw Coking Plant, Treforest.

Gauge : 4'8½".

Shunting Locos.

4	(12061)	0-6-0DE	Derby		1949	NgC
5	(12063)	0-6-0DE	Derby		1949	NgC
6	(12071)	0-6-0DE	Derby		1950	NgC
1	(D2139)	0-6-0DM	Sdn		1960	CEC
2	(D2178)	0-6-0DM	Sdn		1962	CEC
		0-6-0DH	AB	494	1964	AP
		0-6-0DH	AB	497	1964	AP

Coke Oven Locos.

	0-4-0WE	GB	2180	1948	NgC
1	0-4-0WE	GB	2690	1950	CwC
2	0-4-0WE	GB	2691	1950	CwC
	0-4-0WE	GB	420174	1969	NgC

THOMAS NESS LTD. (A wholly owned subsidiary company.)

Locomotives are kept at the following locations :-

MID GLAMORGAN

CT   (ST 162864)   Caerphilly Tar Distillation Plant, Caerphilly.

TYNE & WEAR

SAT   (NZ 294633)   St. Anthony's Tar Distillation Plant, Walker. RTC.

Gauge : 4'8½".

No.1

0-4-0ST	OC	P	2142	1953	SAT	OOU
0-6-0DM		HE	5511	1960	CT	

OPENCAST EXECUTIVE

Locomotives listed may be the property of the site operating contractor (listed alongside each location), of the opencast executive, or be hired by the contractor from a third party.

SCOTTISH REGION

Locomotives are kept at the following locations :-

FIFE

Wf   (NT 196983)   Westfield Disposal Point, Kinglassie.

STRATHCLYDE

Bb   (NS 443083)   Benbain Disposal Point, Waterside.

Gauge : 4'8½".

WL No.1		4wDH	S	10012	1959	Bb	
WL No.2		4wDH	RR	10268	1967	Wf	
		4wDH	RR	10269	1967	Wf	

NORTHERN REGION

Locomotives are kept at the following locations :-

NORTHUMBERLAND

Wd   (NZ 237957)   Widdrington Disposal Point, Widdrington. (Derek Crouch (Contractors) Ltd.)

TYNE & WEAR

Sw   (NZ 205625)   Swalwell Disposal Point, Swalwell. (Johnsons (Chopwell) Ltd.)

Gauge : 4'8½".

68078	L 2	0-6-0ST	IC	AB	2212	1946	Wd	
(12052)	MP 228	0-6-0DE		Derby		1949	Wd	
12074		0-6-0DE		Derby		1950	Sw	
12088		0-6-0DE		Derby		1951	Sw	
(12093)	MP 229	0-6-0DE		Derby		1951	Wd	
	DEREK CROUCH	0-6-0DH		EEV	D1201	1967	Wd	
		0-4-0DH		HE	6678	1968	?	OOU

CENTRAL (EAST) REGION

Locomotives are kept at the following locations :-

DERBYSHIRE

Ox   (   )   Oxcroft Disposal Point.

Sm   (SK 451738)   Seymour Disposal Point, Woodthorpe. (Hargreaves (West Riding) Ltd.)

WH   (SK 445424)   West Hallam Disposal Point, Mapperley. (Hargreaves (West Riding) Ltd.)

LEICESTERSHIRE

SL   (SK 432118)   Coalfield Farm Disposal Point, (South Leicester Colliery), Hugglescote. (Shepherd, Hill & Co Ltd.)

NOTTINGHAMSHIRE

So   (SK 471441)   Shilo Disposal Point, Ilkeston. (Hargreaves (West Riding) Ltd.)

WEST YORKSHIRE

BR   (SE 399285)   Bowers Row Disposal Point, Astley.(Hargreaves (West Riding) Ltd.)

BO   (SE 300164)   British Oak Disposal Point, Crigglestone. (Hargreaves (West Riding) Ltd.)

Gauge : 4'8½".

12099	0-6-0DE	Derby	1952	BO	
12122	0-6-0DE	Dar	1952	BO	
08016	0-6-0DE	Derby	1953		

D2258	0-6-0DM	(RSH (DC	7879 1957 2602 1957	So
(D2049)	0-6-0DM	Don		1958	WH
03037	0-6-0DM	Sdn		1959	BO
(D2148)	0-6-0DM	Sdn		1960	BR
1/13	0-6-0DH	HE	7410 1976	SL
	0-6-0DH	HE	8979 1979	Ox
	4wDH R/R	Unilok	2005	Sm

CENTRAL (WEST) REGION

Locomotives are kept at the following locations :-

CLWYD

Gt   (SD 314516)   Gatewen Disposal Point, New Broughton.(Lindley Plant Ltd.)

GREATER MANCHESTER

Ab   (SD 628015)   Abram Disposal Point, Abram.

Gauge : 4'8½".

		4wDH	S	10059 1961	Ab
(D2182)	3/3	0-6-0DM	Sdn	1962	Gt

SOUTH WESTERN REGION

Locomotives are kept at the following locations :-

DYFED

CB   (SN 424059)   Coed Bach Disposal Point, Kidwelly. (Powell Duffryn Fuels Ltd.)

CM   (SN 529125)   Cwm Mawr Disposal Point, Tumble. (Powell Duffryn Fuels Ltd.)

MID GLAMORGAN

Cb   (SO 092060)   Cwmbargoed Disposal Point, Cwmbargoed. (Taylor Woodrow Construction Ltd.)

Lh   (SS 997828)   Llanharan Disposal Point. (Derek Crouch (Contractors) Ltd.)

WEST GLAMORGAN

GCG   (SN 713120)   Gwaun-cae-Gurwen Disposal Point, Gwaun-cae-Gurwen. (Powell Duffryn Fuels Ltd.)

On   (SN 843105)   Onllwyn Disposal Point, Onllwyn. (Derek Crouch (Contractors) Ltd.)

Gauge : 4'8½".

No.	Name		Type	Builder	Works No.	Year	Code
D3019 (D2260) 102	GWYNETH THOMAS HARLING		0-6-0DE	Derby		1953	GCG
			0-6-0DM	(RSH	7890	1957	CM
			0-4-0DM	(DC	2604	1957	
			0-6-0DM	RH	418790	1958	CM
D4092 (07002) (07006) 07012	CHRISTINE		4wDM	RH	421702	1959	Cb
			0-6-0DE	RH	441936	1960	LH
			0-6-0DE	Dar		1962	GCG
			0-6-0DE	RH	480687	1962	CB
			0-6-0DE	RH	480691	1962	CB
468048 MP 201 MP 202	VICTOR	LSSH	0-6-0DM	RH	480697	1962	CM
			0-6-0DH	HC	D1254	1962	CB
			0-6-0DH	RH	468048	1963	GCG
	HILDA		0-6-0DH	EEV	D1202	1967	On
			0-6-0DH	EEV	D1230	1969	On
			0-6-0DH	EEV	3994	1970	Lh

MINING RESEARCH and DEVELOPMENT ESTABLISHMENT

Locomotives are kept at the following location :-

DERBYSHIRE

STS (SK 285193) Swadlincote Test Site, Swadlincote.

Gauge : 3'0".

		+ DE	GMT	1976	STS

+ Experimental 3-car articulated linear motor manriding train.

Other locomotives and personnel carriers here from time to time for tests and evaluation.

SECTION 5

MINISTRY OF DEFENCE
Army Department

MINISTRY OF DEFENCE
Army Department

Locomotives are used at the following locations; those depots marked + normally work their traffic with permanent local stock but also use Army Department locomotives as the occasion demands (see entry under appropriate County). Depot types are :-

C.A.D.	+ Central Ammunition Depot.
C.O.D.	= Central Ordnance Depot.
D.O.E.	= Department of the Environment.
O.E.S.D.	=
P.E.E.	= Proof & Experimental Establishment.
R.A.F.	= Royal Air Force.
R.A.O.C.	= Regional Depot Royal Army Ordnance Corps.
R.E.M.E.	= Royal Electrical & Mechanical Engineers.
R.N.A.D.	= Royal Naval Armament Depot.
R.O.F.	= Royal Ordnance Depot.

AFD		(TQ 993434) R.A.O.C. Ashford, Kent.
ALD		(SU 880514) R.A.O.C. Aldershot, Hants. (Locomotive here occasionally.)
ASH		(SO 932338) Ashchurch, Glos.
BIR		(NZ 267558) R.O.F. Birtley, Tyne & Wear.
BIS		(SP 612176) C.O.D. Bicester Central Workshops, Oxon. Locomotives stored at Graven Hill (SP 581203, 583199) and repaired at Arncott. Also N.G. locos occasionally present.
BMR		(SP 581203) C.O.D. Bicester Military Railway, Oxon.
BRM		(SU 658589) C.A.D. Bramley, Hants.
CHI		(SK 509349) C.O.D. Chilwell, Notts.
CM		(SJ 853324) D.O.E. Quality Testing Unit, Cold Meece, near Swynnerton, Staffs. (N.G. only.)
CMK	+	(ST 982302) R.A.F. No.11 Maintenance Unit, Chilmark & Dinton, Wilts. (Integrated system.)
CWT		(ST 48x90x) R.A.F. Caerwent, Gwent. (Locomotive here occasionally.)
DHL	+	(SU 276266) R.N.A.D. Dean Hill, Wilts.
DON		(SJ 696133) C.O.D. Donnington, Salop.
ER		(NY 246656) R.O.F. East Riggs, Dumfries & Galloway.
ESK		(SD 083927) P.E.E. Eskmeals, Cumbria.
HES		(SE 523540) 322 Engineer Park, Hessay, North Yorks.
INC		(NS 677758) P.E.E. Inchterf, near Kirkintilloch, Strathclyde.
KIN		(SP 373523, 374524, 378518) C.A.D. Kineton, Warwicks.
LEC		(SE 024432) Army School Of Mechanical Transport, R.C.T. Museum, Leconfield, Humberside.
LEE	+	(SE 370345) R.O.F. Leeds, West Yorks.
LEU	+	(NO 453212) R.A.F. Leuchars, Fife.
LM		(SP 153473, 155476) Engineer Resources, Long Marston, near Stratford-Upon-Avon, Warwicks.
LON		(NY 355676) C.A.D. Longtown, Cumbria. (One railcar outstationed at Smalmstown. NY 367687)
LYD	+	(TR 033198) D.O.E. Lydd Gun Ranges, Lydd, Romney Marsh, Kent.(N.G. only.)
MCH		(SU 395103) Marchwood Military Railway, Hants.
MOL		(SO 508467) R.A.O.C. Moreton-on-Lugg, Hereford & Worcester.
RAD	+	(SJ 784545) R.O.F. Radway Green, Alsager, Cheshire.
RUD		(SK 575322) Ordnance Storage & Disposal Depot, Ruddington, Notts.
SHO		(TM 946856) P.E.E. New Ranges, Shoeburyness, Southend-on-Sea, Essex.
STI		(NS 803934) R.A.O.C. Stirling, Central.
THA		(SU 524664) R.A.O.C. Thatcham, Berks.
TID		(SU 261507) Tidworth, Wilts. (Shed at Ludgershall.)
WAR		(ST 885450) Warminster, Wilts.
YC		(SP 837555) O.E.S.D. Yardley Chase, Northants. (Closed)

Gauge : 4'8½".

92		WAGGONER	0-6-0ST	IC		HE	3792	1953	MCH
98		ROYAL ENGINEER	0-6-0ST	IC		HE	3798	1953	LM
110			4wDM			RH	411319	1958	INC
114			4wDM			FH	3913	1959	INC
123	A7	15548	0-4-0DM			Bg/DC	2157	1941	BIR
200			0-4-0DM			AB	358	1941	DHL
201		FROG 61/30289	0-4-0DM			AB	362	1942	RAD
202			0-4-0DM			AB	357	1941	HES
212	A7		0-4-0DM			VF	4861	1942	CHI
						DC	2169	1942	
213			0-4-0DM			VF	4862	1942	AFD
						DC	2170	1942	
221			0-4-0DM			Bicester		c1955	LM
222			0-4-0DM			VF	5256	1945	BIS
						DC	2175	1945	
223			0-4-0DM			VF	5257	1945	ASH
						DC	2176	1945	
224			0-4-0DM			VF	5259	1945	DON
						DC	2178	1945	
225			0-4-0DM			VF	5260	1945	LM
						DC	2179	1945	
226			0-4-0DM			VF	5261	1945	STI
						DC	2180	1945	
227			0-4-0DM			VF	5262	1945	STI
						DC	2181	1945	
228			0-4-0DM			VF	5263	1945	CHI
						DC	2182	1945	
229			0-4-0DM			VF	5264	1945	RUD
						DC	2183	1945	
230			0-4-0DM			VF	5265	1945	AFD
						DC	2184	1945	
231			0-4-0DM			VF	5266	1945	MOL
						DC	2204	1945	
232			0-4-0DM			VF	5267	1945	LM
						DC	2205	1945	
233			0-4-0DM			AB	369	1945	MOL
234			0-4-0DM			AB	370	1945	SHO
235			0-4-0DM			AB	371	1945	BIS
236			0-4-0DM			AB	372	1945	ASH
237			0-4-0DM			VF	4863	1942	ER
						DC	2171	1942	
244			0-4-0DH			TH	130C	1963	
			Rebuild of 0-4-0DM			JF	22971	1942	ER
247			0-4-0DM			AB	342	1940	BIS OOU
249			0-4-0DM			VF	5258	1945	ESK
						DC	2177	1945	
250			0-4-0DM			AB	368	1945	THA
251			0-4-0DM			RH	390772	1956	THA
252			4wDH			TH	270V	1977	BMR
253		CONDUCTOR	4wDH			TH	271V	1977	BMR
254			4wDH			TH	272V	1977	KIN
255			4wDH			TH	273V	1977	KIN
256		MARLBOROUGH	4wDH			TH	274V	1977	KIN
257		TELA	4wDH			TH	275V	1978	KIN
258			4wDH			TH	298V	1981	BIS
259			4wDH			TH	299V	1981	BIS
			4wDH			RR	10242	1966	CWT
			4wDH			RR	10244	1966	CMK
400			0-4-0DH			NBL	27421	1955	RUD
402			0-4-0DH			NBL	27423	1955	KIN
404			0-4-0DH			NBL	27425	1955	BIS
405			0-4-0DH			NBL	27426	1955	LEU
406			0-4-0DH			NBL	27427	1955	TID

No.		Name	Type		Builder	Works No.	Year	Shed	Notes
407			0-4-0DH		NBL	27428	1955	ESK	OOU
408			0-4-0DH		NBL	27429	1955	RUD	
409			0-4-0DH		NBL	27644	1959	ESK	
410			0-4-0DH		NBL	27645	1958	BIS	
411			0-4-0DH		NBL	27646	1959	BIS	
412			0-4-0DH		NBL	27647	1959	TID	
413			0-4-0DH		NBL	27648	1959	TID	
420			0-6-0DH		RH	459515	1961	DON	
421			0-6-0DH		RH	459516	1961	BRM	
422			0-6-0DH		RH	459517	1961	BRM	
423			0-6-0DH		RH	459518	1961	SHO	
424			0-6-0DH		RH	459520	1961	SHO	
425			0-6-0DH		RH	459519	1961	MCH	
426			0-6-0DH		RH	459521	1961	LON	
427			0-6-0DH		RH	466616	1961	DON	
428			0-6-0DH		RH	466617	1961	LON	
429			0-6-0DH		RH	466618	1961	LON	
430			0-6-0DH		RH	466621	1961	DON	
431			0-6-0DH		RH	466622	1962	SHO	
432			0-6-0DH		RH	466623	1962	MCH	
433		THE CITY OF NOTTINGHAM	0-6-0DH		RH	468043	1963	DON	
434			0-6-0DH		RH	468044	1963	DON	
435			0-6-0DH		RH	468045	1963	BRM	
436			0-6-0DH		RH	468046	1963	LON	
440		HASSAN	0-6-0DH		RH	468041	1962	KIN	
610		GENERAL LORD ROBERTSON	0-8-0DH		S	10143	1963	SHO	
620		SAPPER	0-8-0DH		AB	500	1964	BMR	
621		WAGGONER	0-8-0DH		AB	501	1965	KIN	
622		GREENSLEEVES	0-8-0DH		AB	502	1965	BMR	
623		STOREMAN	0-8-0DH		AB	503	1965	BMR	
624		ROYAL PIONEER	0-8-0DH		AB	504	1965	BMR	
625		CRAFTSMAN	0-8-0DH		AB	505	1966	KIN	
9031	6	GAZELLE	0-4-2WT	IC	Dodman		1893	LEC	
9103			2w-2PMR		Wkm	8089	1958	CWT	
9104			2w-2DMR		DC	2323		BMR	DsmT
9105	3		2w-2PMR		Wkm	7397	1956	YC	
9106			2w-2DMR		DC	2325		BRM	DsmT
9107			2w-2DMR		DC	2326		KIN	
9108	2		2w-2DMR		DC	2327		BRM	DsmT
9112			2w-2DMR		DC	2328		KIN	
9113			2w-2PMR		Bg	3538		BRM	DsmT
9114			2w-2DMR		Bg	3539		BRM	DsmT
9115			2w-2DMR		CE	5380/1	1968	KIN	
9116			2w-2DMR		CE	5380/2	1968	KIN	
9117			2w-2DMR		CE	5427	1968	BIS	
9118			4wDMR		BD	3706	1975	SHO	
9119			4wDMR		BD	3707	1975	BIS	
9120			4wDMR		BD	3708	1975	KIN	
9121			4wDMR		BD	3709	1975	BRM	
9122			4wDMR		BD	3710	1975	KIN	
9123			4wDMR		BD	3711	1975	KIN	
9124			4wDMR		BD	3712	1975	KIN	
9125			4wDMR		BD	3713	1975	BRM	
9126	1		4wDMR		BD	3741	1976	BRM	
9127			4wDMR		BD	3742	1976	LON	
9128	2		4wDMR		BD	3743	1976	KIN	
9129			4wDMR		BD	3744	1976	LON	
9150			4wDMR		BD	3745	1976	SHO	
			4wDMR		BD	3746	1976	KIN	
No.1			2w-2PMR		Wkm	7092	1955	KIN	DsmT
No.2			2w-2PMR		Wkm	7093	1955	KIN	DsmT

9222		2w-2PMR	Wkm	8199	1958	KIN	DsmT
9246		2w-2DMR	DC	1894		KIN	DsmT
9247		2w-2DMR	DC	1896		KIN	DsmT
9248		2w-2DMR	DC	1895		SHO	DsmT

Gauge : 60cm.

22	LOD/758173	4wDM	RH	211609	1941	LYD
25		4wDM	RH	222100	1943	CM
AD27	LOD 758141 L.R. No.5	4wDM	RH	229633	1944	LYD
34		4wDM	HE	7009	1971	ER
35		4wDM	HE	7010	1971	ER
36		4wDM	HE	7011	1971	ER
37		4wDM	HE	7012	1971	ER
38		4wDM	HE	7013	1971	ER
AD39	L.R. No.3	4wDM	RH	223696	1944	LYD
	BLACK BARRON	4wDM	MR	8745	1942	ER
	SNOOPY	4wDM	MR			ER

SECTION 6 IRELAND
ULSTER

ANTRIM	273
ARMAGH	273
DOWN	274
FERMANAGH	+
LONDONDERRY	275
TYRONE	275

REPUBLIC OF IRELAND

CARLOW	276	LONGFORD	281
CAVAN	+	LOUTH	282
CLARE	276	MAYO	282
CORK	276	MEATH	+
DONEGAL	277	MONAGHAN	+
DUBLIN	278	OFFALY	283
GALWAY	278	ROSCOMMON	284
KERRY	279	SLIGO	284
KILDARE	280	TIPPERARY	284
KILKENNY	+	WATERFORD	285
LEITRIM	+	WESTMEATH	286
LEIX/LAOIS	281	WEXFORD	286
LIMERICK	+	WICKLOW	286

+ No known locomotives exist.

ULSTER

ANTRIM

BULLRUSH PEAT CO. BALLAGHY.
Gauge : 75cm.

		4wDM		MR	22220	1964
		4wDM		MR	40S307	1968
		4wDM		MR	40S309	1968

THE LORD O'NEILL, SHANES CASTLE, RANDALSTOWN.
Gauge : 3'0".

No.1	TYRONE					
No.2	RORY	0-4-0T	OC	P	1026	1904
No.3	SHANE	4wDM		MR	11039	1956
No.4	NIPPY	0-4-0WT	OC	AB	2265	1949
	NANCY	4wDM		FH	2014	1936
6	COLUMBKILLE	0-6-0T	OC	AE	1547	1908
12		2-6-4T	OC	NW	830	1907
18		0-4-0+4wDMR		WkB/Dundalk		1934
		0-4-0+4wDMR		WkB/Dundalk		1940
		2w-2PMR		Wkm	7441	1956

RAILWAY PRESERVATION SOCIETY OF IRELAND, WHITEHEAD DEPOT.
Gauge : 5'3".

186		0-6-0	IC	SS	2838	1879
171	SLIEVE GULLION	4-4-0	IC	BP	5629	1913
			Rebuilt Dundalk		42	1938
No.3	GUINNESS	0-4-0ST	OC	HC	1152	1919
No.3	R.H.SMYTH	0-6-0ST	OC	AE	2021	1928
4		2-6-4T	OC	Derby		1947
No.27	LOUGH ERNE	0-6-4T	IC	BP	7242	1949
23		4wDM		FH	3509	1951
1		4w-4wDMR		NCC		1933

ARMAGH

IRISH PEAT DEVELOPMENT CO LTD, MAGHERY. (Closed)
Gauge : 3'0".   (Line to Verners Bog, near Annaghmore.)

		4wDM		FH	3719	1954	OOU
		4wDM		CS	1727	1955	OOU

DOWN

BELFAST AND COUNTY DOWN RAILWAY TRUST, BALLYNAHINCH JUNCTION.
Gauge : 5'3".

		0-4-0T	OC	OK	12475 1934
		0-4-0T	OC	OK	12662 1935

BELFAST TRANSPORT MUSEUM, WITHAM STREET, BELFAST 4.
Gauge : 5'3".

No.1		0-6-0ST	OC	RS	2738 1891
93		2-4-2T	IC	Dundalk	16 1895
30		4-4-2T	IC	BP	4231 1901
74	DUNLUCE CASTLE	4-4-0	IC	NB	23096 1924
No.85	MERLIN	4-4-0	3C	BP	6733 1932 +
800	MAEVE	4-6-0	3C	Inchicore	1939
8178		2w-2DMR		Dundalk	1932

+ Currently under renovation at Harland & Wolff Engine Works, Belfast.

Gauge : 3'0".

2		0-4-0Tram	OC	K	T84 1883
2	KATHLEEN	4-4-0T	OC	RS	2613 1887
2		0-4-0T	OC	P	1097 1906
2	BLANCHE	2-6-4T	OC	NW	956 1912
11	PHOENIX	4wDM		AtW	114 1928
			Rebuilt	Dundalk	1932
1		2-2-0PMR		Alldays & Onions	1906
3		2-4w-2PMR		DC	1926 +
10		0-4-0+4wDMR		WkB	1932

+ Now unmotorised.

Gauge : 1'10".

20		0-4-0T	IC	Spence	1905

CONTAINER REFURBISHING CO LTD, NEWRY.
Gauge : 3'0".

	4wDM	L	36745 1951

NORTHERN IRELAND RAILWAYS LTD, GREAT VICTORIA STREET DEPOT, BELFAST.
Gauge : 5'3".

	4wDM	R/R	Unilok	A114

ULSTER FOLK MUSEUM, CULTRA.
Gauge : 2'0".

	4wPM	MR	246 1916
	4wDM	HE	3127 1943
	4wDM	MR	9202 1946
	4wDM	FH	

LONDONDERRY

<u>J.McGILL, EGLINTON</u>.
Gauge : 1'10".

4wDM		FH		

<u>C.TENNANT (N.I.) LTD, GLENCONWAY, DUNGIVEN</u>.
Gauge : 3'0".

4wDM		RH	218030	1942

TYRONE

<u>DR. R.COX, STRABANE STATION</u>.
Gauge : 3'0".

2-6-4T	OC	NW	828	1907	OOU
2-6-4T	OC	NW	829	1907	OOU

REPUBLIC OF IRELAND

CARLOW

COMHLUCHT SIUICRE EIREANN TEO, CARLOW SUGAR FACTORY.
Gauge : 5'3".   RTC.

	4wDM			RH	382827	1955	OOU

CLARE

CORAS IOMPAIR EIREANN.
Ennis Depot.
Gauge : 5'3".

416A	2w-2PMR			Wkm	8919	1962	

Ennis Station.
Gauge : 3'0".

5	0-6-2T	OC	—	D	2890	1892	Pvd

CORK

COMHLUCHT SIUICRE EIREANN TEO, MALLOW SUGAR FACTORY.
Gauge : 5'3".

1	4wDM			RH	305322	1951	
2	4wDM			RH	312425	1951	

CORAS IOMPAIR EIREANN.
Cork Depot.
Gauge : 5'3"

415A	2w-2PMR			Wkm	8918	1962	

Cork (Kent) Station.
Gauge : 5'3".

36	2-2-2	IC		Bury		1848	Pvd

Mallow Station.
Gauge : 5'3".

90		0-6-0T	IC		Inchicore	1875	Pvd

IRISH STEEL HOLDINGS LTD, HAULBOWLINE ISLAND, COBH.
Gauge : 5'3".

	4wDM	R/R	Unilok?
	4wDM	R/R	Unilok?

MARTIN O'KEEFFE, ROSTELLAN, near MIDLETON.
Gauge : 3'0".

C42	2w-2PMR		Wkm	7129	1955	Pvd

ROSMINIAN FATHERS, UPTON, INNISHANNON.
Gauge : 2'0".

	4wDM	RH	264244	1949

WINN TECHNOLOGY LTD, KILBRITTAIN.
Gauge : 1'10".

	0-4-0T	IC	Spence	1912	Pvd

DONEGAL

BORD NA MONA, GLENTIES // NA GLEANNTA.

Site is 2 miles south west of Glenties on the Adara road.
For loco details see separate Section.

DONEGAL PEAT DEVELOPMENT CO, BELLANAMORE, near FINTOWN. (Closed)
Gauge : 2'0".

	4wPM	MR	7944	1943	Dsm

DUBLIN

CORAS IOMPAIR EIREANN.
Connolly Depot.
Gauge : 5'3".

2	414A	2w-2PMR		Wkm	8920 1962

Inchicore Works, Dublin.
Gauge : 5'3".

9093 ZL	4wDM	Ford	1973

ARTHUR GUINNESS, SON & CO (DUBLIN) LTD, ST. JAMES GATE BREWERY, DUBLIN.
Gauge : 1'10".   RTC.

17	0-4-0T	IC	Spence	1902	Pvd
	4wDM		FH		Dsm
	4wDM		FH		OOU
	4wDM		FH		OOU
	4wDM		FH		OOU

FH's are four of 3068/1947; 3255/1948 and 3444, 3446, 3447, 3449/1950.

P.T.O'HALLORAN (METALS) LTD, KNOCKLYON LANE, TEMPLEOGUE, DUBLIN. (Closed)
Gauge : 2'0".

No.1	4wDM	OK	OOU
No.2	4wDM	OK	OOU

? locos still on site.

TRINITY COLLEGE, DUBLIN.
Gauge : 1'9".

	0-6-0	IC	Kennon	1855

GALWAY

BORD NA MONA.
Attymon // Ath Tiomain.

      Site is 6 miles east of Athenry.
      For loco details see separate Section.

Clonkeen // Cluain Chaoin.

      A sub-shed of Attymon.
      For loco details see separate Section.

COLD CHON (GALWAY) LTD // GALWAY CONCRETE LTD, ORANMORE.
Gauge : 5'3".

JZA 979	4wDM	Scammell	1960	

COMHLUCHT SIUICRE EIREANN TEO.
Gowla Farm, Ballyforan, near Ballinasloe.
Gauge : 2'0".

4	4wDM	RH	226281	1944
5	4wDM	RH	256799	1948
	4wDM	FH	3989	1962
	4wDM	RH	497771	1963

Tuam Sugar Factory.
Gauge : 5'3".

G 613	0-4-0DM	RH	395302	1956
	4wDH	Dtz	57226	1962

GALWAY METAL CO, ORANMORE.
Gauge : 5'3".

K 801	0-8-0DH	MAK	800028	1954 +

    + In use as a stationary power unit.

PRIORITY DRILLING LTD, PLANT DEPOT, KILLIMOR.
Gauge : 2'0".

LM 2	4wDM	HE	1878	1938	Dsm
	4wDM	HE	4473	1955	Dsm

KERRY

BORD NA MONA.
Barna // Bearna, near Ballydesmond.

    Site is 10 miles south east of Castleisland.
    For loco details see Separate Section.

Carrigcannon // Carraig Cheanna.

    Site is 7 miles north of Castleisland on the road to Listowel.
    For loco details see separate Section.

KILDARE

<u>BORD NA MONA.</u>
<u>Ballydermot</u> // Baile Dhiamarda.

     Site is 1½ miles north of Rathnagan on road from Kildare to Edenderry.
     Serves B.S.L. Allenwood and Irish C.E.C.A.
     For loco details see separate Section.

<u>Kilberry</u> // Cill Bheara.

     Site is 3 miles north west of Athy on Monasterevin road.
     Serves Kilberry Fertiliser Factory.
     For loco details see separate Section.

<u>Lullymore</u> // Loilgheach Mor.

     Site is 3 miles from Allenwood on Edenderry road.
     Serves Lullymore Briquetting Factory.
     For loco details see separate Section.

<u>Timahoe</u> // Tigh Mochua.

     Site is 3 miles west of Prosperous on the Clane to Edenderry road.
     Serves B.S.L. Allenwood.
     For loco details see separate Section.

<u>BORD SOLATHAIR AN LEACTREACHAIS, ALLENWOOD SOD-PEAT BURNING POWER STATION, NAAS.</u>
Gauge : 3'0".

4wDM		RH	300518	1950
4wDM		RH	314222	1951
4wDM		RH	314223	1951
4wDM		RH	326051	1952

<u>CORAS IOMPAIR EIREANN, KILDARE.</u>
Gauge : 5'3".

572	RZI 997	4wDM	Massey Ferguson	1966

<u>IRISH C.E.C.A., ALLENWOOD CARBON FACTORY.</u>
Gauge : 3'0".

4wDM		HE	6075	1961

LEIX/LAOIS

<u>BORD NA MONA, COOLNAMONA // CÚIL NA MONÁ.</u>

Site is 3½ miles south of Portlaoise on the road to Abbeyleix.
Serves the Coolnamona Fertiliser Factory.
For loco details see separate Section.

<u>IRISH STEAM PRESERVATION SOCIETY, STRADBALLY HALL, STRADBALLY.</u>
Gauge : 5'3".

	2w-2PMR		DC	1495 1927 +

+ Used as a store and office.

Gauge : 3'0".
No.2	0-4-0WT	OC	AB	2264 1949
	4wDM		HE	2280 1941
C39	2w-2PMR		Wkm	6861 c1954

Gauge : 1'10".
21	0-4-0T	IC	Spence	1905 +

+ Currently under renovation elsewhere.

<u>STRADBALLY STEAM MUSEUM, STRADBALLY.</u>
Gauge : 1'10".
15	0-4-0T	IC	Spence	1912 +

+ Carries plate dated 1895, in error.

LONGFORD

<u>BORD NA MONA, MOUNTDILLON // CNOC DIOLUIN.</u>

Site is 1½ miles north of Lanesborough on the Strokestown road, and consists
of the Mountdillon, Derryaroge, Begnagh, Derryad, Derryaghan & Corlea Bogs.
Serves B.S.L. Lanesborough.
For loco details see separate Section.

<u>BORD SOLATHAIR AN LEACTREACHAIS, LANESBOROUGH PEAT BURNING POWER STATION,
near LOUGHREE.</u>
Gauge : 3'0".
	4wDM		RH	422566 1958
	4wDM		RH	422567 1958

<u>RIO TINTO FINANCE & EXPLORATION LTD, KEEL, near COLEHILL.</u> (Closed)
Gauge : 2'0".

| | | 0-4-0BE | WR | G7179 1967 OOU |

LOUTH

<u>CORAS IOMPAIR EIREANN, DUNDALK STATION.</u>
Gauge : 5'3".

| 131 | | 4-4-0 | IC | NR | 5757 1901 Pvd |

MAYO

<u>BORD NA MONA, TIONNSCA ABHAINN EINNE.</u>

Site is 9 miles west of Crossmolina on the road to Belmullet. This, the
Oweniny River Project, is always spoken of as "T.A.E."
Serves B.S.L. Bellacorick and also supplies locos to Bangor Erris Bog when
required.
For loco details see separate Section.

<u>CORAS IOMPAIR EIREANN, CLAREMORRIS.</u>
Gauge : 5'3".

| 875 ARI | ACE 9 | 2-2wDM | | Ford |

OFFALY

BORD NA MONA.
Blackwater // Uisce Dubh.

Site is 1½ miles east of Shannonbridge on the road to Cloghan, and consists
of the Blackwater, Kilmacshane, Garryduff, Lismanny & Culliaghmore Bogs.
Serves B.S.L. Shannonbridge.
For loco details see separate Section.

Boora // Buarach.

Site is 6 miles east of Cloghan on the Tullamore road, and consists of the
Noggusboy, Derries, Turraun, Pollagh, Oughter, Boora, Derrybrat, Drinagh &
Clongawney More Bogs.
Serves B.S.L. Ferbane and Derrinlough Briquetting Plant.
For loco details see separate Section.

Clonsast // Cluain Sosta.

Site is 3 miles north of Portarlington on the road to Rochfortbridge, and
consists of the Clonsast, Derrylea, Derryounce & Garryhinch Bogs.
Serves B.S.L. Portarlington.
For loco details see separate Section.

Derrygreenagh // Doire Dhraigneach.

Site is 2 miles south of Rochfortbridge on the Rhode road, and consists of
the Derryhinch, Drumman, Derryarkin, Ballybeg, Cavemount, Esker, Mount Lucas,
Ballycon, Derrycricket & Cloncreen Bogs.
Serves B.S.L. Rhode and Croghan Briquetting Plant.
For loco details see separate Section.

Lemanaghan // Liath Manchain.

Site is at Ferbane, off the Birr to Athlone road.
For loco details see separate Section.

BORD SOLATHAIR AN LEACTREACHAIS, PORTARLINGTON SOD-PEAT BURNING POWER STATION.
Gauge : 3'0".

1	4wDM	RH	249526 1947
2	4wDM	RH	249525 1947
3	4wDM	RH	279604 1949
4	4wDM	RH	326052 1952

ERIN PEAT PRODUCTS LTD.
Site near Birr, on L 113 road.
Gauge : 2'0".

4wDM	FH	

Site near Birr, on T 41 road to Tullamore.
Gauge : 2'0".

4wDM	RH	193974 1938
4wDM	FH	2306

SHAMROCK MACHINE TURF CO, SHANE VALLEY.
Gauge : 2'0".

4wDM	R&R	84	1938
4wDM	MR	7919	1940

ROSCOMMON

WESTERN INDUSTRIES (BOYLE) LTD, KEELOGUES LIMESTONE QUARRY, BOYLE.
Gauge : 2'0".   RTC.

4wDM	FH	OOU

SLIGO

McTIERNAN BROS, GLEN BALLINSHEE COLLIERY, GEEVAGH.
Gauge : 2'0".

0-4-0BE	WR

TIPPERARY

BORD NA MONA.
Littleton (Ballydeath) // Baile Dhaith.

    Site is 1½ miles south of Littleton, near Thurles.
    For loco details see separate Section.

Templetuohy. // Teampall Tuaithe.

    Site is 3 miles south east of Templetuohy on the Templemore to Urlingford road.
    For loco details see separate Section.

COMHLUCHT SIUICRE EIREANN TEO. THURLES SUGAR FACTORY.
Gauge : 5'3".

		4wDM	RH	252843	1948
G 611		4wDM	RH	312424	1951
G 615		4wDH	Dtz	57225	1962
G 616		4wDH	Dtz	57223	1962
G 617		4wDH	Dtz	57227	1962
		4wDH	Dtz	57229	1962

MOGUL OF IRELAND LTD, SHALEE SILVER MINES, near NENAGH.
Gauge : 2'6".

(9)		4wDH	MR	115U093	1970	OOU
(10)		4wDH	MR	115U094	1970	OOU
11		4wDH	CE	5879/2	1971
12		4wDH	CE	5879/1	1971
13		4wDH	CE	5952A	1972
14		4wDH	CE	5952B	1972
15		4wDH	CE	5960A	1973
16		4wDH	CE	5960B	1973
17		4wDH	CE	5960C	1973
18		4wDH	CE	5960D	1973

POPE BROS LTD, PEAT WORKS, near URLINGFORD.
Gauge : 2'0".

No.1	AZA 455	4wDM	MR	8749	1942	OOU
3		4wDM	MR	9219	1946	Dsm
No.4	E.I.B. 39	4wDM	MR	7361	1939
5		4wDM	MR	9239	1946
		4wDM	HE	2659	1942	OOU
		4wDM	KKM	312	1961	OOU

WATERFORD

CORAS IOMPAIR EIREANN, WATERFORD.
Gauge : 5'3".

5	417A	2w-2PMR	Wkm	8917	1962

TRAMORE MINIATURE RAILWAY, TRAMORE.
Gauge : 1'3".

278		2-8-0PH S/O	SL	22 1973

WESTMEATH

<u>BORD NA MONA.</u>
<u>Ballivor // Baile Iomhair.</u>

     Site is 6 miles west of Ballivor on road from Trim to Mullingar.
For loco details see separate Section.

<u>Coolnagan // Cuil Na Glon.</u>

     Site is 5 miles west of Castlepollard on the road to Granard.
For loco details see separate Section.

<u>CORAS IOMPAIR EIREANN, ATHLONE.</u>
Gauge : 5'3".

1				2w-2PMR	Wkm	8916 1962
589	2285 ZO	ACE 8		4wDM	Ford	1973
				4wDM		

<u>RAILWAY PRESERVATION SOCIETY OF IRELAND, MULLINGAR DEPOT.</u>
Gauge : 5'3".

No.184		0-6-0	IC	Inchicore	1880
15		2-6-0	IC	BP	6112 1922

WEXFORD

<u>WALLACE BROS LTD, COURTOWN BRICK & TILE WORKS, COURTOWN.</u> (Closed)
Gauge : 1'8".

	4wDM	RH	264237 1948 OOU

WICKLOW

<u>AVOCA MINES LTD, AVOCA.</u>
Gauge : 2'0".

BE	WR	OOU

Bord na Mona   Irish Turf Board

The Bord operates rail systems on peat bogs throughout the country, and locomotives are kept at the locations listed below.

Gauge : 3'0" (except those marked + which are 2'0".)

Locations : (For more details see the relevant County.)

A	Attymon, Co. Galway.
Ba	Barna, Co. Kerry.
Bd	Ballydermot, Co. Kildare.
Bi	Ballivor, Co. Westmeath.
Bl	Blackwater, Co. Offaly.
Bo	Boora, Co. Offaly.
Ca	Carrigcannon, Co. Kerry.
Cg	Coolnagan, Co. Westmeath.
Ck	Clonkeen, Co. Galway.
Cm	Coolnamona, Co. Laois.
Cs	Clonsast, Co. Offaly.
Dg	Derrygreenagh, Co. Offaly.
Gl +	Glenties, Co. Donegal.
K +	Kilberry, Co. Kildare.
Le	Lemanaghan, Co. Offaly.
Li	Littleton, Co. Tipperary.
Lu	Lullymore, Co. Kildare.
M	Mountdillon, Co. Longford.
TAE	Tionnsca Abhainn Einne, Co. Mayo.
Te	Templetuohy, Co. Tipperary.
Ti	Timahoe, Co. Kildare.

LM 12							
LM 13 D			4wDM		Wcb	40331 1945	Ba
LM 14 D			4wDM		RH	198251 1939	Cs
LM 15			4wDM		RH	198290 1940	Cs
LM 16 B			4wDM		RH	198326 1940	Li
LM 17 G			4wDM	+	RH	200075 1940	K
LM 18 G			4wDM		RH	242901 1946	Lu
(LM 19)			4wDM		RH	242902 1946	TAE
LM 20 A			4wDM		RH	243386 1946	Cs Dsm
LM 21 A			4wDM	+	RH	243387 1946	Gl
LM 22			4wDM	+	RH	243392 1946	K
			4wDM		RH	243395 1946	
			rebuilt 4wDMR		BnM		Ti
LM 23 E			4wDM		RH	244788 1946	Cs
LM 24 E			4wDM		RH	244870 1946	Cs
LM 25 E	7		4wDM		RH	244871 1946	Cs
LM 26 E			4wDM	+	RH	248458 1946	Gl
LM 27 E	8		4wDM		RH	249524 1947	Ti
LM 28 G	12		4wDM		RH	249543 1947	Ti
LM 29 G			4wDM		RH	249544 1947	M
LM 30 G	11		4wDM		RH	249545 1947	Ti
LM 31 E			4wDM		RH	252232 1947	Cg
LM 32 E	4		4wDM		RH	252233 1947	Ti
LM 33 E	2		4wDM		RH	252234 1947	Ti
LM 34 E	1		4wDM		RH	252239 1947	A
LM 35 E			4wDM		RH	252240 1947	Bd
LM 36 E	2		4wDM		RH	252241 1947	M
LM 37 E	3		4wDM		RH	252245 1947	M
LM 38 E			4wDM		RH	252246 1947	A
LM 39 E			4wDM		RH	252247 1947	M
LM 40 E			4wDM		RH	252251 1947	M
LM 41 E			4wDM		RH	252252 1947	A
LM 42 C			4wDM	+	RH	252849 1947	K
LM 46			4wDM		RH	259184 1948	Bd

LM 47	F		4wDM	RH	259185	1948	Bd
LM 48	F		4wDM	RH	259186	1948	Bd
LM 49	F		4wDM	RH	259189	1948	TAE
LM 50			4wDM	RH	259190	1948	Bd
LM 51	F		4wDM	RH	259191	1948	Cg
LM 52	F	15	4wDM	RH	259196	1948	Ti
LM 53	F		4wDM	RH	259197	1948	Bd
LM 54	F		4wDM	RH	259198	1948	Cm
LM 55	F		4wDM	RH	259203	1948	Bd
LM 56	F	7	4wDM	RH	259204	1948	Ti
LM 57	F		4wDM	RH	259205	1948	Bd
LM 58	F		4wDM	RH	259206	1948	Li
LM 59			4wDM	RH	259737	1948	Cs
LM 60	F		4wDM	RH	259738	1948	Cs
LM 61	F		4wDM	RH	259739	1948	Bi
LM 62	F		4wDM	RH	259743	1948	Lu
LM 63	F		4wDM	RH	259744	1948	Bi
LM 64	F		4wDM	RH	259745	1948	Cs
LM 65			4wDM	RH	259749	1948	Te
LM 66		5	4wDM	RH	259750	1948	M
LM 67	F	17	4wDM	RH	259751	1948	Ti
LM 68			4wDM	RH	259752	1948	Li
LM 69	F		4wDM	RH	259755	1948	Bi
LM 70	F		4wDM	RH	259756	1948	Cg
LM 71	F		4wDM	RH	259757	1948	M
LM 72	F		4wDM	RH	259758	1948	M
LM 73		8	4wDM	RH	259759	1948	M
LM 74	F	6	4wDM	RH	259760	1948	Ti
LM 75	R		4wDM	RH	326047	1952	Lu
LM 76			4wDM	RH	326048	1952	Cm
LM 77	H		4wDM	RH	329680	1952	Bd
LM 78	H		4wDM	RH	329682	1952	Cs
LM 79	H	10	4wDM	RH	329683	1952	Ti
LM 80	H	13	4wDM	RH	329685	1952	Ti
LM 81	H		4wDM	RH	329686	1952	Cs
LM 82			4wDM	RH	329688	1952	Bd
LM 83		1	4wDM	RH	329690	1952	Ba
LM 84	J	14	4wDM	RH	329691	1952	Ti
LM 85	J		4wDM	RH	329693	1952	M
LM 86	J		4wDM	RH	329695	1952	M
LM 87	J		4wDM	RH	329696	1952	Bl
LM 88	J	16	4wDM	RH	329698	1952	Ti
LM 89	J		4wDM	RH	329700	1952	Cg
LM 90	J		4wDM	RH	329701	1952	Bi
LM 91			4wDM	RH	371962	1954	Bo
LM 92	L		4wDM	RH	371967	1954	Bo
LM 93	T		4wDM	RH	373376	1954	Bd
LM 94	T		4wDM	RH	373377	1954	Bd
LM 95	T		4wDM	RH	373379	1954	Lu
LM 96	L		4wDM	RH	375314	1954	TAE
LM 97	T		4wDM	RH	375332	1954	Lu
LM 98	T		4wDM	RH	375335	1954	Cs
LM 99	U		4wDM	RH	375336	1954	Lu
LM 100			4wDM	RH	375341	1954	Cs
LM 101	U		4wDM	RH	379059	1954	Bo
LM 102			4wDM	RH	379076	1954	Bo
LM 103			4wDM	RH	375318	1954	Bo
LM 104		8	4wDM	RH	375322	1954	TAE
LM 105	U		4wDM	RH	375344	1954	Bl
LM 106	U		4wDM	RH	375345	1954	M
LM 107	U		4wDM	RH	379055	1954	Bo
LM 108	U		4wDM	RH	379061	1954	TAE
LM 109	U		4wDM	RH	379064	1954	Bo
LM 110	U		4wDM	RH	379066	1954	Bo

LM						No.	Year	
LM 111			4wDM		RH	379079	1954	Bl
LM 112			4wDM	+	RH	375699	1954	K
LM 113	U		4wDM		RH	379068	1954	Dg
LM 114	U		4wDM		RH	379070	1954	Bo
LM 115	U		4wDM		RH	379073	1954	Cs
LM 116	M	15	4wDM		RH	379077	1954	TAE
LM 117	U		4wDM		RH	379910	1954	Bo
LM 118	V		4wDM		RH	379913	1954	Bo
LM 119	V		4wDM		RH	379916	1954	TAE
LM 120	V		4wDM		RH	379917	1954	Lu
LM 121	V		4wDM		RH	379922	1954	Dg
LM 122	V		4wDM		RH	379923	1954	Bo
LM 123			4wDM		RH	379925	1954	TAE
LM 124	N		4wDM		RH	379081	1954	Cs
LM 125	O		4wDM		RH	379084	1954	Cs
LM 126	W		4wDM		RH	379927	1954	Bo
LM 127	W		4wDM		RH	379928	1954	Bo
LM 128	X		4wDM		RH	383260	1955	Bl
LM 129	X		4wDM		RH	383264	1955	M
LM 130	P		4wDM		RH	382812	1955	TAE
LM 131			4wDM		RH	379086	1955	Bo
LM 132	P		4wDM		RH	382809	1955	Cs
LM 133	P		4wDM		RH	379090	1955	Dg
LM 134	Q		4wDM		RH	382811	1955	Dg
LM 135	Q		4wDM		RH	382814	1955	Dg
LM 136	Q		4wDM		RH	382815	1955	Dg
LM 137	Q	6	4wDM		RH	382817	1955	TAE
LM 138	Q		4wDM		RH	382819	1955	TAE
LM 139	Q		4wDM		RH	392137	1955	Te
LM 140	Q		4wDM		RH	392139	1955	Bi
LM 141	Q		4wDM		RH	392142	1955	Bo
LM 142	Q		4wDM		RH	392145	1955	Bo
LM 143			4wDM		RH	392148	1956	Bl
LM 144	Q		4wDM		RH	392149	1956	Dg
LM 145	X		4wDM		RH	394023	1956	Dg
LM 146			4wDM		RH	394024	1956	Lu
LM 147	X		4wDM		RH	394025	1956	Lu
LM 148	X		4wDM		RH	394026	1956	Dg
LM 149	X		4wDM		RH	394028	1956	Dg
LM 150	X		4wDM		RH	394027	1956	Dg
LM 151	Q		4wDM		RH	392150	1956	Bl
LM 152	Q		4wDM		RH	392151	1956	Bi
LM 153	Q		4wDM		RH	394029	1956	Dg
LM 154	X		4wDM		RH	394030	1956	Bl
LM 155	X		4wDM		RH	394031	1956	M
LM 156	X		4wDM		RH	394032	1956	Dg
LM 157	X		4wDM		RH	394033	1956	Dg
LM 158	X		4wDM		RH	394034	1956	Dg
LM 159	X		4wDM		RH	402174	1956	Dg
LM 160	X		4wDM		RH	402176	1956	Dg
LM 161	X		4wDM		RH	402175	1956	Bo
LM 162	X		4wDM		RH	402177	1956	TAE
LM 163	X		4wDM		RH	402178	1956	Bo
LM 164	Q		4wDM		RH	392152	1956	A
LM 165			4wDM		RH	402179	1956	Dg
LM 166	Q		4wDM		RH	402977	1956	Bi
LM 167	Q		4wDM		RH	402978	1956	Cm
LM 168	Q		4wDM		RH	402980	1956	Dg
LM 169	Q		4wDM		RH	402981	1956	Dg
LM 170	Q		4wDM		RH	402982	1956	Te
LM 171	Q		4wDM		RH	402983	1956	Le
LM 172	Q		4wDM		RH	402984	1956	Ca
LM 173	Q		4wDM		RH	402985	1957	Le
LM 174	Q		4wDM		RH	402986	1957	TAE

LM 175		0-4-0DM		RH	420042	1958	Bo	
LM 176		0-4-0DM		BnM		1961	Dg	
LM 177	5	4wDM		RH	218037	1943	Li	
18		4wDM		RH	211687	1941	Lu	Dsm
		4wDM		RH	421428	1958	Lu	
LM 178		0-4-0DM		Dtz	57120	1960	Bo	
LM 179		0-4-0DM		Dtz	57121	1960	Bo	
LM 180		0-4-0DM		Dtz	57122	1960	Bo	
LM 181		0-4-0DM		Dtz	57123	1960	Bo	
LM 182		0-4-0DM		Dtz	57126	1960	TAE	
LM 183		0-4-0DM		Dtz	57127	1960	Bo	
LM 184		0-4-0DM		Dtz	57130	1960	Bl	
LM 185		0-4-0DM		Dtz	57131	1960	Bo	
LM 186		0-4-0DM		Dtz	57132	1960	Bo	
LM 187		0-4-0DM		Dtz	57133	1960	Bo	
LM 188		0-4-0DM		Dtz	57124	1960	Dg	
LM 189	2	0-4-0DM		Dtz	57125	1960	Dg	
LM 190		0-4-0DM		Dtz	57128	1960	Dg	
LM 191		0-4-0DM		Dtz	57129	1960	Dg	
LM 192		0-4-0DM		Dtz	57134	1960	Dg	
LM 193		0-4-0DM		Dtz	57135	1960	Dg	
LM 194	7	0-4-0DM		Dtz	57136	1960	Bl	
LM 195		0-4-0DM		Dtz	57137	1960	TAE	
LM 196		0-4-0DM		Dtz	57138	1960	TAE	
LM 197		0-4-0DM		Dtz	57139	1960	Dg	
LM 198		4wDM	+	RH	398076	1956	Gl	
LM 199		0-4-0DM		HE	6232	1962	Dg	
LM 200		0-4-0DM		HE	6233	1962	Bo	
LM 201		0-4-0DM		HE	6234	1962	TAE	
LM 202		0-4-0DM		HE	6235	1962	TAE	
LM 203		0-4-0DM		HE	6236	1962	Bo	
LM 204		0-4-0DM		HE	6237	1963	Dg	
LM 205		0-4-0DM		HE	6238	1963	TAE	
LM 206		0-4-0DM		HE	6239	1963	Bo	
LM 207		0-4-0DM		HE	6240	1963	TAE	
LM 208		0-4-0DM		HE	6241	1963	Dg	
LM 209		0-4-0DM		HE	6242	1963	TAE	
LM 210		0-4-0DM		HE	6243	1963	Dg	
LM 211		0-4-0DM		HE	6244	1963	Dg	
LM 212		0-4-0DM		HE	6245	1963	TAE	
LM 213		0-4-0DM		HE	6246	1963	Dg	
LM 214		0-4-0DM		HE	6247	1963	TAE	
LM 215		0-4-0DM		HE	6248	1963	Dg	
LM 216		0-4-0DM		HE	6249	1963	Bo	
LM 217		0-4-0DM		HE	6250	1963	Bl	
LM 218		0-4-0DM		HE	6251	1963	Dg	
LM 219		0-4-0DM		HE	6252	1963	Bo	
LM 220		0-4-0DM		HE	6253	1963	Bl	
LM 221		0-4-0DM		HE	6254	1963	Dg	
LM 222		0-4-0DM		HE	6255	1963	Bl	
LM 223		0-4-0DM		HE	6256	1963	Dg	
LM 224		4wDM	+	RH	375317	1954	K	
LM 225		0-4-0DM		HE	6304	1964	Dg	
LM 226		0-4-0DM		HE	6305	1964	M	
LM 227		0-4-0DM		HE	6306	1964	Bo	
LM 228		0-4-0DM		HE	6307	1964	Bo	
LM 229		0-4-0DM		HE	6308	1964	Bl	
LM 230		0-4-0DM		HE	6309	1964	Bl	
LM 231		0-4-0DM		HE	6310	1964	Bl	
LM 232		0-4-0DM		HE	6311	1964	Dg	
LM 233		0-4-0DM		HE	6312	1965	Bo	
LM 234		0-4-0DM		HE	6313	1965	Bl	
LM 235		0-4-0DM		HE	6314	1965	Bo	
LM 236		0-4-0DM		HE	6315	1965	Dg	

LM 237		0-4-ODM		HE	6316 1965	Cm
LM 238		0-4-ODM		HE	6318 1965	Bd
LM 239		0-4-ODM		HE	6317 1965	Dg
LM 240		0-4-ODM		HE	6319 1965	Bo
LM 241		0-4-ODM		HE	6320 1965	Bd
LM 242		0-4-ODM		HE	6321 1965	M
LM 243		0-4-ODM		HE	6322 1965	Bo
LM 244		0-4-ODM		HE	6323 1965	Bl
LM 245		0-4-ODM		HE	6324 1965	Bo
LM 246		0-4-ODM		HE	6325 1965	Bo
LM 247		0-4-ODM		HE	6326 1965	M
LM 248		0-4-ODM		HE	6328 1965	Bd
LM 249		0-4-ODM		HE	6327 1965	M
LM 250		0-4-ODM		HE	6329 1965	M
LM 251		0-4-ODM		HE	6330 1965	M
LM 252		0-4-ODM		HE	6331 1965	M
LM 253		0-4-ODM		Dtz	57834 1965	Bl
LM 254		0-4-ODM		Dtz	57835 1965	Bl
LM 255		0-4-ODM		Dtz	57838 1965	Cm
LM 256		0-4-ODM		Dtz	57837 1965	Bo
LM 257		0-4-ODM		Dtz	57836 1965	Dg
LM 258		0-4-ODM		Dtz	57839 1965	M
LM 259		0-4-ODM		Dtz	57840 1965	Bo
LM 260		0-4-ODM		Dtz	57841 1965	Bo
LM 261		0-4-ODM		Dtz	57842 1965	Bl
LM 262		0-4-ODM		Dtz	57843 1965	M
LM 263		4wDM	+ RH		7002/0600-1 1968	Gl
LM 264		4wDM	+ RH		371535 1954	K
LM 265		4wDM	+ RH		375696 1954	K
LM 266		0-4-ODM		HE	7232 1971	Bo
LM 267		0-4-ODM		HE	7233 1971	M
LM 268		0-4-ODM		HE	7234 1971	Li
LM 269		0-4-ODM		HE	7235 1971	Bo
LM 270		0-4-ODM		HE	7237 1971	M
LM 271		0-4-ODM		HE	7236 1971	Ti
LM 272		0-4-ODM		HE	7239 1971	TAE
LM 273		0-4-ODM		HE	7246 1972	Bo
LM 274		0-4-ODM		HE	7238 1971	M
LM 275		0-4-ODM		HE	7240 1972	Cm
LM 276		0-4-ODM		HE	7241 1972	Lu
LM 277		0-4-ODM		HE	7242 1972	Bl
LM 278		0-4-ODM		HE	7243 1972	Bl
LM 279		0-4-ODM		HE	7244 1972	Bd
LM 280		0-4-ODM		HE	7245 1972	Bl
LM 281		0-4-ODM		HE	7247 1972	Bl
LM 282		0-4-ODM		HE	7248 1972	Bl
LM 283		0-4-ODM		HE	7250 1972	Ti
LM 284		0-4-ODM		HE	7249 1972	Bd
LM 285		0-4-ODM		HE	7253 1972	Bl
LM 286		0-4-ODM		HE	7254 1972	Bl
LM 287		0-4-ODM		HE	7255 1972	Bl
LM 288		0-4-ODM		HE	7256 1972	Bl
LM 289	LM 1	0-4-ODM		HE	7252 1972	Dg
LM 290	2	0-4-ODM		HE	7251 1972	Dg
LM 292		0-4-ODM		HE	8529 1977	Bl
LM		0-4-ODM		HE	8530 1977	
LM		0-4-ODM		HE	8531 1977	
LM		0-4-ODM		HE	8532 1977	
LM 296		0-4-ODM		HE	8534 1977	Bl
LM 297		0-4-ODM		HE	8533 1977	Bo
LM		0-4-ODM		HE	8535 1977	
LM		0-4-ODM		HE	8536c1977	
LM		0-4-ODM		HE	8537c1977	
LM		0-4-ODM		HE	8538c1977	

ID	Type	Rebuilt	Builder	Number	Year	Code1	Code2
LM	0–4–0DM		HE	8539c	1978		
LM	0–4–0DM		HE	8540c	1978		
LM	0–4–0DM		HE	8541c	1978		
LM	0–4–0DM		HE	8542c	1978		
LM	0–4–0DM		HE	8543	1978		
LM	0–4–0DM		HE	8544	1978		
LM	0–4–0DM		HE	8545	1978		
LM 308	0–4–0DM		HE	8546	1978	Cs	
LM	0–4–0DM		HE	8547	1978		
LM	0–4–0DM		HE	8548	1978		
LM 313	0–4–0DM		HE	8549	1977	Bl	
LM 312	0–4–0DM		HE	8550	1977	Bl	
LM	0–4–0DM		HE	8551	1978		
LM 315	0–4–0DM		HE	8922	1979	Bl	
LM 322	0–4–0DM		HE	8923	1979	Bl	
LM 323	0–4–0DM		HE	8924	1979	Bl	
LM	0–4–0DM		HE	8925	1979		
LM	0–4–0DM		HE	8926	1979		
LM	0–4–0DM		HE	8927	1979		
LM	0–4–0DM		HE	8928	1979		
LM	0–4–0DM		HE	8929	1979		
LM 311	0–4–0DM		HE	8930	1980	Bl	
LM 324	0–4–0DM		HE	8931	1980	Dg	
LM	0–4–0DM		HE	8932	1980		
LM 341	4wDM		SMH	60SL740	1980	Cm	
LM	4wDM		SMH	60SL741	1980		
LM	4wDM		SMH	60SL742	1980		
LM	4wDM		SMH	60SL743	1980		
LM	4wDM		SMH	60SL744	1980		
LM	4wDM		SMH	60SL745	1980		
LM	4wDM		SMH	60SL746	1980		
LM	4wDM		SMH	60SL747	1980		
LM	4wDM		SMH	60SL748	1980		
LM	4wDM		SMH	60SL749	1980		
LM	4wDM		SMH	60SL750	1980		
LM	4wDM		SMH	60SL751	1980		
C 11	2w-2PMR		BnM			Cs	
(C 12)	2w-2PMR		BnM		1966	Cs	Dsm
C 16	2w-2PMR		+ Wkm	4806	1948	Gl	Dsm
C 17	2w-2PMR		Wkm	4807	1948	Ba	OOU
(C 27)	2w-2PMR		Wkm	4817	1948	Cs	Dsm
(C 28)	2w-2PMR		Wkm	4818	1948	Lu	Dsm
(C 32)	2w-2PMR		Wkm	4824	1949		
		Rebuilt	BnM			Bi	Dsm
(C 33)	2w-2PMR		Wkm	4821	1949		
		Rebuilt	BnM			Cs	Dsm
C 34	2w-2PMR		Wkm	4822	1949	Ti	
C 35	2w-2PMR		BnM	1		Ti	
(C 36)	2w-2PMR		BnM	2		Bo	
(C 38)	2w-2PMR		BnM			Bd	Dsm
C 40	2w-2PMR		Wkm	7127	1955	Bo	Dsm
(C 43)	2w-2PMR		Wkm	7130	1955	Bl	
C 45	2w-2PMR		Wkm	7132	1955	Dg	
C 47	2w-2PMR		BnM	3		Cs	
C 48	2w-2PMR		BnM	4		Cg	
C 49	4wPMR		BnM	5		M	
C 50	2w-2PMR		BnM	6		Bd	
C 51	2w-2PMR		BnM	7	1958	Bi	
C 52	2w-2PMR		BnM	8		TAE	
C 53	2w-2PMR		BnM	9		Dg	
C 54	2w-2PMR		BnM	10		Bl	
C 55	2w-2PMR		Wkm	7680	1957	Ck	
C 56	2w-2PMR		Wkm	7681	1957	Te	

C 57	2w-2PMR	Wkm	7682 1957	Cg	Dsm
C 58	2w-2PMR	Wkm	8730 1960	Bl	
(C 59)	2w-2PMR	Wkm	8631 1960	Bi	
(C 60)	2w-2DMR	BnM	c1960	Bd	Dsm
(C 61)	2w-2PMR	BnM		Li	
(C 62)	2w-2DMR	BnM	1960	M	
C 63	2w-2PMR	BnM	1972	A	
C 64	2w-2PMR	BnM	1972	Li	
C 65	2w-2PMR	BnM	1972	M	
C 66	2w-2PMR	BnM	1972	Cm	
C 67	2w-2PMR	BnM	1972	Dg	
C 68	2w-2PMR	BnM	1972	Bd	
C 69	2w-2PMR	BnM	1972	Ti	
C 70	2w-2PMR	BnM	1972	TAE	
C 71	2w-2PMR	BnM	1972	M	
C 72	2w-2PMR	BnM	1972	Bi	
C 73	2w-2PMR	BnM	1972	Cg	
C 74	2w-2PMR	BnM	1972	Bo	
C 75	2w-2PMR	BnM	1972	Bl	
C 76	2w-2PMR	BnM	1972	Dg	
C 77	2w-2PMR	BnM	1972	Cm	
C 78	2w-2PMR	BnM	1972	Bo	
C 79	2w-2PMR	BnM	1972	Bd	
C 80	2w-2PMR	BnM	1972	Bl	

SECTION 7 BRITISH RAILWAYS DEPARTMENTAL STOCK

This list is in accordance with our records at January, 1982 and those held by Roger Butcher of the R.C.T.S.

PART ONE

BRITISH RAIL ENGINEERING LTD.
BRITISH RAIL RESEARCH AND DEVELOPMENT DIVISION.

PART TWO

BATTERY LOCOMOTIVES.
DE-ICING UNITS.
DIESEL SHUNTING LOCOMOTIVES.
ENGINEERS MOTOR TROLLEYS.
INSPECTION SALOONS.
MISCELLANEOUS POWERED DEPARTMENTAL VEHICLES.
ROUTE LEARNING UNITS.
TRAIN HEATING UNITS.

PART ONE

BRITISH RAIL ENGINEERING LTD.
Crewe Works, Cheshire.
Gauge : 4'8½".   (SJ 691561)

7042		4wDM R/R	S&H	7511	1969
7158		4wDH R/R	NNM	77501	1980

Eastleigh Works, Hampshire.
Gauge : 4'8½".   (   )

D2991		0-6-0DE	RH	480692 1962 +

+ In use as a stationary generator.

Glasgow Works, Strathclyde.
Gauge : 4'8½".   (NS 605665)

1179		4wPM R/R	Unilok	404
1185		4wDM R/R	Unilok	1815 1973
1193		4wDM R/R	Unilok	1861

Litchurch Lane Carriage Works, Derby.
Gauge : 4'8½".   (SK 364345)

2337		0-4-0WE	Derby C&W	1922	
8		0-4-0WE	Derby C&W	1935	
3335	3	0-4-0WE	Derby C&W	1935	
(4953)		0-4-0WE	Derby C&W	1956	
1656		0-4-0WE	Derby C&W	1913	OOU
No.4		0-4-0WE	Derby C&W	(1913?)	

Wolverton Works, Buckinghamshire.
Gauge : 4'8½".   (SP 812413)

DB 975426	(Sc 79098)	4w-4DMR	Sdn		1956 + OOU
		4wDM	SMH	103GA078	1978

+ Sold to Kings, Snailwell for scrap. Awaiting removal from site.

BRITISH RAIL RESEARCH AND DEVELOPMENT DIVISION.

Although most of the vehicles detailed in this section can be seen at the locations listed, they go where they are required and so will of course be found at other locations as well.

Derby Technical Centre.
Gauge : 4'8½".   (   )

The C.M. & E.E. also possess some Derby-based departmental vehicles which are therefore conveniently included here.

97201	EXPERIMENT	(24061)	4w-4wDE	Crewe	1960
84003			4w-4wWE	NB	27795 1960 + OOU
ADB 968021		(84009)	4w-4wWE	NB	27801 1960
DRC 730J			4wDM R/R	Unimog	1970
RDB 975010	IRIS	(M 79900)	4w-4DMR	Derby	1956
RDB 975089		(M 50396)	4w-4DMR	PR	1957
RDB 975385	HYDRA	(M 55997)	4w-4DMR	Cravens	1958
ADB 975430		(S 64300)	4w-4wRE	York	1971
ADB 975432		(S 64301)	4w-4wRE	York	1971
RDB 975634	PC 3		4w-4w-4warticGE	Derby	1972
RDB 975635	PC 4		4w-4w-4warticGE	Derby	1972

ADB 975812		(W 43000)	4w–4wDE	Crewe		1972
ADB 975813		(W 43001)	4w–4wDE	Crewe		1972
ADB 975844	057	(S 64305)	4w–4wRE	York		1971
ADB 975845	056	(S 62427)	4w–4wRE	York		1971 ++
ADB 975846	056	(S 62428)	4w–4wRE	York		1971 ++
ADB 975847	056	(S 64302)	4w–4wRE	York		1971 ++
ADB 975848	056	(S 64303)	4w–4wRE	York		1971 ++
ADB 975849	057	(S 62426)	4w–4wRE	York		1971
ADB 975850	057	(S 62429)	4w–4wRE	York		1971
ADB 975851	057	(S 64304)	4w–4wRE	York		1971
RDB 975874			4wDMR	Leyland		1978
RDB 977020			4wDMR	Leyland/BREL		1981
RDB 999507			4wDMR	Wkm	8025	1958

+ Source of spares for ADB 968021.
++ Currently stored in Derby St. Mary's yard.

Mickleover, Derbyshire.
Gauge : 4'8½".   (   )

PWM 3949			2w–2PMR	Wkm	6934	1955 +
PWM 3956	RDB 3956		2w–2PMR	Wkm	6941	1955
97801	PLUTO	(08267)	0–6–0DE	Derby		1957
RDB 975003	GEMINI	(Sc 79998)	4w–4wBER	Derby/Cowlairs		1958

+ Privately owned - temporarily stored here.

Old Dalby, Leicestershire.
Gauge : 4'8½".   (   )

DB 998900		4wDMR	DC/Bg	2267	1950 + OOU
RDB 998901		4wDMR	DC/Bg	2268	1950
7076		4w–4wDH	BPH	7980	1963 ++
7096		4w–4wDH	BPH	8000	1963 ++

+ Source of spares for RDB 998901.
++ Used as a dead load vehicle. (Non operable.)

PART TWO

BATTERY LOCOMOTIVES.

Former Watford Line motor vehicles converted to battery locomotives.

B = Birkenhead North, Merseyside.
C = Cricklewood T.M.D. London.
H = Hornsey T.M.D. London.

Gauge : 4'8½".

DB 975178	(M 61136)	4w-4wBE		Afd/Elh	1958	
			Rebuilt	Wolverton	1974	B
DB 975179	(M 61139)	4w-4wBE		Afd/Elh	1958	
			Rebuilt	Wolverton	1974	B
97703	(M 61182)	4w-4wBE		Afd/Elh	1958	
			Rebuilt	Don	1980	C
97704	(M 61185)	4w-4wBE		Afd/Elh	1958	
			Rebuilt	Don	1980	C
97705	(M 61184)	4w-4wBE		Afd/Elh	1958	
			Rebuilt	Don	1980	C
97706	(M 61189)	4w-4wBE		Afd/Elh	1958	
			Rebuilt	Don	1980	C
LDB 97707	(M 61166)	4w-4wBE		Afd/Elh	1958	
			Rebuilt	Don	1975	H
LDB 97708	(M 61173)	4w-4wBE		Afd/Elh	1958	
			Rebuilt	Don	1975	H
97709	(M 61172)	4w-4wBE		Afd/Elh	1958	
			Rebuilt	Don	1975	H
97710	(M 61175)	4w-4wBE		Afd/Elh	1958	
			Rebuilt	Don	1975	H

DE-ICING UNITS.

Gauge : 4'8½".

LONDON MIDLAND REGION.

These units are currently based at the following depots :-

K = Kirkdale, Merseyside.
S = Southport, Merseyside.

M 28319		4w-4wRE		Derby	c1939	S
M 28334		4w-4wRE		Derby	c1939	S
ADB 977017	(M 28354)	4w-4wRE		Derby	c1939	
			Rebuilt	Hor	1981	K
ADB 977018	(M 28357)	4w-4wRE		Derby	c1939	
			Rebuilt	Hor	1981	K

SOUTHERN REGION.

These units are currently based at the following depots. Each autumn some units are temporarily reallocated for de-leafing duties.

BM = Bournemouth.
BR = Brighton.
CL = Chart Leacon.
EH = Eastleigh.
FR = Fratton.
GI = Gillingham.
RE = Ramsgate.
SU = Selhurst.
WP = Wimbledon Park.

The unit number is given first, followed by the two cars forming it.

001	ADS 70268 (S 10726 S)	4w-4wRE	Rebuilt	Stewarts Lane	1967	
	ADS 70273 (S 10500 S)	4w-4wRE	Rebuilt	Stewarts Lane	1969	BM
002	ADB 975594 (S 12658 S)	4w-4wRE	Rebuilt	Selhurst	1978	
	ADB 975595 (S 10994 S)	4w-4wRE	Rebuilt	Selhurst	1978	SU
003	ADS 70270 (S 10497 S)	4w-4wRE	Rebuilt	Stewarts Lane	1968	
	ADS 70272 (S 10499 S)	4w-4wRE	Rebuilt	Stewarts Lane	1968	EH
004	ADB 975586 (S 10907 S)	4w-4wRE	Rebuilt	Selhurst	1977	
	ADB 975587 (S 10908 S)	4w-4wRE	Rebuilt	Selhurst	1977	CL
005	ADB 975588 (S 10981 S)	4w-4wRE	Rebuilt	Selhurst	1977	
	ADB 975589 (S 10982 S)	4w-4wRE	Rebuilt	Selhurst	1977	CL
006	ADB 975590 (S 10833 S)	4w-4wRE	Rebuilt	Selhurst	1978	
	ADB 975591 (S 10834 S)	4w-4wRE	Rebuilt	Selhurst	1978	BR
007	ADB 975592 (S 10993 S)	4w-4wRE	Rebuilt	Selhurst	1978	
	ADB 975593 (S 12659 S)	4w-4wRE	Rebuilt	Selhurst	1978	GI
008	ADB 975596 (S 10844 S)	4w-4wRE	Rebuilt	Selhurst	1979	
	ADB 975597 (S 10987 S)	4w-4wRE	Rebuilt	Selhurst	1979	WP
009	ADB 975598 (S 10989 S)	4w-4wRE	Rebuilt	Selhurst	1980	
	ADB 975599 (S 10990 S)	4w-4wRE	Rebuilt	Selhurst	1980	WP
010	ADB 975600 (S 10988 S)	4w-4wRE	Rebuilt	Selhurst	1980	
	ADB 975601 (S 10843 S)	4w-4wRE	Rebuilt	Selhurst	1980	BR
011	ADB 975602 (S 10991 S)	4w-4wRE	Rebuilt	Selhurst	1980	
	ADB 975603 (S 10992 S)	4w-4wRE	Rebuilt	Selhurst	1980	RE
012	ADB 975604 (S 10939 S)	4w-4wRE	Rebuilt	Selhurst	1981	
	ADB 975605 (S 10940 S)	4w-4wRE	Rebuilt	Selhurst	1981	FR

DIESEL SHUNTING LOCOMOTIVES.

EASTERN REGION.

Chesterton Junction Central Materials Depot, Cambridge.
Gauge : 2'0".   (TL 477613)

85049		4wDM	RH	393325 1956
85051		4wDM	RH	404967 1957

Lowestoft C.C.E. Plant and Machinery Depot, Suffolk.
Gauge : 4'8½".   (   )

041399	(DB 988525)	2w-2DM		+

+ Formerly the power unit of a Viaduction Inspection Unit.

SCOTTISH REGION.

Polmadie Carriage Sidings, Glasgow, Strathclyde.
Gauge : 4'8½".   (   )

P.O.1.	(08173)	0-6-0DE	Dar	1956

SOUTHERN REGION.

Gauge : 4'8½".

SA = Sandown, Isle of Wight.
SL = Slade Green, Kent.

97800	IVOR	(08600)	0-6-0DE	Derby	1959	SL
97803		(05001)	0-6-0DM	HE	4870 1956	SA

Civil Engineer's Locomotives.

Gauge : 4'8½".

They are normally only permitted to work on lines completely occupied by the Chief Civil Engineer or in engineering department sidings. They are based at the following depots but normally at weekends they will be working wherever the Chief Civil Engineer requires them.

N	(SO 792495)	Newland P.A.D., Malvern, Hereford & Worcester.
Ra	(ST 137800)	Radyr P.A.D., South Glamorgan.
Re	(SU 706740)	Reading C.C.E. Yard, Berkshire.
Sw	(SU 138844)	Swindon Newburn P.A.D., Wiltshire.
T	(ST 210257)	Taunton Fairwater P.A.D., Somerset.

97650	0-6-0DE	RH	312990	1952	Sw
97651	0-6-0DE	RH	431758	1959	+
97652	0-6-0DE	RH	431759	1959	T
97653	0-6-0DE	RH	431760	1959	Ra
97654	0-6-0DE	RH	431761	1959	N

+ Currently being overhauled in Swindon Works.

Reading Signal Works, Berkshire.

Gauge : 4'8½".   (SU 716739)

97020		4wDM	RH	408493	1956	
06002		0-4-0DM	AB	434	1959	+ OOU
97804	(06003)	0-4-0DM	AB	435	1959	+

+ Currently at Reading T.M.D.

ENGINEERS MOTOR TROLLEYS.

The location shown for each trolley is its "Base Location" but trolleys go where they are required and so can be found working elsewhere.

Gauge : 4'8½".

EASTERN REGION.

AY = Ainderby.	B = Battersby.
BI = Billingham.	C = Castleton.
F = Ferryhill.	G = Grosmont.
HE = Hexham.	HO = Horsforth.
LG = Low Gates,	M = Marsden.
Northallerton.	TY = Tyne Yard.

BA = Bishop Auckland.	
DBT = Darlington Bank Top.	
H = Haltwhistle.	
L = Leyburn.	
N = Northallerton.	
W = Whitby.	

900476	DX 68011	2w-2PMR	Wkm	413	1931	N
900392		2w-2PMR	Wkm	672	1932	C OOU
900468		2w-2PMR	Wkm	675	1932	B OOU
(900420)		2w-2PMR	Wkm	730	1932	H DsmT
		2w-2PMR	Wkm	1548	1934	M DsmT
DE 900853		2w-2PMR	Wkm	6684	1953	BA Dsm
DB 965045		2w-2PMR	Wkm	7073	1955	AY
DB 965065		2w-2PMR	Wkm	7580	1956	H
DB 965071		2w-2PMR	Wkm	7586	1957	LG
DB 965073		2w-2PMR	Wkm	7588	1957	B
DB 965075		2w-2PMR	Wkm	7590	1957	F DsmT
DB 965078		2w-2PMR	Wkm	7593	1957	F
DB 965079		2w-2PMR	Wkm	7594	1957	G
DB 965080		2w-2PMR	Wkm	7595	1957	C
DB 965082		2w-2DMR	Wkm	7597	1957	B
DB 965083	DE 320501	2w-2PMR	Wkm	7598	1957	W
DB 965085		2w-2PMR	Wkm	7600	1957	L
(DB 965086)		2w-2PMR	Wkm	7601	1957	H DsmT
(DB 965087)		2w-2PMR	Wkm	7602	1957	H DsmT

```
DB 965092                              2w-2PMR          Wkm    7607 1957 TY
DB 965096                              2w-2PMR          Wkm    7611 1957 BA
DB 965097                              2w-2PMR          Wkm    7612 1957 F
DB 965099                              2w-2PMR          Wkm    7614 1957 BA
DB 965102                              2w-2PMR          Wkm    7617 1957 F
DB 965104                              2w-2PMR          Wkm    7619 1957 DBT
DB 965949    68/005                    2w-2PMR          Wkm   10645 1972 HO
DB 965950    68/006                    2w-2PMR          Wkm   10646 1972 H
DB 965951    68/007                    2w-2PMR          Wkm   10647 1972 M
DB 965952    68/008                    2w-2PMR          Wkm   10648 1972 HE
DB 965987    68/010                    2w-2PMR          Wkm   10731 1974 F
                                       2w-2PMR          Wkm              BI OOU
```

LONDON MIDLAND REGION.

```
B   = Bangor.              BM = Barmouth.        BO = Borth.
C   = Caersws.            H  = Hanwood.         HL = Harlech.
L   = Llanbrynmair.       M  = Machynlleth.     N  = Newtown.
VOR = Vale of Rheidol line.                     W  = Welshpool.
```

```
TR 1     A1M                                     2w-2PMR           Wkm    6872 1954 H  +
TR 6                                             2w-2PMR           Wkm    6901 1954 L  +
TR 11    A155W    PWM 2187                        2w-2PMR           Wkm    4164 1948 HL +
TR 13             PWM 2189                        2w-2PMR           Wkm    4166 1948 M  +
TR 16    B25      PWM 3951                        2w-2PMR           Wkm    6936 1955 W
TR 18             PWM 4301                        2w-2PMR           Wkm    7504 1956    BO
TR 19             PWM 4302                        2w-2PMR           Wkm    7505 1956 B  +
TR 20    B49      PWM 4310   DB 965563            2w-2PMR           Wkm    7513 1956 C  +
(TR 21)  B50      PWM 4311                        2w-2PMR           Wkm    7514 1956 HL
TR 22    B51      PWM 4312                        2w-2PMR           Wkm    7515 1956 BM +
TR 23    B52      PWM 4313                        2w-2PMR           Wkm    7516 1956 B  +
TR 26             PWM 2214                        2w-2PMR           Wkm    4131 1947 VOR @
TR 34                       DB 965566            2w-2PMR           Wkm    8272 1959 H
TR 36    A14W     PWM 2786                        2w-2PMR           Wkm    6885 1954 BM +
TR 37    A25M     PWM 2787                        2w-2PMR           Wkm    6896 1954 N  +
TR 38             PWM 3764                        2w-2PMR           Wkm    6643 1953 M  +
TR 39             PWM 4306   DB 965564            2w-2PMR           Wkm    7509 1956 C
TR 40             PWM 4314   DB 965565            2w-2PMR           Wkm    7517 1956 B  +
TR 41    A159M    PWM 2191                        2w-2PMR           Wkm    4168 1948 BO +
         72107    DX 68025   DB 965392            2w-2DMR           Matisa  D8 006 1971
                                                                   Rebuilt Crewe      1980 W
         72109    DX 68026   DB 965491            2w-2DMR           Matisa  D8 012 1972
                                                                   Rebuilt Crewe      1980 N
         72012    DX 68027   DB 965473            2w-2DMR           Matisa  D8 002 1970
                                                                   Rebuilt Crewe      1980 N
```

```
          + Sold for preservation. Awaiting removal from site.
          @ Gauge 1'11½".
```

SCOTTISH REGION.

```
C  = Crianlarich.          G  = Georgemas Jct.    H   = Halkirk Ballast Tip.
K  = Kilmarnock.          KL = Kyle of Lochalsh Branch.
SB = Spean Bridge.        ST = Stranraer.         WHL = West Highland Line.
```

```
                                       2w-2PMR          Wkm               H  +  OOU
                                       2w-2PMR          Wkm    1405 1934 KL @ DsmT
                                       2w-2PMR          Wkm    3707 1945 KL @ DsmT
DB 965135     68001                    2w-2PMR          Wkm    7847 1957 K     OOU
(DB 965136)                            2w-2PMR          Wkm    7848 1957 G ++ DsmT
(DB 965139)                            2w-2PMR          Wkm    7851 1957 G ++ DsmT
DB 965331     68003                    2w-2PMR          Wkm   10179 1968 C
```

DB 965919		2w-2DMR	Matisa			WHL
DB 966025		2w-2DMR	Schoma	4016	1974	C
DB 966027		2w-2DMR	Schoma	4017	1974	G
	68028	2w-2DMR	Matisa	D8 004		
		Rebuilt	Kilmarnock		1977	C
	68029	2w-2DMR	Matisa	D8 011		
		Rebuilt	Kilmarnock		1978	SB
	68030	2w-2DMR	Matisa	D8 005		
		Rebuilt	Kilmarnock		1979	G
	68031	2w-2DMR	Matisa	D8 014		
		Rebuilt	Kilmarnock		1981	ST

+ In 1981 this trolley was partially concealed in the undergrowth.
@ The body of one of these is in Strathcarron Station Yard.
++ The bodies of these trolleys are at 6, Glamis Road, Wick and Newfield Farm, Thurso Road, Wick.

SOUTHERN REGION.

A varying number of trolleys on this region are normally in use by the building department for work on tunnels, etc. Their allocations are therefore liable to frequent change.

CD = Central Division.	DP = Dover Priory.	EW = Earlswood.	
FW = Folkestone Warren.	H = Hastings.	HG = Hither Green.	
PP = Preston Park.	PU = Purley.	S = Southampton.	
TWC = Tunbridge Wells Central.			

DS 52	2w-2PMR	Wkm	7031	1954	H DsmT
DS 3232	2w-2PMR	Wkm	6471	1952	PP DsmT
DS 3304	2w-2PMR	Wkm	7823	1957	PU DsmT
DS 3317	2w-2PMR	Wkm			S + DsmT
DB 965143	2w-2PMR	Wkm	7974	1958	EW DsmT
DS 965336	2w-2DMR	Wkm	10343	1969	PP
DB 965990	2w-2DMR	Wkm	10705	1974	TWC +
DB 965991	2w-2DMR	Wkm	10707	1974	PU
DB 965992	2w-2DMR	Wkm	10708	1974	TWC +
DB 965993	2w-2DMR	Wkm	10706	1974	H
DB 966030	2w-2DMR	Plasser	419	1975	CD
DB 966031	2w-2DMR	Wkm	10839	1975	S +
DB 966033	2w-2DMR	Wkm	10841	1975	DP
DB 966034	2w-2DMR	Wkm	10842	1975	FW
DB 966035	2w-2DMR	Wkm	10843	1975	HG +
DB 966036	2w-2DMR	Wkm	10844	1975	EW

+ Currently in use by the building department.

WESTERN REGION.

EJ = Exmouth Jct.	N = Neath.	P = Plymouth, Valletort Road.
T = Taunton.	Y = Yeovil.	

PWM 1932	2w-2PMR	Wkm	3366	1943	P DsmT
PWM 2177	2w-2PMR	Wkm	4154	1948	P DsmT
PWM 2814	2w-2PMR	Wkm	4992	1949	P DsmT
PWM 2824	2w-2PMR	Wkm	5002	1949	P DsmT
PWM 2831	2w-2PMR	Wkm	5009	1949	N
PWM 3960	2w-2PMR	Wkm	6945	1955	T OOU
PWM 4303	2w-2PMR	Wkm	7506	1956	EJ
PWM 4305	2w-2PMR	Wkm	7508	1956	Y

<u>INSPECTION SALOONS.</u>

The vehicles detailed below are the power vehicles of the two twin-car civil engineer's inspection saloons based at York, but for use as required anywhere on the Eastern Region.

DB 975349	(E 51116)	4w-4DMR	GRCW	1957	
DB 975664	(Sc 51122)	4w-4DMR	GRCW	1957	

<u>MISCELLANEOUS POWERED DEPARTMENTAL VEHICLES.</u>

	Sc 79971		4wDMR	PR	1958	a
022	ADS 70315	(S 10731 S)	4w-4wRE			b
	ADS 70318	(S 10787 S)	4w-4wRE			b
023	ADS 70316	(S 10742 S)	4w-4wRE			b
	ADS 70317	(S 10760 S)	4w-4wRE			b
024	ADB 975250	(S 10829 S)	4w-4wRE			b
	ADB 975251	(S 10830 S)	4w-4wRE			b
	DB 975007	(E 79018)	4w-4DMR	Derby	1954	c
	ADB 975027	(M 61162)	4w-4wRE	Afd/Elh	1958	d
	ADB 975319	(S 10919 S)	4w-4wRE			e
	ADB 975322	(S 10920 S)	4w-4wRE			e

a : Grounded at Millerhill Yard - in use as a mess hut.
b : Stores Units for use between various S.R. depots.
c : Ultrasonic Unit based at Old Oak Common.
d : C.M. & E.E. trials from Strawberry Hill.
e : S.R. Instruction Unit based at Selhurst.

<u>ROUTE LEARNING UNITS.</u>

These units work wherever required on the region to which they are allocated.

The W.R. units are based for maintenance at Reading, whilst the L.M.R. units are normally maintained at Tyseley and Carlisle Kingmoor. The E.R. units are normally maintained at York or Lincoln.

TDB 975023	(W 55001)	4w-4DMR	GRCW	1958	WR
TDB 975042	(M 55019)	4w-4DMR	GRCW	c1958	LMR
TDB 975227	(M 55017)	4w-4DMR	GRCW	c1958	LMR
TDB 975309	(M 55008)	4w-4DMR	GRCW	1958	ER
TDB 975310	(M 55010)	4w-4DMR	GRCW	1958	ER
TDB 975540	(W 55016)	4w-4DMR	GRCW	c1958	WR
TDB 975659	(W 55035)	4w-4DMR	Pressed Steel	1960	WR
TDB 975994	(Sc 55014)	4w-4DMR	GRCW	c1958	ER
TDB 975998	(Sc 55013)	4w-4DMR	GRCW	c1958	ER

<u>TRAIN HEATING UNITS.</u>

Former Capital Stock locomotives converted to E.T.H. pre-heating units.

NA = Newton Abbot.   NO = Norwich T.M.D.   ST = Stratford T.M.D.
SW = Maliphant Sidings, Swansea.   TO = Toton T.M.D.   Y = Yarmouth.

ADB 968000	(8243)	4w-4wDE	BTH	1961	TO
ADB 968001	(8233)	4w-4wDE	BTH	1131 1959	ST
ADB 968002	(8237)	4w-4wDE	BTH	1960	TO
ADB 968008	(24054)	4w-4wDE	Crewe	1959	NA
ADB 968009	(24142)	4w-4wDE	Derby	1960	SW
ADB 968013	(31013)	2w-2-2w+2w-2-2wDE	BT	84 1958	Y
ADB 968014	(31002)	2w-2-2w+2w-2-2wDE	BT	73 1957	NO
ADB 968015	(31014)	2w-2-2w+2w-2-2wDE	BT	85 1958	Y
ADB 968016	(31008)	2w-2-2w+2w-2-2wDE	BT	79 1958	ST